PRODUCTION MANAGER'S HANDBOOK OF FORMULAS AND TABLES

PRODUCTION MANAGER'S HANDBOOK OF FORMULAS AND TABLES

Lewis R. Zeyher, CMC

Prentice-Hall, Inc.
Englewood Cliffs, N. J.

Prentice-Hall International, Inc., *London*
Prentice-Hall of Australia, Pty. Ltd., *Sydney*
Prentice-Hall of Canada, Ltd., *Toronto*
Prentice-Hall of India Private Ltd., *New Delhi*
Prentice-Hall of Japan, Inc., *Tokyo*

Library of Congress Cataloging in Publication Data

Zeyher, Lewis R
  Production manager's handbook of formulas and
tables.

  1. Industrial engineering--Handbooks, manuals,
etc. I. Title.
T56.23.Z48   658.4'03'0212   72-93
ISBN 0-13-724427-4

Printed in the United States of America

Dedication

To managers everywhere, at all levels, who each day conscientiously endeavor to improve their knowledge, technical skills and human relations practices. This improved competence serves to better lead their subordinates to the production of a superior quality product, shipped on schedule, and with a minimum of cost.

About the Author

Lewis R. Zeyher, CMC

Lewis R. Zeyher has been President of Zeyher Associates, Management Consultants, Jenkintown, Pennsylvania, for seventeen years. His total experience covers over thirty years in eighty different plants, and he has held a wide variety of positions from factory manager to vice president of manufacturing. During one period he was head office staff industrial engineer for a multi-plant company.

Mr. Zeyher has conducted industrial management seminars for business leaders in this country and abroad, and he has been active as a lecturer and teacher on management subjects. He is a Fellow of the Society for Advancement of Management and he holds the Society's *Advancement of Management Award* for his two outstanding books in the operations management field: *Cost Reduction in the Plant* and *Production Manager's Desk Book*, both published by Prentice-Hall. He is also a Founding Member of the Institute of Management Consultants.

He received his B. S. degree in Mechanical Engineering from Lafayette College, and he has completed graduate studies at Pennsylvania, Columbia, and Rutgers Universities.

A Word from the Author

Over the years many managers have asked me why there wasn't one source they could go to to find the formulas and tables they use frequently in their industrial operations. These comments led me to make a review of my own experience. There immediately came to mind situations where references had to be made to three or four sources before the proper formula or table could be found. On other occasions, I knew a solution of a problem could easily be resolved if only the necessary formula could be remembered, but I didn't know where to start to look for it. There were also many cases where I was fortunate to locate a formula but didn't know how to solve it, because no instructions were provided.

Obviously, a single source was urgently needed. With this in mind, I undertook the compilation of this handbook. It brings together, in one volume, material that is ordinarily widely scattered. It can also be a great time saver. It is convenient to use—the topical arrangement and the index make for ease and speed of reference. Since every formula in this handbook illustrates a live problem and its solution, it will quickly refresh your memory or demonstrate a new approach. In many cases, all that need be done to solve your problem is to substitute the data in the formula and then do the simple mathematical operation. In addition, a review of the Table of Contents will provide fresh ideas and suggest better techniques for making your management more effective.

This handbook will serve the needs not only of the Production Manager, but also of anyone engaged in industrial and operations management. Some practical features that this handbook offers are listed here:

(a) Provides scores of ready-to-use formulas, tables, charts, and timesaving shortcuts for swift, sure handling of every phase of your management work.

(b) Gives dozens of worked-out examples depicting how these formulas are applied to typical operating problems.

- Discusses inventory turnover calculations, economical lot size ordering, and the manufacturing time method of production scheduling. (See Chapter 1.)
- Describes the four internationally accepted methods for measuring work. (See Chapters 2 and 3.)
- Tells how to establish materials handling labor ratios and costs. Methods of forecasting profitability of new equipment are also presented. (See Chapter 5.)
- Depicts use of various warehousing formulas, i.e., handling time standards, space-utilization standards, performance-control reports, planning standards, cost controls, charge rates for company warehouses, and development of direct-labor and space cost standards. (See Chapter 6.)
- Outlines proper utilization of individual relations formulas, i.e., labor turnover, slope of wage trend-line, personnel performance index, standard injury frequency rate, and the standard injury severity rate. (See Chapter 7.)
- States how to make capital investment decisions through quantitative analysis. Also describes how to reach a decision by using probability concepts, decision tree analysis, and by the simulation (Monte Carlo) technique. (See Chapter 8.)
- Specifies how to employ operation research techniques to solve complex problems—either by the model system or directly by differential calculus. (See Chapter 9.)
- Relates, in detail, how to apply the queuing theory to business problems. This discusses methods of calculating machine interference time for one crew servicing two, three and four machines. (See Chapter 10.)
- Acquaints you with various inventory formulas—estimating inventory by the gross profit method and the retail inventory method. (See Chapter 11.)
- Informs you about the use of cost and production formulas, i.e., equivalent units of production, breakeven point, normal burden rate (Job Order Cost), volume variation, budget variation, quantity variation and related techniques. (See Chapter 12.)
- Advises you about the use of statistics by both the sampling and correlation methods. (See Chapters 13 and 14.)
- Discusses the six formulas for calculating depreciation, i.e., straight line, units of production, sum of the year's digit method, declining balance (real), 200% declining balance, and composite rate. (See Chapter 15.)

I am indebted to many individuals who have given generously of their time and efforts to aid me in the development of this book. I am especially grateful to Professor Charles M. Merrick of Lafayette College and Professor James M. Apple of Georgia Institute of Technology. The following business executives also were most cooperative in contributing ideas and providing advice: Richard

P. Towne, Donald Rose, Richard Burdick, Phil Carroll, John R. Kennel, Joseph A. Panico, John B. Taggart, James P. O'Brien and John J. Wilkinson.

The following companies also kindly cooperated in assisting me: Eaton Yale & Towne, Inc.; The Maytag Company; Atlantic Richfield Co.; Victor F. Weaver, Inc.; Work-Factor Company—Division of Science Management Corporation; Maynard Training Center, Columbus, Ohio; H. B. Maynard & Company; and Methods-Time Measurement Association for Standards and Research.

Lewis R. Zeyher

How to Use This Handbook

It is assumed that the reader has sufficient knowledge of production, industrial engineering, accounting theory, mathematics, industrial relations, management controls, and statistics to be able to recognize a problem and know how to use the proper formula in order to solve it. This book is not intended to be a substitute for such knowledge. It is to be used as a reference tool to aid in problem solving. Every formula is illustrated with a simple problem and its solution. In a number of instances all one needs to do to solve the problem is merely substitute the actual data in the formula and perform the indicated operations.

Included in the Appendix is a section on *Make or Buy*. Most companies frequently make these decisions and each may use a different approach in reaching a verdict. A practical criteria has been developed in this section which can serve as a blueprint for further exploration and application of this and other major areas of decision making.

Also included in the Appendix is a section on *Warehousing*. This covers pallet terminology and sizes with definitions and drawings depicting the various types in use and their construction. In addition, a procedure for comparative rating for fork lift trucks and a method of calculating a load with a length other than that specified by the manufacturer is provided. This is added as a supplement to the chapter on warehousing formulas (Chapter 6).

The section on *Planning and Control* is addressed to the use of charting, describing both PERT (Program Evaluation and Review Techniques) and the use of the Gantt chart. Production managers and related operating executives are usually kept busy attending to their daily operating problems and have a minimum amount of time to devote to reviewing their budgetary performance and similar reports. An assortment of graphical forms of reporting are depicted. Any number of these can be used as guidelines and can be tailored to fit your specific needs. The use of these charts will keep the busy manager well informed about his important operating areas with a minimum expenditure of his limited time.

Other special charts covering important and useful data are made available in the Appendix. Six tables used frequently by responsible operating personnel are provided in the back of the book in Appendix D.

If one is seeking a specific formula, he needs only turn to the index to see if the formula is included in the handbook. When he locates the page on which the formula is presented, a quick glance over the formula and the example-problem will be sufficient to determine if the formula is the one he is seeking. This exercise will be of great help in developing a solution to his own problem.

The next step is to acquaint himself with the symbols used in the formula. Then it is important to assemble all the facts relevent to his particular problem. This data is then substituted in the formula and the calculations performed, preferably on a calculator. If practicable, the results should be checked by a competent associate. It also is helpful to examine the solution to the problem and decide if it appears reasonable. Occasionally a minus quantity may indicate a mistake, or a decimal point may be in the wrong place.

Following are some typical examples of the types of problems which a production manager encounters and how this handbook can help solve them.

Example

A manager suspects that one of his production supervisors has too many operators (a crew of three) assigned to one operation on a conveyor line which is running at a speed of 60 units per minute, with a production standard of 0.0495 standard hours per 100 units. After referring to the index for *crew determinations,* the following formula is found:

$$N = P \times S$$

where

N = Number of crew required
P = Production required by crew in units
S = Production standard in hours per 100 units

The facts show the following:

N = Unknown
P = 60 units per minute
S = 0.0495 standard hours per 100 units

Substituting in formula:

$$N = \frac{60 \times 60}{100} \times 0.0495$$

$$N = 36 \times 0.0495$$
$$N = 1.782 \text{ operators (use two operators)}$$

The crew is over-manned, actually requiring only two operators instead of the present crew of three.

A review of Chapter 4, "Production Line Techniques," and particularly the section "Chart of Standards and Crew Determinations," will prove very helpful.

Example

A production manager is often responsible for the operation of warehousing and inventory turnover. Suppose a plant manager is informed by his superior that his cost of carrying inventory is too high, the cost being $3.83 per unit. He is asked to reduce this cost to less than $2.00 per unit. He knows the easiest and quickest way to accomplish this is by increasing his inventory turnover. First, he determines that his current turnover figure is 2.0 times a year. (Refer to the index for inventory turnover and find Chapter I, page 26, section on "Inventory Turnover Calculations.") Next he determines the turnover he must achieve to reach a unit cost of less than $2.00. (Refer to the index for warehousing formulas, Chapter 6, section on "Warehousing Evaluation and Requirements," page 113.) The following calculations are then made:

The unit cost of carrying inventory when related to turnover is determined by using the following formula:

$$\text{Unit cost} = A + \frac{B}{\text{turnover/year}}$$

where: A = Warehousing cost related to *activity* expressed in dollars per unit.

   B = Warehousing cost related to *storing* expressed in dollars per unit.

(See pages 112 and 113 for inventory cost elements to consider.)
The facts show the following:

$$A = \$1.50/\text{unit}$$
$$B = \$5.10/\text{unit}$$

Examples of how turnover affects inventory carrying cost per unit are shown below:

Turnover per Year	Equation	Inventory Carrying Cost per Unit
1.0	$\$1.50 + \dfrac{\$5.10}{1.0}$	$7.65
2.0	$\$1.50 + \dfrac{\$5.10}{2.0}$	$3.83
► 4.0	$\$1.50 + \dfrac{\$5.10}{4.0}$	$1.91
6.0	$\$1.50 + \dfrac{\$5.10}{6.0}$	$1.28

The production manager reviews these figures and decides that a turnover of 4.0 times a year will provide an inventory carrying cost per unit of $1.91 which is less than the $2.00 per unit his superior requested.

The next question the manager might ask is how should he go about improving his turnover figure in order to reach the 4.0 times a year required. That question, of course, is beyond the scope of this book. For further information on this subject, refer to the Prentice-Hall publication *Cost Reduction in the Plant*, Lewis R. Zeyher, Chapter 17, Production Control and Inventory Turnover, pages 183–193.

Example

Suppose the production manager purchased a new machine that cost $150,000. From past experience and further projections it is estimated that the asset will have a salvage value of $30,000 at the end of its useful life. Engineering estimates indicate that with adequate maintenance the machine will probably produce 800,000 units during its efficient life. During the eighth year of its life the machine produced 40,000 units. He then wants to know the annual charge for depreciation by the *units of production method*.

Referring to the index under the heading "Depreciation," we find the formula for "Depreciation—units of production," located on pages 209 and 210.

The formula is:

$$D_j = C - S\frac{U_x}{U_n}$$

where

D_j = Depreciation for particular year
C = Cost
S = Salvage value
U_x = Units produced in particular year
U_n = Estimated number of units that asset will produce during its life.

The facts are:

D_j = Unknown
C = $150,000
S = $30,000
U_x = 40,000 units
U_n = 800,000 units

Solution:

$$D_j = (\$150,000 - \$30,000)\frac{40,000}{800,000}$$
$$D_j = \$120,000 \times 0.05$$
$$D_j = \$6,000$$

Note: Units may be interpreted to be units, labor hours, machine hours, etc.

Table of Contents

1

Production Control and Inventory Turnover

ECONOMICAL LOT SIZE ORDERING[1]

Formula

$$Q' = \sqrt{\frac{2\,AR}{CU}}$$

where

Q' = Economic lot size order in units
A = Annual requirements in units
R = Restocking cost in dollars
C = Annual carrying cost ratio
U = Unit factory cost in dollars

Example

You are required to determine the economical lot size for scheduling the production order of a product element with annual requirements of 480 units, restocking costs of $25.00, annual carrying cost ratio of 21%, and a unit factory cost of $15.00.

Solution

$$Q' = \sqrt{\frac{2\,AR}{CU}}$$

$$Q' = \sqrt{\frac{2 \times 480 \times \$25.00}{0.21 \times \$15.00}}$$

$$Q' = \sqrt{\frac{\$24,000}{\$3.15}}$$

---

[1] From "Modern Approaches to Production and Control," by Robert A. Pritzker and Robert A. Gring, © 1960 by The American Management Association, Inc. Reprinted by permission of the American Management Association, Inc. P. 198.

$$Q' = \sqrt{7619}$$
$$Q' = 87 \text{ units}$$

INVENTORY TURNOVER CALCULATIONS

Rate of Return Formula[2]

$$R = \frac{P}{S} \times \frac{S}{I}$$

where

R = Rate of return
P = Profit
S = Sales
I = Investment

In this formula, frequency is brought into the measure by the multiplier called turnover of investment. The second fraction is the number of times capital was circulated.

Example

Determine the actual rate of return of a company with gross sales of $2,000,000, profit of $240,000, and plant investment of $500,000.

$$R = \frac{P}{S} \times \frac{S}{I}$$
$$R = \frac{\$240,000}{\$2,000,000} \times \frac{\$2,000,000}{\$500,000}$$
$$R = .12 \times 4$$
$$R = .48 \text{ or approximately } 48\%$$

The actual management performance here is approximately 48% and represents a rate of return on investment.

[3] Du Pont's formula for figuring return on investment is neither secret nor unique. The calculation is made by taking gross profit as a per cent of sales, and multiplying that by turnover. Assuming no changes in price, an improvement in profit indicates success in cutting costs; higher turnover would point to harder working of capital.

All assets are cranked into a formula on an original cost basis. If plant investment was depreciated each year, return on investment would continually rise, thus failing to reveal actual management performance.

Calculation of Inventory Turnover

Formula

$$T = \frac{S}{I}$$

---

[2] From *Practical Production and Inventory Control,* by Phil Carroll. Copyright © 1966 by Phil Carroll. Used by permission of McGraw-Hill Book Company, New York. Pp. 4–5.

[3] Reprinted from November 9, 1963 issue of *Business Week* by special permission. Copyright © 1963 by McGraw-Hill, Inc. P. 144.

where

T = Number of inventory turns per year
S = Yearly sales volume (manufactured cost)
I = Average monthly inventory (manufactured cost)

Example

Determine the inventory turnover of a metalworking plant with a yearly sales volume (manufactured cost) of $24,000,000 and an average monthly inventory (manufactured cost) of $8,000,000.

$$T = \frac{S}{I}$$

$$T = \frac{\$24,000,000}{\$8,000,000}$$

$$T = 3.0 \text{ turns per year}$$

Comment: This plant should have had an average inventory of approximately $3,000,000 or a turnover of 8.0 times a year. The difference between $8,000,000 and $3,000,000 is $5,000,000. At 6.0% this reflects an annual loss of interest alone of $300,000, with additional losses not calculated here.

Cost of Inventory

Development of formula for cost of carrying inventory, using the classical *EOQ* formula:

$$\frac{Q'}{2}CU = \frac{A}{Q'}R$$

Then solving for C—multiplying both sides of equation by 2 gives us:

$$Q'CU = \frac{2\ AR}{Q'}$$

Then dividing both sides of equation by $Q'U$ gives us:

$$C = \frac{2\ AR}{(Q')^2\ U}$$

where

C = Annual carrying cost ratio
A = Annual requirements in units
R = Restocking costs in dollars
Q' = Economic lot size of order in units
U = Unit factory cost in dollars

Example

Determine the annual carrying cost ratio in per cent for a product element with annual requirements of 480 units, restocking costs of $25.00, unit factory cost of $15.00, and the economic lot size order of 87 units.

$$C = \frac{2\,AR}{(Q')^2\,U}$$

$$C = \frac{2 \times 480 \times \$25.00}{(87)^2 \times \$15.00}$$

$$C = \frac{\$24,000}{\$1135.4}$$

$$C = 21\%$$

Annual carrying cost ratio is 21%.

DETERMINATION OF INVENTORY PARKING RATIO

Explanation

Carrying and maintaining inventory, whether raw materials, in-process or finished stock, is costly. The application of this formula to your operations will highlight weaknesses in your material flow. Detection should lead to a reduction in the time factor by improving your material's handling, planning, routing, scheduling, warehousing and shipping operations. A reduction of one day in your cycle time of a $100,000 inventory could save you $60.00 per day or $22,000 per year. Delivery schedules would be shortened, improving customer relations.

Formula[4]

$$R = \frac{C}{T} - 1$$

where

R = Parking ratio (sitting time)
C = Customer cycle time
T = Operation time per lot

Example

Determine the parking ratio of a product that has a customer cycle time of 25 days (600 hours) for a lot of 200 units consuming a total of 14 operations, of which 4 are productive and 10 unproductive times. The unproductive operations include inspection, temporary storage, transportation and delays. The productive operations took 133 hours and the unproductive 467 hours.

Solution

$$R = \frac{C}{T} - 1$$

$$R = \frac{600}{133} - 1$$

---

[4] From *Practical Production and Inventory Control,* Phil Carroll. Copyright © 1966 by Phil Carroll. Used by permission of McGraw-Hill Book Company, New York. P. 225.

$$R = 4.5 - 1$$
$$R = 3.5$$

Our answer tells us that the product sits or is parked 3.5 times as long as it takes us to make it.

PRODUCTION SCHEDULING—MANUFACTURING TIME METHOD[5]

(for use in plant's operations on the piecework type of incentive system)

Formula

$$MT = \frac{N + D + \left(\dfrac{QL + S}{AE}\right)}{7}$$

where

N = Number of operations
D = Number of departments
L = Total direct labor, dollars per piece
S = Total normal setup, in dollars
Q = Order quantity
AE = Average earnings per hour per man
7 = Effective working hours per man per day
MT = Total manufacturing time in man-days

Example

You are required to determine the number of man-days of manufacturing time, given the following data:

$N = 5$ operations $\qquad D = 4$ departments
$Q = 1,000$ pieces $\qquad L = \$0.008$ per piece
$S = \$5.00$ $\qquad AE = \$3.00$

$$MT = \frac{N + D + \left(\dfrac{QL + S}{AE}\right)}{7}$$

$$MT = \frac{5 + 4 + \left(\dfrac{1,000 \times \$0.008 + \$5.00}{\$3.00}\right)}{7}$$

$$MT = 1.91 \text{ man-days} \quad (\text{use } 2.0 \text{ man-days})$$

The following slightly different formula may be used in plants or divisions that have wage-incentive systems using *standard hours* as a basis.

$$MT = \frac{N + D + \left[\dfrac{(Q/L) + S}{AE}\right]}{7}$$

[5] Thomas M. Landy, *Production Planning and Control, Industrial Engineering Handbook,* First Edition, Ed. H. B. Maynard (New York: McGraw-Hill Book Co. Copyright © 1956), pp. 6–16, 17.

where

AE = Per cent efficiency
L = Total direct labor, pieces per hour
S = Total normal setup, standard hours

and the other symbols are used in the piecework formula.

Example

You are required to determine the number of man-days of manufacturing time, where *standard hours* are used as a basis for a wage-incentive system, given the following data:

$$AE = 0.90 \qquad N = 5 \qquad Q = 500 \qquad L = 75 \qquad S = \text{standard hours}$$

$$MT = \frac{4 + 3 + \left[\dfrac{(500/75) + 3}{0.90}\right]}{7}$$

$$MT = 2\ 1/2 \text{ man-days of manufacturing time}$$

Note: Items N, number of operations, and D, number of departments, may be ignored completely in both formulas where the department is conveyorized and all operations are performed within the department. However, in plants operating with the colony-style layouts these items will play a very important part.

2

Work Measurement: Time Study Standards and Work Sampling

Time Study Standards

Formula

$$S = TR\,(A + 1)$$

where

- S = Standard time in standard minutes
- T = Observed time in actual minutes
- R = Rating—an estimated factor indicating the effective effort of the operator being observed, in per cent
- A = Allowances in per cent, for personal needs, fatigue, and unavoidable delays

Example

Determine the standard time for Element "A," given the following data:

$$T = 0.10 \text{ minutes}$$
$$R = 1.10 \text{ per cent}$$
$$A = 0.15 \text{ per cent}$$

Solution

$$S = TR\,(A + 1.0)$$
$$S = 0.10 \times 1.10\,(0.15 + 1.0)$$
$$S = 0.10 \times 1.10 \times 1.15$$
$$S = 0.127 \text{ standard minutes}$$

Example

Determine the production standard, per 1,000 pieces, for an operation consisting of six elements, with the following data:

Elements	Standard Minutes/Piece
A	0.127
B	0.532
C	3.040
D	2.210
E	0.980
F	5.111
	12.000

$$\text{Standard hours per piece} = \frac{12.00}{60} = 0.20 \text{ standard hours}$$

$$100 \times 0.20 = 20 \text{ standard hours/100 pieces}$$

The Standard Hour Plan

Formula

$$P = \frac{SU}{H}$$

where

P = Production performance in per cent
S = Standard hours per unit of measure
H = Actual hours worked on the operation
U = Running yards produced

Example

Determine the production performance of a weaver in a mill manufacturing Wilton carpet on a 20/4 width loom producing a quality of a given color, style and manufacturing specifications, with the following data:

$S = 0.575$ Standard hours per running yard
$H = 8.000$ Actual working hours
$U = 14$ Running yards of 20/4 carpet produced

Solution

$$P = \frac{SU}{H}$$

$$P = \frac{0.575 \times 14}{8.00}$$

$$P = \frac{8.05}{8.00}$$

$$P = 100.6 \text{ per cent (production performance for one day)}$$

Example

Determine the production performance of a weaver covering a five-day week in the same mill mentioned above, working on different quality of carpet each day, as follows:[1]

DAY	QUALITY NO.	DAILY PRODUCTION IN RUNNING YARDS		PRODUCTION STANDARD		TOTAL ST'D HRS.
Monday	216	14	×	0.575	=	8.05
Tuesday	328	11	×	0.695	=	7.65
Wednesday	622	18	×	0.452	=	8.14
Thursday	815	12	×	0.725	=	8.70
Friday	153	16	×	0.525	=	8.40
					Total	40.94

Number of actual hours worked on standards was 40.00 hours.
The weaver's production performance for the week was:

Solution

$$\text{Production Performance} = \frac{\text{Production Hours Produced}}{\text{Actual Hours Worked}}$$

$$\text{Production Performance} = \frac{40.94}{40.00} = 102\%$$

Standard Data

(Weaving Wilton broadloom carpet)

Formula

$$S = \frac{(M + WD)\,(A + 1.0)}{60}$$

where

$S =$ Production standard expressed in standard hours per running yard.

$D =$ Process delay time (allowed), expressed in normal minutes per square yard.

$W =$ Width of carpet in feet.

$A =$ Personal and fatigue allowance in per cent.

$M =$ Machine-time expressed in actual minutes per running yard.

Example

Determine the production standard "S" for weaving Wilton broadloom carpet using the standard data indicated in the chart for 3/42 yarn count (Exhibit 2–1).

---

[1] Lewis R. Zeyher, *Cost Reduction in the Plant* (Englewood Cliffs, N. J.: Prentice-Hall, Inc. Copyright © 1965), pp. 168, 172.

YARN Count	PROCESS DELAYS NORMAL MINUTES PER SQUARE YARD	ALLOWANCES			TOTAL ALLOWANCES
		Personal Attention	Unavoid-able	Fatigue	
2/55	0.375	3.0%	5.0%	5.0%	13.0%
2/45	0.480	3.0%	5.0%	5.0%	13.0%
3/42	0.520	3.0%	5.0%	5.0%	13.0%
3/36	0.580	3.0%	5.0%	5.0%	13.0%
3/37	0.625	3.0%	5.0%	5.0%	13.0%
3/30	0.650	3.0%	5.0%	5.0%	13.0%
4/25	0.700	3.0%	5.0%	5.0%	13.0%
4/22 ½	0.730	3.0%	5.0%	5.0 %	13.0%
	(D)				(A)

EXHIBIT 2-1

Standard Data Chart

(Used for calculating standards for various yarn counts of Wilton broadloom carpet)

NOTE: The machine time (M) can be determined by observation or by use of charts usually available from loom manufacturers.

No extra allowances for cut-wire or for height of wire. These considerations were built into the above process delays (D).

For skein dyed yarn add another 3.5% to the total allowances, some consideration having previously been built into above standards.

Process delays (D) was determined by taking a number of time studies of all yarn counts and plotting them on a graph.

Solution

$$S = \frac{(M + WD)\ (A + 1.0)}{60}$$

$$S = \frac{[22.7 + (15 \times 0.520)]\ [0.13 + 1.0]}{60}$$

$$S = \frac{30.5 \times 1.13}{60} = \frac{34.47}{60}$$

$S = 0.575$ Standard hours per running yard.

Method of Least Squares[2]

(to determine the slope of a line)

---

[2] From *How to Chart Timestudy Data,* Phil Carroll. Copyright © 1950 by Phil Carroll. Used by permission of McGraw-Hill Book Company, New York. Pp. 157–160.

Formula

$$S = C + (D \times V)$$

where

S = Standard time
C = Constant
D = Dimension in square inches
V = Value per

Example

Determine the slope of a line in graph (a), given eight figures for time values and eight for area values as follows:

POINT	TIME	AREA	EQUATION
1	.15	10	$.15 = 10A + C$
2	.20	15	$.20 = 15A + C$
3	.20	25	$.20 = 25A + C$
4	.25	35	$.25 = 35A + C$
5	.35	45	$.35 = 45A + C$
6	.35	55	$.35 = 55A + C$
7	.40	60	$.40 = 60A + C$
8	.45	70	$.45 = 70A + C$

The mathematical procedure will permit us to work these eight equations into two we can handle. The procedure requires that we do some multiplying. Two multipliers are used. This will give us two sets of equations. The first multiplier is that of the C term. It is unity (1) in our equations. Therefore, each of the equations will be the same afterward as it is now.

We now must obtain two equations with the second multiplier the factor with the A term. The result will be quite unlike the present equations. (See Exhibit 2–2).

Equation	FIRST SET		SECOND SET	
Number	Multiplier	Equation	Multiplier	Equation
1	1	$.15 = 10A + C$	10	$1.50 = 100A + 10C$
2	1	$.20 = 15A + C$	15	$3.00 = 225A + 15C$
3	1	$.20 = 25A + C$	25	$5.00 = 625A + 25C$
4	1	$.25 = 35A + C$	35	$8.75 = 1{,}225A + 35C$
5	1	$.35 = 45A + C$	45	$15.75 = 2{,}025A + 45C$
6	1	$.35 = 55A + C$	55	$19.25 = 3{,}025A + 55C$
7	1	$.40 = 60A + C$	60	$24.00 = 3{,}600A + 60C$
8	1	$.45 = 70A + C$	70	$31.50 = 4{,}900A + 70C$
TOTALS	Eq. (9)	$2.35 = 315A + 8C$	Eq. (10)	$108.75 = 15{,}725A + 315C$

EXHIBIT 2-2

These two equations we can solve simultaneously as in our other examples. We can cancel the C term first. So multiply (Eq. 10) $\times$ 8:

$$\ldots\ldots\ldots\ldots\ldots\ldots\ldots\ldots\ldots \quad 870.00 = 125{,}800A + 2{,}520C$$

(Eq. 9) × 315: $\ldots\ldots\ldots\ldots \quad 740.25 = 99{,}225A + 2{,}520C$

Subtracting: $\qquad\qquad\qquad 129.75 = 26{,}575A$

$$A = 129.75/26{,}575$$
$$A = .00488 \text{ min./sq. in.}$$

Substituting the value for A in Eq. (9) we can determine C:

$$2.35 = 315A + 8C$$
$$2.35 = (315 \times .00488) + 8C$$
$$2.35 = 1.5372 + 8C$$
$$-\,8C = -2.35 + 1.5372$$
$$-\,8C = -0.8128 = 0.8128/8$$
$$C = 0.1016 \text{ minutes.}$$

Drawing the line

The value of "C" is our constant time for "O" Area, and the time answer for "A" is the slope of the line. This is .0048 minutes per square inch. Hence, our equation will read as:

Standard time $= 0.10$ min. $+ (.0049$ min. sq. in. $\times$ sq. in.) To draw in the line, we should compute some points. We have one—our constant of 0.10 minutes. But to check, let's figure two more. At 20 square inches, the answer to our equation is:

Standard time $= .10$ min. $+ (.0049$ min./sq. in. $\times$ 20 sq. in.)
$$= .10 \text{ min.} + .098 \text{ min.}$$
$$= .198 \text{ min.}$$

At 65 square inches the standard time is:

Standard time $= .10$ min. $+ (.0049$ min./sq. in. $\times$ 65 sq. in.)
$$= .10 \text{ min.} + .3185 \text{ min.}$$
$$= .4185 \text{ min.}$$

With these two points plotted X on the chart in Exhibit 2–3, we see that our arithmetic checks. The line continued will pass through 0.10 minute, our constant. Also, the line bears a close relationship to the points originally plotted.

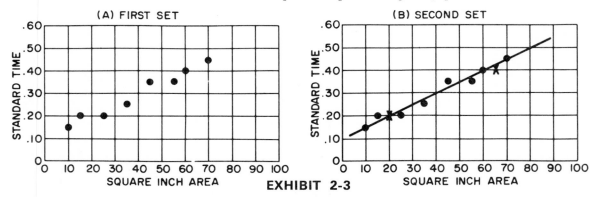

EXHIBIT 2-3

With these several applications of equations, we can conclude. They are representative of the types you need to know for time study work. Others are useful, of course.

Machine-time Calculation:[3]

Formula

$$T = \frac{\pi}{12} \times \frac{DL}{SF}$$

where

T = Cutting time in minutes

π = 3.1416 or $\frac{\pi}{12}$ = 0.2618

D = Diameter in inches

L = Length in inches

S = Surface speed in surface feet per minute

F = Feed in inches per minute

Example

Determine the cutting time in minutes, given the following data:

$\frac{\pi}{12}$ = 0.2618 inches

D = 5 inches

L = 9 inches

S = 80 feet per minute

F = .015 inches per minute

Solution

$$T = \frac{\pi}{12} \times \frac{\text{Diameter} \times \text{Length}}{\text{Surface Speed} \times \text{Feed}}$$

$$T = .262 \times \frac{5 \times 9}{80 \times .015}$$

$$T = 9.83 \text{ Minutes cutting time}$$

What Quantity of Pieces Should Be Produced When a Change Is Made from One Method to Another?

The fundamental rule to remember is that "Two straight-line equations are equal to each other at the point where their lines cross." Applying this rule to the above question to determine "what quantity?" you can simply set down the two equations equal to each other.

Formula

$$S = Su + (PQ)$$

where

S = Standard Time

Su = Setup minutes

P = Process or operation minutes

Q = Quantity, in pieces

---

[3] From *How to Chart Timestudy Data*, Phil Carroll. Copyright © 1950 by Phil Carroll. Used by permission of McGraw-Hill Book Company, New York. Pp. 212, 216, 148–151.

Solution

Given

Lathe	Setup Minutes	Operations Minutes
Engine	20.0	4.0
Turret	50.0	1.9

EXHIBIT 2-4

The equations of these two straight lines would be:

(Engine lathe) $S = Su + (PQ) = 20.0 + (4.0\ Q)$

(Turret lathe) $S = Su + (PQ) = 50.0 + (1.9\ Q)$

The quantity we want to determine is the economical number of pieces where we change method. It is the same Quantity in both equations. It is where the two lines cross each other. The total Standard Time for the lot produced either way would be identical at this point. For that reason we can set these down equal to each other; with Q standing for the Quantity, we have:

$20.0 + 4.0Q = 50.0 + 1.9Q$

collecting like terms we get:

$4.0Q + 20.0 = 1.9Q + 50.0$

$4.0Q - 1.9Q = 50.0 - 20.0$

$2.1Q = 30.0$

$Q = 30/2.1$

$Q = 14.3$ pieces

Short Element Times

Another use for equations is in the breakdown of timestudy for data. Two applications are commonly required. First, there is the necessity for sub-dividing watch readings too short to observe and to record. You may have run into this in fast punch-press operations.

Example

A study is made of an operation made up of four elements. We can call them a, b, c, and d. Our data requirements make it necessary to get separate element standards for each one. Observations should be taken in several ways on each study. You also should get over-all readings. . . . Let's take some watch readings to work out a sample problem.

Elements		*Minutes*	*Observations*
a + b + c	=	.045	(1)
b + c + d	=	.045	(2)
c + d + a	=	.035	(3)
d + a + b	=	.055	(4)
Total cycle	=	.060	(5)

Solution

Arranging the equations so we can add them, we get:

a + b + c $\quad$ = $\quad$ .045	(1)
$\quad$ b + c + d $\quad$ = $\quad$ .045	(2)
a $\quad$ + c + d $\quad$ = $\quad$ .035	(3)
a + b $\quad$ + d $\quad$ = $\quad$ .055	(4)

$$3a + 3b + 3c + 3d \; = \; .180$$

Dividing by 3, the result is:

$$a + b + c + d \quad = \quad .060 \qquad (5)$$

Next we solve for our unknowns:

a + b + c + d $\quad$ = $\quad$ .060	(5)
$\quad$ b + c + d $\quad$ = $\quad$ .045	(2)

Subtracting:

$$a \quad = \quad .015$$

a + b + c + d $\quad$ = $\quad$ .060	(5)
a $\quad$ + c + d $\quad$ = $\quad$ .035	(3)

$$b \quad = \quad .025$$

a + b + c + d $\quad$ = $\quad$ .060	(5)
a + b $\quad$ + d $\quad$ = $\quad$ .055	(4)

$$c \quad = \quad .005$$

a + b + c + d $\quad$ = $\quad$ .060	(5)
a + b + c $\quad$ = $\quad$ .045	(1)

$$d \quad = \quad .015$$

Man-Machine Charting

All jobs can essentially be broken down into:
(1) Get ready
(2) Do
(3) Put away

The "get ready" operation in the machining of a casting in a turret lathe would include picking-up casting, placing in jaws of chuck, tightening and positioning tool; the "do" part would consist of the actual cutting of metal; and the "put away," stopping machine, removing casting from chuck, and placing aside in tote-box. Many machines are semi-automatic and the machine cycle will run without attention after being started, stopping automatically at the completion of a fixed cycle.

Exhibit 2–5 depicts a typical man-machine chart.

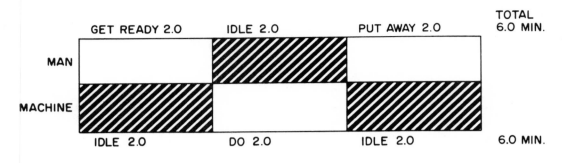

EXHIBIT 2-5

This situation, which is not unusual in many factories, brings up the commonly used term of *machine utilization*. The formula can be expressed as follows:

Formula

$$MU = \frac{D}{T}$$

where

MU = Machine utilization ratio
D = The "do" element in minutes
T = The total cycle time in minutes

Example (A): Determine the machine utilization ratio of a semi-automatic milling machine that actually cuts metal for two minutes, and the total cycle time is six minutes.

Solution

$$MU = \frac{2.0}{6.0} = .333$$

The machine utilization ratio is 33%.

Operator Effectiveness Ratio

Another analytical tool is the formula for calculating *operator effectiveness,* represented as follows:

Formula

$$OE = \frac{R + D}{T}$$

where

OE = Operator effectiveness ratio
R = Get-ready time in minutes
T = The total cycle time in minutes
D = The "do" element in minutes

Example (B): Determine the *operator effectiveness ratio* of an operator who takes two minutes to "get ready" and two minutes to "put away," with the total "cycle time" of six minutes.

Solution

$$OE = \frac{2 + 2}{6} = .667$$

The operator effectiveness ratio is 67%.

Example (C): Determine graphically (Exhibit 2–6) if an operator can operate two machines instead of one, using the following data:

Get ready	=	1.5 minutes
Do	=	1.5 minutes
Put away	=	1.5 minutes
Total cycle time	=	4.5 minutes (Machine No. 1)
Total cycle time	=	4.5 minutes (Machine No. 2)

Solution

Man	Put away and get ready, Machine No. 1		Put away and get ready, Machine No. 2		6.0 Min
Mach. No. 1	Idle 3.0 Mins.		Do 1.5 Mins.	Idle 1.5 Mins.	6.0 Min
Mach. No. 2	Do 1.5 Mins	Idle 1.5 Mins	Idle 3.0 Mins.		6.0 Mins

EXHIBIT 2-6

Costing Out the Man-Machine Cycle

The relative cost of the man and the machine is important in deciding how many machines a man should run. Suppose we illustrate this with the following example:

$$C = \frac{100 \, NTL}{60}$$

where

Man Cost
$\begin{cases} C = \text{Man cost in dollars per 100 pieces} \\ N = \text{Number of machines} \\ T = \text{Cycle time in minutes} \\ L = \text{Labor cost per hour} \end{cases}$

Machine Cost
$\begin{cases} C = \text{Machine cost in dollars per 100 cycles} \\ N = \text{Number of machines} \\ T = \text{Total cycle time in minutes} \\ L' = \text{Machine cost per hour} \end{cases}$

Example (D): Compare the cost of one man, one machine against one man and two machines. The machine is a semi-automatic rubber molding press, has a curing cycle of three minutes, takes two minutes to load and two minutes to unload. In addition, the man cost is $3.00 per hour, and the machine cost, on a one-shift basis, is $1.00 per hour.

One Man, One Machine

Man cost: $\qquad 1 \times \dfrac{7}{60} \times \$3.00 \times 100 = \$35.00/100$ units

Machine cost: $\qquad 1 \times \dfrac{7}{60} \times \$1.00 \times 100 = \$11.70/100$ units

Total $46.70/100 units

One piece produced per cycle: $\qquad \dfrac{\$46.70}{1} = \$46.70/100$ units

One Man, Two Machines

Man cost: $\qquad 1 \times \dfrac{8}{60} \times \$3.00 \times 100 = \$40.00/100$ units

Machine cost: $\qquad 2 \times \dfrac{8}{60} \times \$1.00 \times 100 = \$26.67/100$ units

Total $66.67/100 units

Two pieces produced per cycle: $\qquad \dfrac{\$66.67}{2} = \$33.34/100$ units

The answer, of course, is that it is more economical to have one operator run two machines.

Example (E)

Table of random sampling numbers (see Exhibit 2-7[4]).

WORK SAMPLING

Statistical Validity:[5]

Accuracy may be defined as the degree to which a given value or set of data is a true measure of the quantity observed or evaluated. However, all measurements contain some degree of imperfection; and this is the reason statistics enters the picture. The science of statistics permits the analysis of numbers or data to ascertain the amount of range of departure of the numbers from the true value which would be obtained by measuring the entire population for the quantity in question. It also helps predict from imperfect data what the true value is within stated probability limits. Such techniques can be applied equally well to basic investigations and to particular data such as that resulting from a work measurement study.

---

[4] Reprinted from A. Hald, *Statistical Tables,* John Wiley & Sons, Inc., New York, N. Y. Copyright © 1952.

[5] Delmar W. Karger and Franklin H. Bayha, *Engineered Work Measurement,* Second Edition, Industrial Press, Inc., New York, N. Y. 10016, Copyright © 1966, pp. 540–541.

Random Sampling Numbers

90	78	82	54	47	20	83	80	10	41	35	22	23	03	98	79	74	41	35	05	78	73	95	47	83
78	58	68	87	41	11	08	81	29	89	71	23	10	01	79	25	06	00	45	80	64	70	95	34	29
51	42	21	03	88	20	05	35	93	00	68	12	09	55	09	36	54	95	22	82	48	30	09	56	87
93	15	07	60	86	67	37	94	24	35	82	44	19	92	96	21	84	29	04	29	83	32	05	10	48
27	12	31	66	62	09	54	17	31	23	27	30	37	36	79	75	50	39	57	12	67	23	22	09	33
79	44	83	55	47	96	50	93	56	82	58	16	35	18	87	64	08	22	47	93	86	43	43	30	17
89	73	43	91	03	57	91	35	40	64	13	61	94	37	16	09	93	96	25	87	30	23	42	54	31
29	30	90	00	58	15	99	93	33	67	80	08	59	21	66	13	54	56	85	25	05	32	03	52	52
97	33	17	26	25	04	73	18	10	05	34	40	32	65	07	28	68	29	31	97	89	57	55	16	
07	15	44	92	47	28	50	93	03	53	37	70	19	68	59	95	39	87	90	46	98	64	46	24	71
82	50	35	50	80	23	67	81	25	02	83	08	12	70	00	25	31	33	80	06	19	86	14	59	27
59	21	86	16	30	27	85	16	26	34	50	15	87	22	69	71	36	95	90	76	90	99	79	63	21
04	19	60	33	05	29	02	33	74	56	38	84	21	07	35	93	54	70	18	47	14	62	75	45	02
96	91	44	09	94	06	89	50	88	83	82	50	11	82	51	30	68	91	06	28	86	65	17	45	20
31	71	03	53	38	94	02	52	72	15	44	49	53	42	43	00	36	97	67	64	12	27	46	00	18
03	70	22	67	59	98	10	64	68	08	79	06	89	48	41	85	72	10	87	24	96	04	20	68	00
08	45	79	46	89	74	73	67	60	15	70	37	61	44	07	67	89	81	54	26	57	17	63	27	74
37	80	05	75	64	48	51	68	68	27	71	75	45	32	27	76	35	26	58	88	67	74	48	90	94
90	63	56	69	37	19	74	48	63	31	52	36	84	40	66	72	66	03	41	87	65	29	12	36	64
22	69	38	02	88	89	71	43	01	87	41	79	42	99	29	41	08	47	32	19	45	29	59	69	90
05	79	69	67	64	36	14	82	65	26	40	51	63	42	48	85	48	34	12	04	33	26	52	26	52
48	91	53	03	82	64	24	06	31	03	97	44	82	24	89	88	48	66	54	10	41	27	09	11	61
94	64	97	27	25	62	23	94	40	54	56	32	97	78	90	58	86	41	75	19	42	90	85	36	68
15	85	82	52	08	52	96	26	92	88	93	11	03	23	52	78	23	57	85	43	53	90	42	22	22
09	81	37	66	56	99	08	59	19	48	29	69	21	64	95	12	08	15	24	45	59	25	22	76	96
43	83	99	02	76	12	16	45	52	66	35	70	93	09	52	75	40	34	35	62	65	42	27	20	59
31	98	09	80	62	75	26	64	57	26	46	41	47	90	97	99	46	10	51	42	73	28	98	89	91
81	35	42	62	84	37	02	59	78	16	17	96	05	71	39	88	05	34	05	92	22	43	89	66	89
97	95	56	39	75	65	47	61	86	33	14	88	55	33	69	70	87	79	94	46	17	61	72	27	01
37	63	35	93	23	17	30	14	51	51	17	28	21	74	67	12	11	57	19	27	38	70	73	82	92
39	22	96	00	48	52	49	62	09	40	08	30	27	54	70	96	06	52	12	80	36	12	38	68	05
61	29	84	34	51	60	19	77	82	16	64	45	02	27	04	65	55	90	95	04	20	39	29	96	28
38	84	18	10	29	19	09	66	06	78	37	09	60	50	21	52	72	01	52	70	29	65	05	37	16
64	29	48	04	08	55	72	25	25	77	54	26	27	24	39	66	67	06	40	00	99	35	70	69	58
64	02	32	99	63	62	42	89	32	20	81	14	08	40	45	22	15	37	49	38	96	51	19	08	27
13	83	39	51	30	31	49	94	83	66	02	50	95	18	98	58	84	90	58	81	00	40	91	12	46
83	30	90	09	35	41	12	87	93	66	85	96	20	65	34	23	13	05	41	01	91	48	95	59	45
46	63	53	97	63	18	86	37	56	20	35	62	66	11	37	30	91	89	97	51	64	78	06	95	65
54	43	40	02	41	55	70	52	96	87	02	82	61	21	88	60	65	98	42	09	03	61	20	83	01
27	18	65	62	01	97	45	79	51	37	74	47	20	11	48	97	93	73	86	50	46	61	95	01	24
45	42	16	13	20	34	51	08	71	52	39	17	71	39	84	97	27	72	49	42	81	62	32	87	22
35	92	97	02	34	93	32	95	81	13	92	05	40	70	95	71	66	61	24	08	77	32	73	66	79
60	55	35	57	24	52	95	84	90	64	38	39	72	70	17	98	42	85	96	67	41	11	83	17	78
43	17	21	09	60	58	86	12	31	11	66	61	43	96	00	93	97	00	15	20	37	96	73	56	63
07	85	74	58	28	38	74	68	32	61	87	14	71	83	47	90	11	96	70	08	67	04	34	46	08
33	00	29	08	87	42	59	40	24	97	44	99	13	56	87	95	02	47	97	89	23	51	45	37	83
97	14	00	42	23	72	03	19	02	41	11	23	36	98	32	19	91	42	03	58	62	23	74	45	06
68	58	32	80	82	40	49	71	83	37	93	49	99	60	72	88	14	26	88	95	48	69	35	40	63
39	87	38	16	06	82	92	62	32	75	67	64	50	49	39	29	55	53	92	97	04	48	60	53	90
37	73	01	84	87	42	88	30	93	75	01	18	34	73	30	28	44	28	18	01	00	38	26	38	57

EXHIBIT 2-7   Table of Random Sampling Numbers[4]

With a sufficiently large sample for any valid measurement of a given property, the data obtained will plot into a normal distribution curve such as shown in Exhibit 2–8. Here the number of observations or occurrences are plotted against appropriate units on the vertical scale, with the measured or calculated value of the variable being tested for distribution on the horizontal scale. The sigma limits extend equally on both sides of the average value $(\overline{X}')$.

With smaller samples, as usually found in practice, statistical analysis provides the only means by which the distribution of a complete lot may be predicted without actually measuring each item in the lot. If the true distribution of a lot is known, the average value for all the measurements in the high point of the curve and the dispersion of the curve obtained from a plot of the data will give clues as to the validity of the data itself and will also suggest what corrections might be needed to make the data valid. The normal distribution curve with attendant statistical measure is therefore a practical tool if correctly applied to real problems.

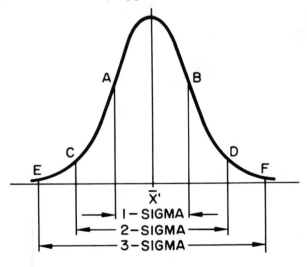

EXHIBIT 2-8   Normal Distribution Curve

Statistical Background of Work Sampling

Concepts and Techniques Applied in Work Sampling
  Measures Used in Work Sampling
  In Work Sampling we assume the existence of a normal probability curve. This being the case, there are six measures that we need to know and use. They will be dealt with separately.
  a. *Arithmetic Mean ($\overline{X}'$)*
    The calculated average of all the occurrences. Already dealt with.
  b. *Standard Deviation (σ)*
    A measure of the dispersion or variability of data. Already dealt with.

c. *Proportion of the Population (p)*

In order to decide upon the size of the sample, it is necessary to estimate the proportion of the most important and/or the smallest event being measured to the whole population. This is expressed as a decimal. Check can be made during the sampling to determine if the proportion is as estimated, or if the sample size must be altered.

d. *Range (R)*

The characteristics of an unbiased random sample differ very little from the characteristics of the population, but they *do* differ, since the sample is a *measure* of the whole, and not *actually* the whole.

If a percentage (or a proportion) has been computed from a sample drawn at random from a larger population, it is subject to sampling variations. The measure of these variations is called Range, and measures the range of p. It is expressed as a ratio of p, and, when applied to the value of p, indicates the upper and lower limits of p. It is determined mathematically as follows:

$$R = \sqrt{\frac{\sigma^2 (1 - p)}{Np}}$$

e. *Size of the Sample (N) Expressed as a Number*

The formula for determining the correct sample size will be given in (g) below.

f. *Level of Confidence*

The degree of reasonable expectation which one has concerning a certain conclusion, rather than an exact mathematical probability.

It is determined mathematically as follows:

$$\sigma = \sqrt{\frac{NR^2 p}{1 - p}}$$

Substituting values to determine σ:

$$N = 400 \qquad R = .10 \qquad p = .20$$

Then:

$$\sigma = \sqrt{\frac{400 \ (.10)^2 \ (.20)}{1.00 - .20}}$$
$$= \sqrt{\frac{400 \ (.01) \ (.20)}{.80}}$$
$$= \sqrt{\frac{.80}{.80}}$$
$$= \sqrt{1}$$
$$= 1$$

In this example, since the value of σ equals one, we can say that we are 68 per cent confident that the true value of p is 20 per cent plus or minus 2 per cent of the total population. If the value of σ had been 2, then we could say that we are 95 per cent confident that the true value of p is 20 per cent plus or minus 2 per cent of the population. (See Exhibit 2–9).

	.00	.01	.02	.03	.04	.05	.06	.07	.08	.09
0.0	.0000	.0040	.0080	.0120	.0159	.0199	.0239	.0279	.0319	.0359
0.1	.0398	.0438	.0478	.0517	.0557	.0596	.0636	.0675	.0714	.0753
0.2	.0793	.0832	.0871	.0910	.0948	.0987	.1026	.1064	.1103	.1141
0.3	.1179	.1217	.1255	.1293	.1331	.1368	.1406	.1443	.1480	.1517
0.4	.1554	.1591	.1628	.1664	.1700	.1736	.1772	.1808	.1844	.1879
0.5	.1915	.1950	.1985	.2019	.2054	.2088	.2123	.2157	.2190	.2224
0.6	.2257	.2291	.2324	.2357	.2389	.2422	.2454	.2486	.2518	.2549
0.7	.2580	.2612	.2642	.2673	.2704	.2734	.2764	.2794	.2823	.2852
0.8	.2881	.2910	.2939	.2967	.2995	.3023	.3051	.3078	.3106	.3133
0.9	.3159	.3186	.3212	.3238	.3264	.3289	.3315	.3340	.3365	.3389
1.0	.3413	.3438	.3461	.3485	.3508	.3531	.3554	.3577	.3599	.3621
1.1	.3643	.3665	.3686	.3718	.3729	.3749	.3770	.3790	.3810	.3830
1.2	.3849	.3869	.3888	.3907	.3925	.3944	.3962	.3980	.3997	.4015
1.3	.4032	.4049	.4066	.4083	.4099	.4115	.4131	.4147	.4162	.4177
1.4	.4192	.4207	.4222	.4236	.4251	.4265	.4279	.4292	.4306	.4319
1.5	.4332	.4345	.4357	.4370	.4382	.4394	.4406	.4418	.4430	.4441
1.6	.4452	.4463	.4474	.4485	.4495	.4505	.4515	.4525	.4535	.4545
1.7	.4554	.4564	.4573	.4582	.4591	.4599	.4608	.4616	.4625	.4633
1.8	.4641	.4649	.4656	.4664	.4671	.4678	.4686	.4693	.4699	.4706
1.9	.4713	.4719	.4726	.4732	.4738	.4744	.4750	.4758	.4762	.4767
2.0	.4773	.4778	.4783	.4788	.4793	.4798	.4803	.4808	.4812	.4817
2.1	.4821	.4826	.4830	.4834	.4838	.4842	.4846	.4850	.4854	.4857
2.2	.4861	.4865	.4868	.4871	.4875	.4878	.4881	.4884	.4887	.4890
2.3	.4893	.4896	.4898	.4901	.4904	.4906	.4909	.4911	.4913	.4916
2.4	.4918	.4920	.4922	.4925	.4927	.4929	.4931	.4932	.4934	.4936
2.5	.4938	.4940	.4941	.4943	.4945	.4946	.4948	.4949	.4951	.4952
2.6	.4953	.4955	.4956	.4957	.4959	.4960	.4961	.4962	.4963	.4964
2.7	.4965	.4966	.4967	.4968	.4969	.4970	.4971	.4972	.4973	.4974
2.8	.4974	.4975	.4976	.4977	.4977	.4978	.4979	.4980	.4980	.4981
2.9	.4981	.4982	.4983	.4984	.4984	.4984	.4985	.4985	.4986	.4986
3.0	.4987	.4987	.4987	.4988	.4988	.4988	.4989	.4989	.4989	.4990

EXHIBIT 2-9   Areas Under the Normal Probability Curve[6]

[6] Adopted from "Direct Table of the Normal Integral" from *First Course in Probability and Statistics,* J. Neyman. Copyright © 1959 by Holt, Rinehart & Winston, Inc. Adopted and reproduced by permission of Holt, Rinehart & Winston, Inc.

At any given time during the course of a Work Sampling study the level of confidence in the data already available can be determined by substituting values in the above formula, and looking at Exhibit 2–9.

For example, σ equals 1.56. On Exhibit 2–9, follow the left hand column down to 1.5. Follow that line horizontally to the column labeled .06. Multiply the figure there by 2 and divide by 100. The resulting figure will be the level of confidence expressed as a percentage (88 per cent).

g. *Determining the Sample Size*

In order to determine the size of the sample, the following things must be done first:

(1) Estimate p.
(2) Decide upon the range of p (R).
(3) Decide upon the level of confidence desired (σ).

This is expressed mathematically as follows:

$$N = \frac{\sigma^2 (1 - p)}{R^2 p}$$

For example, suppose:
Level of Confidence is 80 per cent. From Exhibit 2–9

$$\sigma = 1.28 \quad p = .20 \quad \text{and} \quad R = .10, \quad \text{then:}$$

$$N = \frac{(1.28)^2 (1.00 - .20)}{(.10)^2 (.20)}$$

$$= \frac{(1.64) (.80)}{(.01) (.02)}$$

$$= \frac{1.312}{.002}$$

$$= 656$$

Example of Use (Interference Delay Summary)[7]

(a) Formula for Standard Deviation: 2–Sigma (95% limits) to determine per cent error—percentage of element tolerance.

$$2\sigma = \sqrt{\frac{p (1 - p)}{N}}$$

---

[7] Delmar W. Karger and Franklin H. Bayha, *Engineered Work Measurement*, Second Edition, Copyright © 1966, The Industrial Press, New York, N. Y., p. 565.

Example

Large Drill Press Section Holbore Corporation	G. X. Smith—Analyst Industrial Engineering
Purpose of Study:	To guide the planning of a new layout for the drill press area, so as to minimize operator interference.
Problem Studied:	Determine per cent of operator time lost due to traffic in aisles with the present layout.
Plan of Study:	Ten random observations per shift to be completed in twelve days. There are three 8–hour shifts of 20 operators each.
Results of Study:	Per following tabulation and calculation.

(Solution)

ACTIVITY	CODE	OBSERVATIONS TAKEN	PER CENT OF TOTAL
Productive work	*A*	5,726	81.3
Delays not specified in study	*B*	998	14.2
Interference delays due to:			
Pedestrian traffic in aisles	*C*	163	2.3
Manual material handlers	*D*	17	0.2
Truck material handlers	*E*	4	0.1
Inspection delay	*F*	130	1.9
Totals		7,038	100.0%

Total Interference Delays (C through F)   314   4.5%

$$\text{Error} = \pm 2\sqrt{\frac{p(1-p)}{N}} = \pm 2\sqrt{\frac{0.045(1-0.045)}{7,038}}$$

$$\textit{Error} = \pm 0.0049 \quad \text{or} \quad \pm 0.49\%$$

Answer: Interference delay is 4.5% ± 0.5%. This means that the interval from 4.0 to 5.0% has aflout a 95% chance of containing the true value. It also shows the percentage of element tolerance to be:

$$\pm \frac{0.49\%}{4.50\%} = 0.109 = 10.9\%$$

The Binomial Theorem—Basic Analysis Tool in Work Sampling[8]

Introduction:

In order to use Work Sampling properly and to draw inferences from samples correctly, it is necessary to have some knowledge of "probability theory." We will attempt here to distill much of the theory which applies to Work Sampling.

In the simplest type of estimation or prediction situation in which probability theory is used, such as in the prediction of the outcome of a flip of a coin, the various ways in which the event can happen are known in advance. The tossing of a coin may result, for instance, in one of three things happening:

(1) The coin may come to rest head up, tail down.

---

[8] Robert E. Heiland and Wallace J. Richardson, *Work Sampling* (New York: McGraw-Hill Co. Copyright © 1957), pp. 81–82, 84–85.

(2) The coin may come to rest tail up, head down.

(3) The coin may come to rest on edge, neither face up.

The fact that no other result may happen leads to the statement that we have defined all the ways in which the event may take place, and we define the summation of the three ways as certainty, i.e., we are certain that one of these three events will happen.

This total probability is always defined as unity, or 1.000. The probability that either event 1, 2, or 3 will happen is expressed as a proportion of unity, or something less than 1.000 . . ., and is usually expressed as a decimal amount, e.g., 0.50, or as a fraction, 1/4 or 1/2, etc. The summation of these individual probabilities is unity, or 1.000.

In our coin flipping, for example, we may have probabilities of each of the three events expressed as, say, 0.4999, 0.4999, and 0.0002, totaling 1.000.

In a bridge game, the probability of being dealt any one card may be considered as one in fifty-two, assuming a thorough shuffle and straight deal. Thus the probability of drawing an ace of spades is 1/52 or 0.01923077, etc. The probability is the same for each other card.

The Binomial Theorem:

Mathematicians have supplied us with a very simple way to express these long-run probabilities where an event may happen in one or two ways. In order to explore the binomial theorem we must adopt some shorthand notations first:

> Let p = the probability of result A
> Let q = the probability of result B
> Let n = the number of sample items drawn.

Assume: A and B are the only ways in which the event can occur, and are mutually exclusive; i.e., if A happens, B cannot, and vice versa.

Assume: Random sampling conditions prevail, with no change in p and q throughout the sampling process.

Therefore:

$$p + q = 1.000 \text{ and } (p + q)^1 = 1.000$$

The binomial theorem is:

$$(p + q)^n = 1.000$$

This equation can be expanded by the use of simple algebra.

> If $n = 2$, then
> $$(p + q)^2 = p^2 + 2pq + q^2 = 1.000 \tag{1}$$
> If $n = 3$, then
> $$(p + q)^3 = p^3 + 3p^2q + 3pq^2 + q^3 = .1000 \tag{3}$$
> If $n = 4$, then
> $$(p + q)^4 = p^4 + 4p^3q + 6p^2q^2 + 4pq^3 + q^4 = 1.000 \tag{3}$$

Now let us assume this is the coin-tossing experiment developed previously. Now $p = 0.50$ and $q = 0.50$, being the respective probabilities of heads and tails. "Plugging in" these values of p and q in formulas (1) to (3), above, yields:

$$0.5^2 + 2(0.5)(0.5) + 0.5^2 = 1.000 \qquad (1)$$
$$0.25 \;+\; 0.50 \;+\; 0.25 = 1.000$$

$$0.5^3 + 3(0.5)^2(0.5) + 3(0.5)(0.5)^2 + 0.5^3 = 1.000 \qquad (2)$$
$$0.125 \;+\; 0.375 \;+\; 0.375 \;+\; 0.125 = 1.000$$

$$0.5^4 + 4(0.5)^3(0.5) + 6(0.5)^2(0.5)^2 + 4(0.5)(0.5)^3 + (0.5)^4 = 1.000$$
$$0.0625 \;+\; 0.2500 \;+\; 0.3750 \;+\; 0.2500 \;+\; 0.0625 = 1.000 \qquad (3)$$

In formula (1), the probability of two consecutive heads, or two consecutive tails, is 0.25, or one in four, and of one head and one tail is 0.50, or one in two. By the same interpretation in formula (2), the probability of three consecutive heads, or tails, is 0.125, or one in eight; of two heads and one tail is three in eight; etc. In formula (3), the probability of four consecutive heads is 0.0625, or one in sixteen, etc.

Work Sampling—Lube Area Filling and Shipping Study

I *Activity Analysis*

	Minutes	% of Total
Direct Work	8945	62.1
Necessary Delays	631	4.4
Miscellaneous Work	1310	9.1
Leaderman	1135	7.9
Personal Time	831	5.8
Coffee Break, Eating on Job, Etc.	548	3.8
Personal Work	237	1.6
Idle Time	531	3.7
No Work Available	144	1.0
Missed	88	.6
TOTAL	14400	100.0

II *Development of Necessary Delay Factor*

$$\frac{\text{(631 Total Necessary Delay Minutes)}}{\text{8945 (Total Direct Work Minutes)}} = .071$$

III *Development of Miscellaneous Work Allowance*

$$\frac{\text{1402 (Miscellaneous Work Minutes + 7\% Personal)}}{\text{5 (Number of Days in Study)}} = \text{280 Min./8 Hr. Day}$$

IV *Development of Leaderman Allowance*

$$\frac{\text{1214 (Indirect Work Minutes + 7\% Personal)}}{\text{5 (Number of Days in Study)}} = \text{.243 Min./8 Hr. Day}$$

SUMMARY SHEET

Code	Activity	*Numbers of Observations* 6/20	6/21	6/22	6/23	6/24	*Total*	*% of Total*	*Minutes/ Activity*
	DIRECT WORK:								
950	Fill 30 or 55 Gallon Drum	103	96	128	118	97	542	20.83	2999.52
951	Decal Drum	44	21	37	29	25	156	6.00	864.00
952	Ship Full Drum to Car—#2	30	27	65	44	12	178	6.84	984.96
953	Ship Full Drum to Car—G.P.	11	20	18	23		72	2.77	398.88
954	Ship Full Drum to Truck	21	8	18	15	8	70	2.69	387.36
955	Ship Full Drum to Trailer	9	31			12	52	2.00	288.00
956	Place Full Drum in Storage	9	18	10	5	32	74	2.84	408.96
957	Tank Gauging	38	44	41	44	50	217	8.34	1200.96
958	Ship 120 # Drum to Truck								

SUMMARY SHEET (Cont.)

Code	Activity	Number of Observations 6/20	6/21	6/22	6/23	6/24	Total	% of Total	Minutes/ Activity
959	Ship Small Packages to Car—Con.								
960	Ship Small Packages to Car—H.J.								
961	Ship Small Packages to Truck	7	4		7	11	29	1.11	159.84
962	Pump Off Leaker	8	12				20	.77	110.88
970	Receive & Store Empty 5 Gal. Bkts.	1					1	.04	5.76
971	Transfer & Fill 5 Gal. Bkts.	11			1		12	.46	66.24
972	Fill Other Packages		8		17	31	56	2.15	309.60
980	Fill & Ship 30 or 55 Gal. Drum	18	27	6	1		52	2.00	288.00
981	Fill & Ship—Small Lots			1			1	.04	5.76
990	Other	16	11	2	16	10	55	2.11	303.84
	TOTAL DIRECT WORK	326	327	326	320	288	1587	60.99	8782.56
	NECESSARY DELAYS:								
660	Give or Receive Instructions	13	10	5	7	10	45	1.73	249.12
661	Search for Material or Equipment		3	1	2	4	10	.38	54.72
662	Wait for Elevator								
663	Clear Work Area								
664	Phone Calls	4	2	1	1	2	10	.38	54.72
665	Wait—Bblg. House Gang	5	6	8	5	3	27	1.04	149.76
666	Other	5	3	3	5	5	21	.81	116.64
667	Flush Lines of Canning Hse.				1		1	.04	5.76
	TOTAL NECESSARY DELAYS	27	24	18	21	24	114	4.38	630.72
	MISCELLANEOUS WORK:								
550	Rearrange, Stock								
551	Ship Miscellaneous Packages					2	2	.08	11.52
552	Repaint Drums								
553	Housekeeping	17	14	31	20	36	118	4.53	652.32
554	Clean Dirty Car								
555	Take Sample	2	7	12	5	7	33	1.27	182.88
556	Other		4	2			6	.23	33.12
557	Clerical Work—Checking Orders	12	13	19	11	3	58	2.23	321.12
558	Clean Strainers	6			5	3	14	.54	77.76
559	Prepare Drum from Stock			4		1	5	.19	27.36
	TOTAL MISCELLANEOUS WORK	37	38	68	41	52	236	9.07	1306.08
	OTHER								
777	Personal Time	34	38	27	22	29	150	5.77	830.88
000	Coffee Break, Eating on Job, Etc.	14	2	34	17	32	99	3.81	548.64
011	Smoking								
022	Idle Time	16	28	25	32	25	126	4.84	696.96
033	Personal Work; Telephone Call, Etc.	7		9	13	14	43	1.65	237.60
044	No Work Available	5		1	10	10	26	1.00	144.00
055	Missed	14	2				16	.61	87.84
066	Other								
100	Union Business, Safety, Etc.								
	TOTAL OTHER	90	70	96	94	110	460	17.68	2545.92
	LEADERMAN ACTIVITIES:								
330	Supervision	19	11	7	12	10	59	2.27	326.88
331	Clerical Work	13	23	17	8	19	80	3.08	443.52
332	Receive Instructions	2	4		7	2	15	.58	83.52
333	Inspect Packages					4	4	.15	21.60
334	Phone Calls		2	1		5	8	.31	44.64
335	Planning Work	5	3	4	5	7	24	.92	132.48
336	Check Inventory								
337	Prepare Sample List		5		2	3	10	.38	54.72
338	Other		2		3		5	.19	27.36
	TOTAL LEADERMAN	39	50	29	37	50	205	7.88	1134.72
	GRAND TOTAL	519	509	537	513	524	2602	100.00	
	STUDY MAN MINUTES	2880	2880	2880	2880	2880	TOTAL MAN MIN. 14400		

In conducting this study all observations taken were performance rated to compensate for fluctuating performance. If standards were developed without performance rating, they would be representative only of the pace used during the study period.

After compensating for fluctuating performance level, the "Idle Time" observations were adjusted to balance out the total study man minutes observed.

Code	Activity	Number of Obser.	Minutes/ Activity	Average Rating	Rated Minutes/ Activity	% of Total
	DIRECT WORK:					
950	Fill 30 or 55 Gallon Drum	542	2999.52	102.30	3068.509	21.31
951	Decal Drum	156	864.00	104.00	898.560	6.24
952	Ship Full Drum to Car—#2 Platform	178	984.96	103.23	1016.774	7.06
953	Ship Full Drum to Car—Grease Plant	72	398.88	101.81	406.100	2.82
954	Ship Full Drum to Truck	70	387.36	100.50	389.297	2.70
955	Ship Full Drum to Trailer	52	288.00	102.60	295.488	2.05
956	Place Full Drum in Storage	74	408.96	99.19	405.647	2.82
957	Tank Gauging	217	1200.96	100.02	1201.200	8.34
961	Ship Small Packages to Truck	29	159.84	99.48	159.009	1.10
962	Pump Off Leaker	20	110.88	100.25	111.157	.77
970	Receive & Store Empty 5 Gal. Bucket	1	5.76	100.00	5.760	.04
971	Transfer & Fill 5 Gal. Bucket	12	66.24	100.83	66.790	.46
972	Fill Other Packages	56	309.60	101.34	313.749	2.18
980	Fill & Ship 30 or 55 Gal. Drum	52	288.00	102.31	294.653	2.05
981	Fill & Ship—Small Lots	1	5.76	100.00	5.760	.04
990	Other	55	303.84	100.73	306.058	2.13
	TOTAL DIRECT WORK	1587	8782.56	101.84	8944.511	62.11
	TOTAL NECESSARY DELAYS	114	630.72	100.00	630.720	4.38
	TOTAL MISCELLANEOUS WORK	236	1306.08	100.28	1309.737	9.10
	OTHER:					
	All activities excluding "022"	334	1848.96	100.00	1848.960	12.84
022	Idle Time	126	696.96	100.00	530.785	3.69
	TOTAL OTHER	460	2545.92	100.00	2379.745	16.53
	TOTAL LEADERMAN	205	1134.72	100.05	1135.287	7.88
	GRAND TOTAL	2602	14400.00		14400.00	100.00

Reliability of Study Data

It was decided an accuracy to within 10% covering 91% (1.7 Sigma) of all occurrences which constitute 10% or more of the total activities, would be desirable.

Using the following formula, 2601 observations would be required.

$$N = \frac{\sigma^2(1-P)}{R^2 P}$$

$N = 2601 =$ Number of observations required

$\sigma^2 = 2.89 =$ Number of sigma squared

$P = 10\% =$ Per cent of activity related to whole

$R = 10\% =$ Reliability required

For example, if Miscellaneous Work Time was 10% of the observations recorded, then the percentage of the time spent for Miscellaneous Work Activities would be 10% ± 10% or between 9.0% and 11.0%.

The reliability of the allowances developed from this study is as follows:

$$R = \sqrt{\frac{\sigma^2 (1 - P)}{NP}}$$

R = Reliability
σ^2 = No. of sigma squared
P = Per cent of activity related to whole
N = Number of observations in study

	Allowance	Reliability Range	
		Low	High
Necessary Delay Factor	.071	.060	.082
Miscellaneous Work	280 Min.	251	309
Leaderman	243 Min.	215	271

3

Work Measurement—Predetermined Time Standards: Methods-Time Measurement and Work Factor System

---

THE MTM ELEMENTAL TIME SYSTEM[1]

What Is MTM?

It is a system of predetermined time values which analyzes any manual operation into the basic motions required to perform the operation and assigns a time value to each motion which is determined by the nature of the motion and the conditions under which it is performed.

Why MTM?

The Methods-Time Measurement (MTM) procedure was developed because Industrial Engineers felt the need for a procedure which would lead more surely to effective methods of performing all types of work than existing time and motion study procedures. It provides a common language and terminology to use among Engineers, Production Management and Operators when discussing methods for jobs.

What Will MTM Do for You?

It will help in the following methods and time activities:
(1) Developing effective methods in advance of production
(2) Improving existing methods
(3) Developing effective tool design
(4) Training you to become more methods-conscious
(5) Establishing time standards

---

[1] Used with permission of the Maytag Company, Newton, Iowa 50208.

(6) Training operators to use the best motion pattern
(7) Establishing labor costs
(8) Developing standard data
(9) Other activities, including selecting effective equipment and guiding product design

How to Use the Data Card[2]

In the introduction it was stated that MTM assigns a time value which is determined by the nature of the motion and the conditions under which it is performed. The MTM Data Card contains time values required for each basic motion providing the variable conditions are known, such as: distance, size, weight, class of fit, symmetry, ease of handling, etc.

All MTM time values on the Data Card are shown in Time Measurement units (TMU's):

$$1 \text{ TMU} = .00001 \text{ hour}$$
$$= .0006 \text{ minute}$$
$$= .036 \text{ second}$$

Exhibit 3–1 shows two examples of how to read the data card:

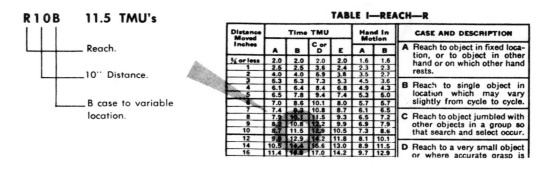

EXHIBIT 3-1

what are BASIC MOTIONS?

We said in our introduction that MTM analyzes any manual operation into the basic motions required to perform the operation. The following is a list of these basic motions with the symbol used for that motion and a brief statement on how to recognize the motion.

DEFINITIONS OF BASIC MOTIONS

*TABLE NO.	BASIC MOTION	SYMBOL	DEFINITION
I	REACH	R	is a motion to move the hand or fingers to an object or location.
II	MOVE	M	is a hand or finger motion to transport an object to a destination.
III	TURN	T	is a motion to rotate (either empty or loaded) the hand, wrist and forearm about the long axis of the forearm.
IV	APPLY PRESSURE	AP	is exertion of additional force to overcome the effects of resistance too great to overcome by a normal move or turn.
V	GRASP	G	is a motion with the hand or fingers to gain control of an object.
VI	POSITION	P	is a motion to align two objects and engage one with another.
VII	RELEASE	RL	is a motion to let go of an object.
VIII	DISENGAGE	D	is a motion to overcome resistance to separate one object from another and is recognized by recoil.
IX	BODY, LEG & FOOT MOTIONS		

* "**Table Number**" refers to table on the MTM data card.

methods improvement examples [3]

MTM Analysis will not improve the job but it will help point out to you where methods improvements can be made. Examine the MTM analysis to get the greatest value from the m e t h o d s improvement possibilities. Here are some possibilities which may lead to methods improvements:

1. **Reduce long reaches and moves.**

2. **Eliminate body motions.**

3. **Have both hands working simultaneously.**

4. **Place as much work as possible within machine cycle.**

5. **Place material so that it can be picked up with simple grasps.**

6. **Eliminate positions or reduce class of fit.**

7. **Eliminate unnecessary inspections.**

8. **Improve design of part to make operation easier to perform.**

9. **Have automatic delivery and/or asiding of parts to eliminate manual motions.**

 **Examples of these possibilities are shown on the following pages.**

---

[3] Used with permission of the Maytag company, Neuton., Iowa 50208

work place layout change
27% improvement

EXAMPLE OF POSSIBILITY NO. 1.

Minor changes in layout can reduce excessive distances of reaches and moves.

OLD METHOD

1. Aside part and obtain next

MOVE part to aside chute	M8B	10.6
RELEASE part	RL1	2.0
REACH to part	R12C	14.2
GRASP part	G4B	9.1
MOVE part to fixture.	M10C	13.5
		49.4

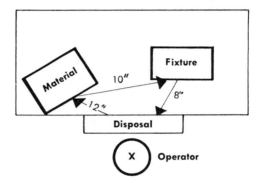

By changing the layout of the work place to have the asiding done while reaching to the next part, reach and move distances can be reduced:

THE IMPROVED METHOD
LOOKS LIKE THIS

1. Aside part and obtain next

MOVE part to aside chute	M3B m	3.6
RELEASE part	RL1	2.0
REACH to part	mR7C	8.0
GRASP part	G4B	9.1
MOVE part to fixture	M10C	13.5
		36.2

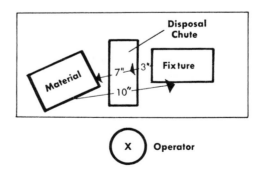

```
    49.4  (OLD)
—   36.2  (NEW)
    13.2  TMU'S Saved or Improved 27%
```

EXHIBIT 3-2

better fixture results in
19% methods improvement

EXAMPLE OF POSSIBILITY NO. 6

Improvement of design of fixtures can make positioning and removing parts easier.

OLD METHOD

1. Move parts to fixture and position.

	LH		RH	
MOVE nozzle near fixture.	(M10B)	15.2	M12C	MOVE bracket to fixture.
REGRASP nozzle during move.	G2		G2	REGRASP bracket during move.
		19.7	⌐ P2SSE	POSITION bracket against spring.
		3.4	M1C	MOVE bracket down in fixture.
MOVE nozzle to fixture.	M2C	5.2	(RL1)	RELEASE bracket.
POSITION nozzle in fixture.	P1SSE	9.1		
RELEASE nozzle.	RL1	2.0		
		54.6		

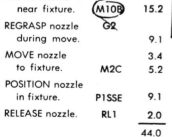

By reducing the size of the fixture spring, thus reducing the tension against the bracket, the <u>P2SSE</u> is reduced to a minimum <u>P1SSE</u>.

IMPROVED METHOD

1. Move parts to fixture and position.

	LH		RH	
MOVE nozzle near fixture.	(M10B)	15.2	M12C	MOVE bracket to fixture.
REGRASP nozzle during move.	G2		G2	REGRASP bracket during move.
		9.1	⌐ P1SSE	POSITION bracket against spring.
MOVE nozzle to fixture.		3.4	M1C	MOVE bracket down in fixture.
	M2C	5.2	(RL1)	RELEASE bracket.
POSITION nozzle in fixture.	P1SSE	9.1		
RELEASE nozzle.	RL1	2.0		
		44.0		

> **54.6** (OLD)
> — **44.0** (NEW)
> **10.6** TMU'S Saved or Improved 19%

EXHIBIT 3-3

part design change
work reduced 32%

EXAMPLE OF POSSIBILITIES NOS. 7 & 8.

Improvement in the design of the part can make the operation easier to perform.

OLD METHOD

1. Move part to fixture and position.

MOVE nozzle toward fixture.	M10B	12.2
REGRASP nozzle during move.	G2	
Select side w/Eye FOCUS	EF	7.3
Regrasp nozzle.	½G2	2.8
MOVE nozzle to fixture.	M2C	5.2
POSITION in fixture.	P1SSE	9.1
RELEASE Part.	RL1	2.0
		38.6

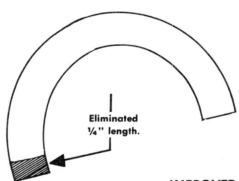

Eliminated ¼" length.

By shortening (1) side of a bent piece of aluminum tubing ¼," the sides are the same length. This eliminates the selection of a slightly longer side before the part was placed in the fixture.

IMPROVED METHOD

1. Move part to fixture and position.

MOVE nozzle to fixture.	M12C	15.2
REGRASP nozzle during move.	G2	
POSITION in fixture.	P1SSE	9.1
RELEASE part.	RL1	2.0
		26.3

38.6	(OLD)
— 26.3	(NEW)
12.3	**TMU'S Saved or Improved 32%**

EXHIBIT 3-4

"aside part" eliminated
100% improvement

EXAMPLE OF POSSIBILITY NO. 9.

Have automatic delivery and/or asiding of parts—eliminate manual motion.

OLD METHOD

Aside Part.

TURN BODY to left.	TBC1	18.6
Toss part to container.	mM8B	7.2
Release part.	RL1	
		25.8

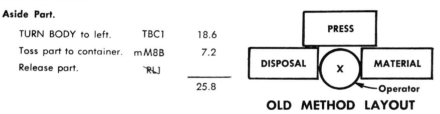

OLD METHOD LAYOUT

**By using automatic air jet to blow
part out of die,
ASIDE Eliminated.**

NEW METHOD

Aside Part.

Part now asided
automatically.

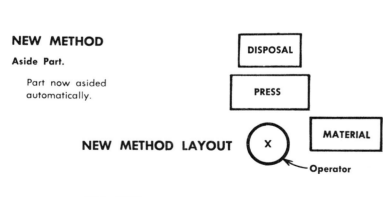

NEW METHOD LAYOUT

25.8 (OLD)
— 0.0 (NEW)
25.8 TMU'S Saved or Improved 100%

EXHIBIT 3-5

METHODS-TIME MEASUREMENT APPLICATION DATA IN TMU[4]
(Exhibits 3–6—3–16)

—REACH—R

Distance Moved Inches	Time TMU				Hand In Motion		CASE AND DESCRIPTION
	A	B	C or D	E	A	B	
¾ or less	2.0	2.0	2.0	2.0	1.6	1.6	**A** Reach to object in fixed location, or to object in other hand or on which other hand rests.
1	2.5	2.5	3.6	2.4	2.3	2.3	
2	4.0	4.0	5.9	3.8	3.5	2.7	
3	5.3	5.3	7.3	5.3	4.5	3.6	
4	6.1	6.4	8.4	6.8	4.9	4.3	**B** Reach to single object in location which may vary slightly from cycle to cycle.
5	6.5	7.8	9.4	7.4	5.3	5.0	
6	7.0	8.6	10.1	8.0	5.7	5.7	
7	7.4	9.3	10.8	8.7	6.1	6.5	
8	7.9	10.1	11.5	9.3	6.5	7.2	**C** Reach to object jumbled with other objects in a group so that search and select occur.
9	8.3	10.8	12.2	9.9	6.9	7.9	
10	8.7	11.5	12.9	10.5	7.3	8.6	
12	9.6	12.9	14.2	11.8	8.1	10.1	
14	10.5	14.4	15.6	13.0	8.9	11.5	**D** Reach to a very small object or where accurate grasp is required.
16	11.4	15.8	17.0	14.2	9.7	12.9	
18	12.3	17.2	18.4	15.5	10.5	14.4	
20	13.1	18.6	19.8	16.7	11.3	15.8	
22	14.0	20.1	21.2	18.0	12.1	17.3	**E** Reach to indefinite location to get hand in position for body balance or next motion or out of way.
24	14.9	21.5	22.5	19.2	12.9	18.8	
26	15.8	22.9	23.9	20.4	13.7	20.2	
28	16.7	24.4	25.3	21.7	14.5	21.7	
30	17.5	25.8	26.7	22.9	15.3	23.2	

Examples

RIOA	Reach, 10 inches, Case A	. . . .	8.7 TMU
RfB	Reach, 3/4 inch or less, Case B	. . . .	2.0 "
mR30C	Reach, 30 inches, Case C, in motion at beginning	. . . . . .	15.3 "
R8Bm	Reach, 8 inches, Case B, in motion at end.	. . . .	7.2 "

EXHIBIT 3-6

TABLE II—MOVE—M

Distance Moved Inches	Time TMU				Wt. Allowance			CASE AND DESCRIPTION
	A	B	C	Hand In Motion B	Wt. (lb.) Up to	Factor	Constant TMU	
¾ or less	2.0	2.0	2.0	1.7	2.5	1.00	0	**A** Move object to other hand or against stop.
1	2.5	2.9	3.4	2.3				
2	3.6	4.6	5.2	2.9	7.5	1.06	2.2	
3	4.9	5.7	6.7	3.6				
4	6.1	6.9	8.0	4.3	12.5	1.11	3.9	
5	7.3	8.0	9.2	5.0				
6	8.1	8.9	10.3	5.7	17.5	1.17	5.6	
7	8.9	9.7	11.1	6.5				
8	9.7	10.6	11.8	7.2				**B** Move object to approximate or indefinite location.
9	10.5	11.5	12.7	7.9	22.5	1.22	7.4	
10	11.3	12.2	13.5	8.6				
12	12.9	13.4	15.2	10.0	27.5	1.28	9.1	
14	14.4	14.6	16.9	11.4				
16	16.0	15.8	18.7	12.8	32.5	1.33	10.8	
18	17.6	17.0	20.4	14.2				
20	19.2	18.2	22.1	15.6	37.5	1.39	12.5	
22	20.8	19.4	23.8	17.0				
24	22.4	20.6	25.5	18.4	42.5	1.44	14.3	**C** Move object to exact location.
26	24.0	21.8	27.3	19.8				
28	25.5	23.1	29.0	21.2				
30	27.1	24.3	30.7	22.7	47.5	1.50	16.0	

EXHIBIT 3-7

[4] Copyright © 1965 by the MTM Association for Standards and Research. No reprint permission without written consent of the MTM Association, 9–10 Saddle Road, Fair Lawn, New Jersey 07410.

TABLE III—TURN AND APPLY PRESSURE—T AND AP

Weight	Time TMU for Degrees Turned										
	30°	45°	60°	75°	90°	105°	120°	135°	150°	165°	180°
Small— 0 to 2 Pounds	2.8	3.5	4.1	4.8	5.4	6.1	6.8	7.4	8.1	8.7	9.4
Medium—2.1 to 10 Pounds	4.4	5.5	6.5	7.5	8.5	9.6	10.6	11.6	12.7	13.7	14.8
Large— 10.1 to 35 Pounds	8.4	10.5	12.3	14.4	16.2	18.3	20.4	22.2	24.3	26.1	28.2
APPLY PRESSURE CASE 1—16.2 TMU.	APPLY PRESSURE CASE 2—10.6 TMU										

EXAMPLES

M5A	Move, 5 inches, Case A, nominal weight	7.3 TMU
M20Bm	Move, 20 inches, Case B, in motion at end, nominal weight	18.2 "
M30C20	Move, 30 inches, Case C, 20 pound effective net weight	38.1 "
T30	Turn, 30°, hand empty	2.8 "
T90S	Turn, 90°, small weight turned	5.4 "
AP1	Apply Pressure, Case 1	16.2 "

TABLE IV—GRASP—G

Case	Time TMU	DESCRIPTION
1A	2.0	Pick Up Grasp—Small, medium or large object by itself, easily grasped.
1B	3.5	Very small object or object lying close against a flat surface.
1C1	7.3	Interference with grasp on bottom and one side of nearly cylindrical object. Diameter larger than ½".
1C2	8.7	Interference with grasp on bottom and one side of nearly cylindrical object. Diameter ¼" to ½".
1C3	10.8	Interference with grasp on bottom and one side of nearly cylindrical object. Diameter less than ¼".
2	5.6	Regrasp.
3	5.6	Transfer Grasp.
4A	7.3	Object jumbled with other objects so search and select occur. Larger than 1" x 1" x 1".
4B	9.1	Object jumbled with other objects so search and select occur. ¼" x ¼" x ⅛" to 1" x 1" x 1".
4C	12.9	Object jumbled with other objects so search and select occur. Smaller than ¼" x ¼" x ⅛".
5	0	Contact, sliding or hook grasp.

EXHIBIT 3-9

TABLE V—POSITION*—P

CLASS OF FIT		Symmetry	Easy To Handle	Difficult To Handle
1—Loose	No pressure required	S	5.6	11.2
		SS	9.1	14.7
		NS	10.4	16.0
2—Close	Light pressure required	S	16.2	21.8
		SS	19.7	25.3
		NS	21.0	26.6
3—Exact	Heavy pressure required.	S	43.0	48.6
		SS	46.5	52.1
		NS	47.8	53.4

EXHIBIT 3-10

*Distance moved to engage—1" or less.

EXAMPLES

G1A	Grasp, Case 1A	2.0 TMU
G2	Regrasp	5.6 "
G4A	Grasp, Case 4A	7.3 "
P1SE	Position, Class 1 fit, symmetrical fit, easy to handle	5.6 "

P2SSE   Position, Class 2 fit, semisymmetrical fit,
         easy to handle .... 19.7 TMU
P3NSD   Position, Class 3 fit, nonsymmetrical fit,
         difficult to handle .... 53.4 "

—RELEASE—RL

Case	Time TMU	DESCRIPTION
1	2.0	Normal release performed by opening fingers as independent motion.
2	0	Contact Release.

TABLE VII—DISENGAGE—D

CLASS OF FIT	Easy to Handle	Difficult to Handle
1—Loose—Very slight effort, blends with subsequent move.	4.0	5.7
2—Close — Normal effort, slight recoil.	7.5	11.8
3—Tight — Considerable effort, hand recoils markedly.	22.9	34.7

EXHIBIT 3-11

—EYE TRAVEL TIME AND EYE FOCUS—ET AND EF

Eye Travel Time $= 15.2 \times \dfrac{T}{D}$ TMU, with a maximum value of 20 TMU.

where T=the distance between points from and to which the eye travels.
D=the perpendicular distance from the eye to the line of travel T.

Eye Focus Time=7.3 TMU.

EXHIBIT 3-12

—BODY, LEG AND FOOT MOTIONS

DESCRIPTION	SYMBOL	DISTANCE	TIME TMU
Foot Motion—Hinged at Ankle.	FM	Up to 4″	8.5
With heavy pressure.	FMP		19.1
Leg or Foreleg Motion.	LM —	Up to 6″	7.1
		Each add'l. inch	1.2
Sidestep—Case 1—Complete when leading leg contacts floor.	SS-C1	Less than 12″	Use REACH or MOVE Time
		12″	17.0
		Each add'l. inch	.6
Case 2—Lagging leg must contact floor before next motion can be made.	SS-C2	12″	34.1
		Each add'l. inch	1.1
Bend, Stoop, or Kneel on One Knee.	B,S,KOK		29.0
Arise.	AB,AS,AKOK		31.9
Kneel on Floor—Both Knees.	KBK		69.4
Arise.	AKBK		76.7
Sit.	SIT		34.7
Stand from Sitting Position.	STD		43.4
Turn Body 45 to 90 degrees—			
Case 1—Complete when leading leg contacts floor.	TBC1		18.6
Case 2—Lagging leg must contact floor before next motion can be made.	TBC2		37.2
Walk.	W-FT.	Per Foot	5.3
Walk.	W-P	Per Pace	15.0

EXHIBIT 3-13

EXAMPLES

RL1   Release, Case 1 .... 2.0 TMU
D2E   Disengage, Class 2 fit, easy to handle .... 7.5 "
EF   Eye Focus .... 7.3 "
LM8   Leg Motion, 8 inches .... 9.5 "
TBC1   Turn Body, Case 1 .... 18.6 "

SIMULTANEOUS MOTIONS

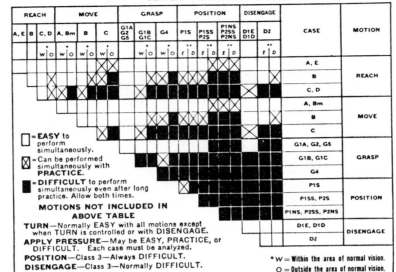

=EASY to perform simultaneously.

☒=Can be performed simultaneously with PRACTICE.

■=DIFFICULT to perform simultaneously even after long practice. Allow both times.

MOTIONS NOT INCLUDED IN ABOVE TABLE

TURN—Normally EASY with all motions except when TURN is controlled or with DISENGAGE.

APPLY PRESSURE—May be EASY, PRACTICE, or DIFFICULT. Each case must be analyzed.

POSITION—Class 3—Always DIFFICULT.

DISENGAGE—Class 3—Normally DIFFICULT.

RELEASE—Always EASY.

DISENGAGE—Any class may be DIFFICULT if care must be exercised to avoid injury or damage to object.

* W = Within the area of normal vision.
O = Outside the area of normal vision.
**E = EASY to Handle.
D = DIFFICULT to Handle.

EXHIBIT 3-14

SUPPLEMENTARY MTM DATA

(POSITION—P (SUPPLEMENTARY DATA)

Class of Fit and Clearance	Case of† Symmetry	Align Only	Depth of Insertion (per ¼")			
			0	2	4	6
21 .150"—.350"	S	3.0	3.4	6.6	7.7	8.8
	SS	3.0	10.3	13.5	14.6	15.7
	NS	4.8	15.5	18.7	19.8	20.9
22 .025"—.149"	S	7.2	7.2	11.9	13.0	14.2
	SS	8.0	14.9	19.6	20.7	21.9
	NS	9.5	20.2	24.9	26.0	27.2
23* .005"—.024"	S	9.5	9.5	16.3	18.7	21.0
	SS	10.4	17.3	24.1	26.5	28.8
	NS	12.2	22.9	29.7	32.1	34.4

*BINDING—Add observed number of Apply Pressure
DIFFICULT HANDLING—Add observed number of G2.

†Determine symmetry by geometric properties, except use S case when object is oriented **prior to** preceding Move.

EXHIBIT 3-15

EXAMPLES

P22SS4   Position, Class 22 fit, semisymmetrical fit, depth of insertion greater than ¾ inch to 1-¼ inches inclusive .... 21.3 TMU

P23SO   Position, Class 23 fit, symmetrical fit, depth of insertion O to ⅛ inch inclusive .... 13.6 "

APPLY PRESSURE—AP(SUPPLEMENTARY DATA)

Apply Force (AF) = 1.0 + (0.3 × lbs.). TMU for up to 10 lb.
= 4.0 TMU max. for 10 lb. and over

Dwell, Minimum (DM) = 4.2 TMU   Release Force (RLF) = 3.0 TMU

AP = AF + Dwell + RLF   APB = AP + G2

EXAMPLES

APO2   Apply pressure, force 2 pounds .... 8.8 TMU

APBO8   Apply pressure, force 8 pounds
Bind occurred .... 66 .... 16.2 "

EXHIBIT 3-16

TABLES (Exhibits 3–17—3–20)

EXHIBIT 3-17   **Time Conversion Table (Based on the defined value of 100,000 TMU per hour)**[5]

Having	Perform This Operation	To Get
Hours	Multiply by 60	Minutes
Hours	Multiply by 3,600	Seconds
Hours	Multiply by 100,000	TMU
Minutes	Divide by 60 (or × 0.0167)	Hours
Minutes	Multiply by 60	Seconds
Minutes	Multiply by 1,667	TMU
Seconds	Divide by 3,600 (or × .000278)	Hours
Seconds	Divide by 60 (or × .0167)	Minutes
Seconds	Multiply by 27.8	TMU
TMU	Multiply by .00001	Hours
TMU	Multiply by .0006	Minutes
TMU	Multiply by .036	Seconds
Pieces per hr.	Take a reciprocal	Hours per piece
Pieces per hr.	Divide into 60	Minutes per piece
Pieces per hr.	Divide into 3,600	Seconds per piece
Pieces per hr.	Divide into 100,000	TMU per piece
Hours per pc.	Take reciprocal	Pieces per hour
Minutes per pc.	Divide into 60	Pieces per hour
Seconds per pc.	Divide into 3,600	Pieces per hour
TMU per piece	Divide into 100,000	Pieces per hour

Example 1

An operator takes five seconds to assemble several parts. Assuming 100% performance, determine the required select minutes and select hourly production for this operation.

Solution

$$\text{Select minutes} = 5 \text{ seconds} \times 0.0167 = 0.0835 \text{ minutes.}$$
$$\text{Hourly production } 3{,}600 \div 5 \text{ seconds/piece} = 720 \text{ pieces/hour.}$$

Example 2

An MTM pattern, with 15% P. F. & D allowance added, shows a task to merit an allowed time standard of 2,060 TMU. Compute the allowed minutes and the allowed time production per hour for this operation.

---

[5] Delmar W. Karger & Franklin H. Bayha, *Engineered Work Measurement,* Second Editon. Copyright © 1966, the Industrial Press, New York, N. Y. 10016. Pp. 577–580.

Solution

Allowed minutes = 2,060 TMU × 0.0006 = 1.2360 minutes.

Allowed rate = 100,000 ÷ 2,060 TMU/pieces = 48.5 pcs/hr.

EXHIBIT 3-18   Conversion of TMU to Minutes and Seconds
(1 minute = 1,666.7 TMU, 1 second = 27.8 TMU)

TMU Range	TMU	Minutes	Seconds
Thousands	10,000	6.000	360.00
	9,000	5.400	324.00
	8,000	4.800	288.00
	7,000	4.200	252.00
	6,000	3.600	216.00
	5,000	3.000	180.00
	4,000	2.400	144.00
	3,000	1.800	108.00
	2,000	1.200	72.00
	1,000	.600	36.00
Hundreds	900	.540	32.40
	800	.480	28.80
	700	.420	25.20
	600	.360	21.60
	500	.300	18.00
	400	.240	14.40
	300	.180	10.80
	200	.120	7.20
	100	.060	3.60
Tens	90	.0540	3.240
	80	.0480	2.880
	70	.0420	2.520
	60	.0360	2.160
	50	.0300	1.800
	40	.0240	1.440
	30	.0180	1.080
	20	.0120	.720
	10	.0060	.360
Units	9	.0054	.324
	8	.0048	.288
	7	.0042	.252
	6	.0036	.216
	5	.0030	.180
	4	.0024	.144
	3	.0018	.108
	2	.0012	.072
	1	.0006	.036
Tenths	0.9	.00054	.0324
	.8	.00048	.0288
	.7	.00042	.0252
	.6	.00036	.0216
	.5	.00030	.0180
	.4	.00024	.0144
	.3	.00018	.0108
	.2	.00012	.0072
	.1	.00006	.0036

EXHIBIT 3-19 Dollar Value of Time Units for Various Labor Rates

TMUs		1,000	2,000	3,000	4,000	5,000	6,000	7,000	8,000	9,000	10,000
MINUTES		.60	1.20	1.80	2.40	3.00	3.60	4.20	4.80	5.40	6.00
HOURS		.01	.02	.03	.04	.05	.06	.07	.08	.09	.10
	1.00	.0100	.0200	.0300	.0400	.0500	.0600	.0700	.0800	.0900	.1000
	1.05	.0105	.0210	.0315	.0420	.0525	.0630	.0735	.0840	.0945	.1050
	1.10	.0110	.0220	.0330	.0440	.0550	.0660	.0770	.0880	.0990	.1100
	1.15	.0115	.0230	.0345	.0460	.0575	.0690	.0805	.0920	.1035	.1150
	1.20	.0120	.0240	.0360	.0480	.0600	.0720	.0840	.0960	.1080	.1200
	1.25	.0125	.0250	.0375	.0500	.0625	.0750	.0875	.1000	.1125	.1250
	1.30	.0130	.0260	.0390	.0520	.0650	.0780	.0910	.1040	.1170	.1300
	1.35	.0135	.0270	.0405	.0540	.0675	.0810	.0945	.1080	.1215	.1350
	1.40	.0140	.0280	.0420	.0560	.0700	.0840	.0980	.1120	.1260	.1400
	1.45	.0145	.0290	.0435	.0580	.0725	.0870	.1015	.1160	.1275	.1475
	1.50	.0150	.0300	.0450	.0600	.0750	.0900	.1050	.1200	.1350	.1500
	1.55	.0155	.0310	.0465	.0620	.0775	.0930	.1085	.1240	.1395	.1550
	1.60	.0160	.0320	.0480	.0640	.0800	.0960	.1120	.1280	.1440	.1600
	1.65	.0165	.0330	.0495	.0660	.0825	.0990	.1155	.1320	.1485	.1650
	1.70	.0170	.0340	.0510	.0680	.0850	.1020	.1190	.1360	.1530	.1700
	1.75	.0175	.0350	.0525	.0700	.0875	.1050	.1215	.1400	.1575	.1750
	1.80	.0180	.0360	.0540	.0720	.0900	.1080	.1260	.1440	.1620	.1800
	1.85	.0185	.0370	.0555	.0740	.0925	.1110	.1295	.1480	.1665	.1850
	1.90	.0190	.0380	.0570	.0760	.0950	.1140	.1330	.1520	.1710	.1900
	1.95	.0195	.0390	.0585	.0780	.0975	.1170	.1365	.1560	.1755	.1950
	2.00	.0200	.0400	.0600	.0800	.1000	.1200	.1400	.1600	.1800	.2000
	2.05	.0205	.0410	.0615	.0820	.1025	.1230	.1435	.1640	.1845	.2050
	2.10	.0210	.0420	.0630	.0840	.1050	.1260	.1470	.1680	.1890	.2100
	2.15	.0215	.0430	.0645	.0860	.1075	.1290	.1505	.1720	.1935	.2150
	2.20	.0220	.0440	.0660	.0880	.1100	.1320	.1540	.1760	.1980	.2200
	2.25	.0225	.0450	.0675	.0900	.1125	.1350	.1575	.1800	.2025	.2250
	2.30	.0230	.0460	.0690	.0920	.1150	.1380	.1610	.1840	.2070	.2300
	2.35	.0235	.0470	.0705	.0940	.1175	.1410	.1645	.1880	.2115	.2350
	2.40	.0240	.0480	.0720	.0960	.1200	.1440	.1680	.1920	.2160	.2400
	2.45	.0245	.0490	.0735	.0980	.1225	.1470	.1715	.1960	.2205	.2450
	2.50	.0250	.0500	.0750	.1000	.1250	.1500	.1750	.2000	.2250	.2500
	2.55	.0255	.0510	.0765	.1020	.1275	.1530	.1785	.2040	.2295	.2550
	2.60	.0260	.0520	.0780	.1040	.1300	.1560	.1820	.2080	.2340	.2600
	2.65	.0265	.0530	.0795	.1060	.1325	.1590	.1855	.2100	.2385	.2650
	2.70	.0270	.0540	.0810	.1080	.1350	.1620	.1890	.2120	.2430	.2700
	2.75	.0275	.0550	.0825	.1100	.1375	.1650	.1925	.2140	.2475	.2750
	2.80	.0280	.0560	.0840	.1120	.1400	.1680	.1960	.2160	.2520	.2800
	2.85	.0285	.0570	.0855	.1140	.1425	.1710	.1995	.2180	.2565	.2850
	2.90	.0290	.0580	.0870	.1160	.1450	.1740	.2030	.2200	.2610	.2900
	2.95	.0295	.0590	.0885	.1180	.1475	.1770	.2065	.2220	.2655	.2950
	3.00	.0300	.0600	.0900	.1200	.1500	.1800	.2100	.2240	.2700	.3000

LABOR RATE IN DOLLARS PER HOUR — DOLLAR VALUES

EXHIBIT 3-20

To Cover the Following Time Ranges in			Shift Decimal in Table Value to the
TMU's	*Minutes*	*Hours*	
1 to 10	0.0006 to 0.006	0.00001 to 0.0001	Left three digits
10 to 100	.006 to .06	.0001 to .001	Left two digits
100 to 1,000	.06 to .6	.001 to .01	Left one digit
10,000 to 100,000	6 to 60	.1 to 1	Right one digit
100,000 to 1,000,000	60 to 600	1 to 10	Right two digits
1,000,000 to 10,000,000	600 to 6,000	10 to 100	Right three digits

Example 3

Convert 912.8 TMU to minutes

Solution

	TMU	MINUTES
	900	0.540
	10	0.0060
	2	0.0012
	0.8	0.00048
Total	912.8	0.54768

Example 4

How many seconds is 247.1 TMU?

Solution

	TMU	SECONDS
	200	7.2000
	40	1.4400
	7	0.2520
	0.1	0.0036
Total	247.1	8.8956

Example 5

A worker earning $1.55 per hour performs a task which the work analyst believes can be retooled to save 420 TMU. What tooling cost will be the breakeven point?

Solution

Referring to the bottom portion of the table, it is necessary to shift the decimal in the table one place to the left for TMU values between 100 and 1,000. A similar step is taken for the 20 TMU value. Then:

	TMU	Dollars
	400	$0.00620
	20	0.00031
Total	420	$0.00651

A saving of $0.00651 per piece will result from retooling. If, say, 180,000 pieces per year are normally produced, the maximum tooling cost that could be paid for out of one year's saving would be:

$$180,00 \text{ pieces/year} \times \$0.00651/\text{pieces} = \$1,171.80$$

WORK MEASUREMENT-WORK FACTOR SYSTEM[6]

The Work-Factor Elemental Time System

The Work-Factor System was designed and first installed more than thirty years ago. Its purpose was to provide an objective measurement technique which would eliminate personal opinion and judgment from the task of establishing work standards. Since the Work-Factor System was developed primarily as a measurement tool, a great deal of research was done in developing the elemental motion time tables. Users of the system claim a high degree of accuracy and consistency for Work-Factor standards. Another important benefit comes from improved work methods. Work-Factor is now widely accepted and is used in almost every industrialized country of the world.

The basic facts about the system are summarized as follows:

1. The Work-Factor System is based on analyzing the motions required to perform an operation.
2. Work is analyzed by dividing an operation into units called Standard Elements. All work, no matter how complex, can be subdivided into the same standard elements.
3. When an operation has been described carefully and analyzed, time values from a standard motion time table are applied to the motions.
4. The total time for all the motions is the time for the job.
5. A stopwatch is used only to determine process time such as drilling time or cutting time which is not controlled by the operator.
6. Only necessary motions are included in the analysis of an operation.

The time that is required for a particular movement is influenced by four major variables, namely, the Body Member that is being moved, the Distance through which it is moved, the Weight that is moved (or the resistance that is overcome), and the Difficulty of the motion. For a given expenditure of energy, the finger moves faster, through a specific distance, than the arm, and the leg moves faster than the trunk. In an operation analysis, the body members that are used can be readily recognized, and separate tables of Work-Factor Time Units are provided for each Body Member that may be involved. In addition, time standards are included for the

---

[6] © 1970 Science Management Corporation, Moorestown, New Jersey. All rights reserved. Tables and illustrations referenced are from Quick, Duncan, Malcolm, *Work-Factor Time Standards,* McGraw-Hill, Inc., N. Y. Copyright © 1962.

special arm motion termed "Forearm Swivel" that is employed, for example, in turning a screw-driver. Typical symbols for body members, and examples of their use, are listed in Exhibit 3–21.

The Work-Factor Symbols for Various Body Members and Typical Operations That They Perform		
Body Member	*Symbol*	*Typical Operations*
Finger	F	Grasping and assembly
Hand	H	
Arm	A	Moving free hand to working position; transporting objects to work location
Trunk	T	Bending to reach objects beyond range of normal arm movements
Leg	L	Walking or stooping; operating pedals
Foot	Ft	Operating treadle-type controls

EXHIBIT 3-21

For a specific Body Member, the time required for a particular motion depends on the distance moved by the member. The length of the motion, in inches, determined at the time that the analysis is made, is employed in selecting the correct time-value from the Motion time tables.

The Four Major Elements of Control, or Difficulty of Motion Employed in the Work-Factor System		
Element	*Symbol*	*Definition*
Steering or Directional Control	S	The control required to direct or steer a body member to a specific target
Precaution or Care	P	The control required to avoid a physical hazard or to prevent damage to a fragile object
Change of Direction	U	The control required to move a body member in a sharply curved path
Definite Stop	D	The control required to terminate a motion at the will of the operator

EXHIBIT 3-22

The effect of weight, or resistance, on the time required to make a movement is not fully recognized by many time study engineers. In the Work-Factor System, due consideration is given to the influence of this important variable, and the necessary time allowances are provided in the tables. These allowances are arranged in multiples of 1 lb. and vary in accordance with the body member that is used.

It will be readily appreciated that the speed of a given movement, and therefore the time required, depends on the degree of control that must be exercised. The variable introduced by the difficulty of, or the degree of control needed for, a particular movement is more difficult to assess than the factors that have been considered so far. From the research on which this system is based, four major elements of control were established as indicated in Exhibit 3–22.

Any movement that involves only a body member, a certain weight, and a distance, is termed a "Basic Motion," a typical example being the action of tossing an object into a box. Control elements, or elements of additional weight, introduce time allowances and the movement ceases to be basic. Combinations of two or three of the variables discussed are quite common, but instances in which all four variables are combined are rare.

Each element of control or weight is termed a "Work-Factor," and to avoid a multiplicity of tables, all the Work-Factors relating to the movement of a specific body member are arranged to be of equal value by carefully defining the rules that govern their application. This concept is made clear by the examples shown in Exhibit 3–23.

DISTANCE MOVED	BASIC	WORK-FACTORS				DISTANCE MOVED	BASIC	WORK-FACTORS			
		1	2	3	4			1	2	3	4
(A) ARM-Measured at Knuckles						(L) LEG—Measured at Ankle					
1″	18	26	34	40	46	1″	21	30	39	46	53
2″	20	29	37	44	50	2″	23	33	42	51	58
3″	22	32	41	50	57	3″	26	37	48	57	65
4″	26	38	48	58	66	4″	30	43	55	66	76
5″	29	43	55	65	75	5″	34	49	63	75	86
6″	32	47	60	72	83	6″	37	54	69	83	95
7″	35	51	65	78	90	7″	40	59	75	90	103
8″	38	54	70	84	96	8″	43	63	80	96	110
9″	40	58	74	89	102	9″	46	66	85	102	117
10″	42	61	78	93	107	10″	48	70	89	107	123
11″	44	63	81	98	112	11″	50	72	94	112	129
12″	46	65	85	102	117	12″	52	75	97	117	134
13″	47	67	88	105	121	13″	54	77	101	121	139
14″	49	69	90	109	125	14″	56	80	103	125	144
15″	51	71	92	113	129	15″	58	82	106	130	149
16″	52	73	94	115	133	16″	60	84	108	133	153
17″	54	75	96	118	137	17″	62	86	111	135	158
18″	55	76	98	120	140	18″	63	88	113	137	161
19″	56	78	100	122	142	19″	65	90	115	140	164
20″	58	80	102	124	144	20″	67	92	117	142	166
22″	61	83	106	128	148	22″	70	96	121	147	171
24″	63	86	109	131	152	24″	73	99	126	151	175
26″	66	90	113	135	156	26″	75	103	130	155	179
28″	68	93	116	139	159	28″	78	107	134	159	183
30″	70	96	119	142	163	30″	81	110	137	163	187
35″	76	103	128	151	171	35″	87	118	147	173	197
40″	81	109	135	159	179	40″	93	126	155	182	206
Weight in Lbs. Male	2	7	13	20	Up	Weight in Lbs. Male	8	42	Up	—	—
Fem.	1	3½	6½	10	Up	Fem.	4	21	Up	—	—

EXHIBIT 3-23 Detailed Work-Factor Motion Time Table

Reading the table and finding an appropriate time value is relatively simple. The distance moved is found at the left of each chart. The times for Basic Motions are found under the basic column immediately next to the distance column. Time values for motions with one of the control variables are found in the column headed 1 Work-Factor; time values for motions with two control variables are found in the 2 Work-Factor column, etc. In all cases, the time for a given motion is found opposite the appropriate motion distance. For example:

A basic arm motion ten inches long is described as A10. The time value—42 Work-Factor Units, is found opposite 10 inches in the Basic column.

A ten-inch arm motion with one control element such as the definite stop Work-Factor, would be described as A10D. The time value for this motion is 61 Work-Factor Units. It is found under the 1 Work-Factor column opposite ten inches.

In the same manner, time values for motions with two Work-Factors are found under column 2, those with three Work-Factors per motion are found under column 3, etc. Thus, it is possible to select from the Arm table an appropriate time value for any arm motion up to 40 inches long with up to four degrees of control. As previously stated, combinations of two or three of the variables are common. Combinations of weight and control with four Work-Factors are rare.

In addition to the tables for the simple movements of individual body members, tabulated data are provided for the selection of times for compound motions such as the elements of assembly, pre-positioning, and grasping objects from random piles. These additional tables greatly facilitate analysis and reduce the time required to establish standards. The times in all Work-Factor tables are in Work-Factor Time Units, each equal to 0.0001 min.* In practice, all analyses are made in these Units and only the final total is converted to minutes by moving the decimal point four places to the left.

For analysis purposes, all work is subdivided into Standard Elements which provide a common denominator for analyzing all types of work. The eight Standard Elements recognized by the Work-Factor System were developed from the Gilbreth therbligs with modification. Each standard element is composed of individual motions. The Work-Factor Standard Elements of work are illustrated in Exhibit 3–24.

In actual practice there are three modifications of the System.

The Detailed Work-Factor System is used for short cycle, highly repetitive, mass production type operations. Detailed Work-Factor is particularly applicable to the small motions relating to micro electronics where microscopes and micromanipulators are used. It is used wherever great accuracy is important. The Work-Factor Time Unit, as mentioned above, is 0.0001 minute.

Ready Work-Factor is a simplified version of the Work-Factor System which is used for job shop work where cycles are longer and the quantities produced are relatively small. The Ready System is based on averages. It provides the same answer as Detailed Work-Factor but, since averages are used, the accuracy of the

---

* Work-Factor Time is defined as that time required for the average Experienced Operator working with good skill and good effort (commensurate with physical and mental well-being) and under standard working conditions to perform one work cycle, or operation, on one unit, or piece, according to prescribed method and specified quality. The Work-Factor Time includes no allowance for personal needs, fatigue, environmental unavoidable delays, or incentive payment.

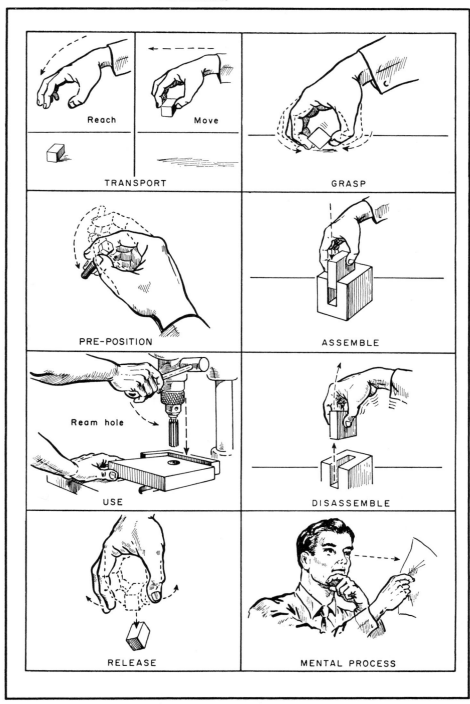

EXHIBIT 3-24 The Work-Factor Standard Elements

individual work elements may not be as great. Ready Work-Factor is widely used because the need for great accuracy and precision is not found in many industries. Ready Work-Factor Time Units are 0.001 minute each, ten times greater than in the Detailed technique.

Abbreviated Work-Factor is a separate and independent system which is a further modification of Detailed Work-Factor. It was developed to fill the need for an estimating tool and for a technique to measure maintenance and machine set-up work. Abbreviated Work-Factor Time Units equal 0.005 minute.

Operation Name & Description

Reproduce One 4-Page Report on Xerox Machine

No	Elemental Description	Hand		Analysis	Time Units		No
		L	R		Elemental	Cumulative	
1	Walk 30' to Xerox Machine 12Pa			12+12 (8)	108	108	1
2	Reach to set knob—place copy on machine	X		1–30	9	117	2
	top simo						
3	Grasp knob	X		0–	1	118	3
4	Turn knob to 1—pick up 1st page simo	X		2–4	3	121	4
5	Release knob	X		0–	1	122	5
6	Reach to flexible cover	X		1–20	7	129	6
7	Grasp cover handle	X		1–	2	131	7
8	Raise cover	X		1–10	5	136	8
9	Move 1st page to window		X	2–20	9	145	9
10	Line up copy to mark		X	OTS–1/8	8	153	10
11	8″ gripping dist.		X	50% (5)	3	156	11
12	Index		X	Ind	4	160	12
13	Release page # 1–lower cover simo		X	0–	1	161	13
14	Reach to start button and hit		X	2–10	6	167	14
15	Machine cycle—timed			M.T.	70	237	15
16	Lift cover— R to page simo R.H.	X		1–10	5	242	16
17	Gr page in machine		X	2–V	3	245	17
18	Remove and turn over on machine top		X	0–20, 1–10	10	255	18
19	Release page 1		X	0–	1	256	19
20	Reach to page 2		X	1–20	7	263	20
21	Grasp page 2		X	3–V	5	268	21
22	Move page 2 to window		X	2–30	11	279	22
23	Place page 2, print and lay aside		X	El 9–18 incl	111	390	23
24	Pick up and print pages 3 and 4		X	(El 19–22) x2	268	658	24
1	Pick up copy from Xerox top			PU 3–20	21	679	1
2	Jog pages to even edges–3 jogs	X	X	6 (1–+1–4)	24	703	2
3	Step to catch tray			3–20	11	714	3
4	Pick up prints from tray	X		P/U 1–20	18	732	4
5	Walk back to desk—30' 12Pa			12+12 (8)	108	840	5
7	Total cycle time					.84 min.	7

EXHIBIT 3-25

While there are three variations of Work-Factor available for use, individual analyses of operations are seldom prepared. Standards men find it more convenient to analyze work by type, than to develop data tables from which standards for operations can be readily established.

A Typical Work-Factor Application

The application of Work-Factor can be demonstrated by the analysis of a simple operation. A clerk is required to walk 30 feet to a Xerox machine and reproduce a copy of a four-page report, then return to the starting point at her desk.

The analysis for this simple operation using the Ready Work-Factor technique is shown in Exhibit 3–25.

Preparation of Standard Date

One of the principal advantages of the Work-Factor system is the ease with which standard data can be prepared. With conventional time study methods, it is generally necessary to undertake many studies in order to establish a pattern or curve from which the variables in a given operation may be determined. By means of the Work-Factor technique, such data can frequently be obtained from a single analysis, since the time required for the different movements can be selected directly from the tables.

If the Xeroxing operation already considered is taken as an illustration, the time values required for standard data are as follows:

Constant Values		Work-Factor Time Units
Walk to Xerox Machine		108
Set counter		14
Pick up copy from machine and jog		45
Step to catch tray and pick up prints		29
Walk back to desk		108
	Total Constants	304
Variables		
Lift cover and print 1st page		134
Pick up 2nd page and print 2nd		134
Pick up and print last 2 pages		268
	Total Variables	536
	Operational Total	840

This reduces to a formula: 0.304 minute (constant) + 0.134 minute per page

Of course, it is readily apparent that a data table can be easily developed to provide a time value for any number of pages or any number of copies of a report. Standards thus developed are more easily "sold," and, once employees gain confidence in Work-Factor times, standards complaints are substantially reduced.

EXHIBIT 3-26 Work Factor Abbreviations and Element Analysis

TRANSPORT

MOTION DISTANCE in Inches	BASIC	WORK-FACTORS				MOTION DISTANCE in Inches	BASIC	WORK-FACTORS			
		1	2	3	4			1	2	3	4
ARM (A): Measured at Knuckles						**LEG (L): Measured at Ankle**					
1	18	26	34	40	46	1	21	30	39	46	53
2	20	29	37	44	50	2	23	33	42	51	58
3	22	32	41	50	57	3	26	37	48	57	65
4	26	38	48	58	66	4	30	43	55	66	76
5	29	43	55	65	75	5	34	49	63	75	86
6	32	47	60	72	83	6	37	54	69	83	95
7	35	51	65	78	90	7	40	59	75	90	103
8	38	54	70	84	96	8	43	63	80	96	110
9	40	58	74	89	102	9	46	66	85	102	117
10	42	61	78	93	107	10	48	70	89	107	123
11	44	63	81	98	112	11	50	72	94	112	129
12	46	65	85	102	117	12	52	75	97	117	134
13	47	67	88	105	121	13	54	77	101	121	139
14	49	69	90	109	125	14	56	80	103	125	144
15	51	71	92	113	129	15	58	82	106	130	149
16	52	73	94	115	133	16	60	84	108	133	153
17	54	75	96	118	137	17	62	86	111	135	158
18	55	76	98	120	140	18	63	88	113	137	161
19	56	78	100	122	142	19	65	90	115	140	164
20	58	80	102	124	144	20	67	92	117	142	166
22	61	83	106	128	148	22	70	96	121	147	171
24	63	86	109	131	152	24	73	99	126	151	175
26	66	90	113	135	156	26	75	103	130	155	179
28	68	93	116	139	159	28	78	107	134	159	183
30	70	96	119	142	163	30	81	110	137	163	187
35	76	103	128	151	171	35	87	118	147	173	197
40	81	109	135	159	179	40	93	126	155	182	206
Weight, lb.:						**Weight, lb.:**					
Male	−2	−7	−13	−20	>20	Male	−8	−42	>42	·	·
Female	−1	−3½	−6½	−10	>10	Female	−4	−21	>21	·	·
TRUNK (T): Measured at Shoulder						**FINGER–HAND (F, H): Measured at Finger Tip**					
1	26	38	49	58	67	1	16	23	29	35	40
2	29	42	53	64	73	2	17	25	32	38	44
3	32	47	60	72	82	3	19	28	36	43	49
4	38	55	70	84	96	4	23	33	42	50	58
5	43	62	79	95	109	**Weight, lb.:**					
6	47	68	87	105	120	Male	−2/3	−2½	−4	>4	·
7	51	74	95	114	130	Female	−1/3	−1¼	−2	>2	·
8	54	79	101	121	139	**FOOT (Ft): Measured at Toe**					
9	58	84	107	128	147	1	20	29	37	44	51
10	61	88	113	135	155	2	22	32	40	48	55
12	66	94	123	147	169	3	24	35	45	55	63
14	71	100	130	158	182	4	29	41	53	64	73
16	75	105	136	167	193	**Weight, lb.:**					
18	80	111	142	173	203	Male	−5	−22	>22	·	·
20	84	116	148	179	209	Female	−2½	−11	>11	·	·
22	88	121	153	185	215	**FOREARM SWIVEL (FS): Measured at Knuckle**					
24	92	125	158	190	220	45°	17	22	28	32	37
26	95	130	163	196	226	90°	23	30	37	43	49
28	99	134	168	201	231	135°	28	36	44	52	58
30	102	139	173	206	236	180°	31	40	49	57	65
Weight, lb.:						**Torque, lb.-in:**					
Male	−11	−58	>58	·	·	Male	−3	−13	>13	·	·
Female	−5½	−29	>29	·	·	Female	−1½	−6½	>6½	·	·

WALK

TYPE	30-Inch PACES		
	1	2	OVER 2
General	Analyze from Table	260	120 + 80/Pace
Restricted		300	120 + 100/Pace
Add 100 for >120° – 180° Turn at Start or Finish of Walk			
Up Steps (8-inch rise, 10-inch flat)	126/step		
Down Steps	100/step		

HEAD TURN (HT): Measured at Nose Tip

Degrees Turn	Distance in Inches	Number of Work-Factors			
		Basic or 1	2	3	4
>22½−45	>2−4	40	51	58	66
−90	−8	60	76	86	99

78

EXHIBIT 3-27

COMPLEX GRASP * FROM RANDOM PILES

SIZE (Major Dimension or Length) (inches)		SOLIDS AND BRACKETS Thickness (inches) >.047 (>3/64) Blind n	Blind s	Visual n	Visual s	THIN FLAT OBJECTS Thickness (inches) –.016 (–1/64) Blind n	Blind s	Visual n	Visual s	–.047 (–3/64) Blind n	Blind s	Visual n	Visual s	CYLINDERS AND REGULAR CROSS-SECTIONED SOLIDS Diameter (inches) –.063 (–1/16) Blind n	Blind s	–.125 (–1/8) Blind n	Blind s	–.188 (–3/16) Blind n	Blind s	–.500 (–1/2) Blind n	Blind s	Visual n	Visual s	>.500 (>1/2) Blind n	Blind s	Visual n	Visual s	Add for Entangled, Nested, or Slippery Objects† n	s
–.063	–1/16	120	172	B	B	–	–	–	–	131	189	B	B	S	S	S	S	S	S	S	S	S	S	S	S	S	S	17	26
–.125	–1/8	79	111	B	B	108	154	B	B	85	120	B	B	85	120	S	S	S	S	S	S	S	S	S	S	S	S	12	18
–.188	–3/16	64	88	B	B	102	145	B	B	74	103	B	B	79	111	74	103	S	S	S	S	S	S	S	S	S	S	12	18
–.250	–1/4	48	64	B	B	72	100	B	B	56	76	B	B	79	111	68	94	64	88	S	S	S	S	S	S	S	S	8	12
–.500	–1/2	40	52	B	B	64	88	B	B	48	64	B	B	62	85	56	76	56	76	44	58	B	B	S	S	S	S	8	12
–1.000	–1	40	52	32	40	64	88	60	82	48	64	44	58	62	85	56	76	48	64	48	64	44	58	40	52	32	40	8	12
–4.000	–4	37	48	20	22	53	72	36	46	45	60	28	34	56	76	48	64	40	52	40	52	36	46	37	48	20	22	8	12
>4.000	>4	46	61	20	22	70	97	44	58	62	85	36	46	56	76	48	64	40	52	40	52	36	46	37	48	20	22	9	14

* Special Grasp conditions should be analyzed in detail
n = Non-simo; s = Simo
B = Use Blind column, since Visual Grasp offers no advantage
S = Use Solid Table

† Add the indicated amounts when objects: (1) are entangled (not requiring two hands to separate); (2) are nested together because of shape or film; (3) are slippery (as from oil or polished surface). When objects both entangle and are slippery, or both nest and are slippery, use double the value in the table.

ASSEMBLE

AVERAGE NUMBER OF ALIGNS (A1S MOTIONS)

TARGET DIMENSION (Inches)	CLOSED TARGETS – RATIO OF PLUG DIAMETER TO TARGET DIMENSION –.225		–.290		–.415		–.900		–.935 †		>.935 ‡		OPEN TARGETS – RATIO OF PLUG DIAMETER TO TARGET DIMENSION –.225		–.290		–.415		–.900		–.935 †		>.935 ‡	
>.875	(D*)	18	(D*)	18	(D*)	18	(¼)	25	(¼)	51	(¼)	59	(D*)	18	(D*)	18	(D*)	18	(D*)	18	(¼)	51	(¼)	59
–.875	(D*)	18	(D*)	18	(SD*)	18	(¼)	25	(¼)	51	(¼)	59	(D*)	18	(D*)	18	(D*)	18	(SD*)	18	(¼)	51	(¼)	59
–.625	(SD*)	18	(SD*)	18	(¼)	25	(½)	31	(½)	57	(½)	65	(SD*)	18	(SD*)	18	(SD*)	18	(½)	31	(½)	57	(½)	65
–.375	(½)	31	(1)	44	(1)	44	(1½)	57	(1½)	83	(1½)	91	(¼)	25	(½)	31	(½)	31	(¾)	38	(¾)	64	(¾)	72
–.225	(1)	44	(1)	44	(1)	44	(1½)	57	(1½)	83	(1½)	91	(½)	31	(½)	31	(½)	31	(¾)	38	(¾)	64	(¾)	72
–.175	(1)	44	(1¼)	51	(1½)	57	(1½)	57	(1½)	83	(1½)	91	(¾)	38	(1)	44	(1)	44	(1)	44	(1)	70	(1)	78
–.124	(2½)	83	(2½)	83	(2½)	83	(2½)	83	(2½)	109	(2½)	117	(1¼)	51	(1¼)	51	(1¼)	51	(1¼)	51	(1¼)	77	(1¼)	85
>.025–.074	(3)	96	(3)	96	(3)	96	(3)	96	(3)	122	(3)	130	(1½)	57	(1½)	57	(1½)	57	(1½)	57	(1½)	83	(1½)	91

*Letters indicate Work-Factors in Move preceding Assemble. †Requires A(X)S Upright for all ratios >.900 (Table value includes A1S Upright and A1 Insert).

‡ Requires A(Y)S Upright and A(Z)P Insert for all ratios >.935 (Table value includes A1S Upright and A1P Insert).

DISTANCE BETWEEN TARGETS

Distance Between Targets (Inches)	Percent Addition to Aligns	Method of Align
– 1	Neg.	Simo
– 2	10	Simo
– 3	30	Simo
– 5	50	Simo
– 7	70	Simo
–15	Align and Insert first Plug, then Assemble* second end.	
>15	Align and Insert first Plug, turn head toward second Plug, React (60 Time Units), Assemble* second Plug.	

*If connected, treat 2nd Assemble as Open Target with no Upright. Index may be required up to 7 inch distances.

GRIPPING DISTANCE

Distance from Grip Point to Align Point (Inches)	Percent Addition to Aligns	Length of Upright Motion (Inches)
– 2	Neg.	1
– 3	10	1
– 5	20	2
– 7	30	2
–10	40	3
–15	60	5
–20	80	6
>20	100	7 and up

BLIND TARGETS

Blind Distance (Inches)	Percent Addition to Aligns: Permanent (Blind at all times)	Temporary (Blind during Assemble)
– ½	20	0
– 1	30	10
– 2	40	20
– 3	70	30
– 5	130	50
– 7	250	70
–10	380	120

GENERAL RULES FOR ASSEMBLE

1. When required, add W and P Work-Factors to all Assemble Motions according to Transport Rules.
2. Reduce number of Aligns by 50 percent when hand is rigidly supported.
3. Where Gripping Distance, two Targets, and Blind Targets are involved, add each percentage to original Align. Do not pyramid percentages.
4. Aligns for Surface Assemble are taken from –.225 column and are A1SD Motions.
5. Index is F1S, A1S, or FS45° S.

POST-DISENGAGE TRAVEL

DISENGAGE RESISTANCE (Pounds)	POST-DISENGAGE TRAVEL (Inches)
– 2	Negligible
– 7	3
–13	6
–20	10

MENTAL PROCESS (MP) – SIMPLE

Focus (Fo)	20
React (Rn)	20
Inspect (I)	30
Mento (Mt)	10

PRE-POSITION

SHAPE, SIZE (INCHES), AND WEIGHT (POUNDS) OF OBJECT[1] [4]		ONE HAND: VERY SMALL V5F1	OPTIMUM V3F1	MEDIUM V4F1		TWO HANDS: MEDIUM 2TrP + VA4D	LARGE 2TrP(W) + VA8D
1. Cylinders – Regular Cross-section[2]	Diameter	–.375	>0 –1.25	>0 1.25		>1.25– 4.50	–
	Major Dimension	–.375	>.375–4.00	>4.00–16.00		>1.25–30.00	–
2. Solids, Thin Flats, etc.[3]	Width	–.375	>0 –1.25	>0 – 1.25	>1.25– 2.50	>2.50–10.00	>10.00–16.00
	Thickness	–.375	>0 –1.25	>0 – 1.25	>0 – 2.50	>0 – 4.50	> 0 – 4.50
	Major Dimension	–.375	>.375–4.00	>4.00–16.00	>1.25–16.00	>2.50–16.00	>10.00–16.00
3. Weight Limits – All Objects (Pounds)	Male	–.667	–.667	–.667		–1.00	–3.50
	Female	–.333	–.333	–.333		–0.50	–1.75

NO. OF POSITIONS SATISFACTORY FOR USE		% PP REQUIRED	WORK-FACTOR TIME UNITS[5] PP–V n	PP–V s	PP–O n	PP–O s	PP–M n	PP–M s	PP–M_2 n	PP–L n
1. From Stack (all wrong way)		100%	80	120	48	72	64	96	70	100
2. One Specific Face Up	1 Edge Only	75%	60	90	36	54	48	72	53	75
	2 Adjacent Edges	63%	50	75	30	45	40	60	44	63
	2 or more Opposite Edges	50%	40	60	24	36	32	48	35	50
3. Two or More Faces Up	1 Edge Only	50%	40	60	24	36	32	48	35	50
	2 Adjacent Edges	25%	20	30	12	18	16	24	18	25
	2 or more Opposite Edges	0%	–	–	–	–	–	–	–	–

Notes:
[1] *Pre-positions requiring one Finger Motion or Wrist Turn can be done Simo with the Move. Other Pre-positions are taken from this Table, or analyzed.*
[2] *Cylinders with a diameter > 4.50 inches or a Major Dimension > 30.00 inches require Special Analyses.*
[3] *Solids and Thin Flats with a width or Major Dimension > 16.00 inches or a thickness > 4.50 inches require Special Analyses.*
[4] *Objects weighing more than weight limits on this Table require Special Analyses.*
[5] *n = Non-Simo; s = Simo.*

Work Factor Abbreviations and Element Analysis

Abbreviations: Abbreviations are also employed wherever practical for various other descriptive terms applied to work. The most common of these are as follows:

Term	Abbreviation	Term	Abbreviation
Align	AL	Left Hand	LH
Approach	AP	Leg	L
Arm	A	Mental Process	MP
Assemble	ASY	Mento	MT
Average	V	Move	M
Balancing Delay	BD	Nested	N
Blind	B	Open Target	OT
Bracket	BR	Plug-Target Ratio	RA
Care (Precaution)	P	Point of Inspection	PT
Change Direction	U	Pre-position	PP
Closed Target	CT	Reach	R
Contact Grasp	CT-GR	React	RE
Cylinder	CYL	Regrasp	R-GR
Definite Stop	D	Release	RL
Direction Control (Steer).	S	Right Hand	RH
Disassemble	DSY	Seat	ST
Distance between Targets	DB	Separate	SEP
Entangled	E	Simultaneous	SIMO
Finger	F	Slippery	SLP
Focus	FO	Solid	SOL
Foot	FT	Transport	TRP
Forearm Swivel	FS	Thin Flat	TF
Grasp	GR	Transfer Grasp	TR-GR
Gripping Distance	GD	Upright	UP
Hand	H	Trunk	T
Head Turn	HT	Use	US
Insert	INS	Weight or Resistance	W
Inspect	INSP	Work Area	WA

EXHIBIT 3-28

4

Production Line Techniques

CALCULATION OF CONVEYOR LINE STANDARDS

1. Definition of Line Standard:[1]

A production standard is defined as a unit of measurement indicating the time necessary for an operation to be performed by an experienced operator, working effectively at a normal pace in a predetermined manner and taking the adequate allowed time for fatigue and personal needs. This is generally expressed in hours or decimal hours per unit. The "normal pace" referred to is that pace accepted throughout industry as being fair and reasonable.

In conveyor line operations where different tasks are performed by various groups of production employees and paced by specified conveyor speeds, crew complements are determined by analysis of their related production standards. The sum of these required man-hours is then employed to calculate the line standard.

It is after the crew assignments have been determined (this can be progressive processing involving a continuous line operation or a motorized conveyor system), the total standard hours for all the crews involved is then divided by the total number of pieces expected to be produced in a given period, resulting in the line standard. The development follows:

2. Line Standard Development:

Formula

$$L = \frac{T}{S \times H} \times 1,000$$

---

[1] Lewis R. Zeyher, *Cost Reduction in the Plant* (Englewood Cliffs, N. J.: Prentice-Hall, Inc. Copyright © 1965), p. 164.

where
L = Line standard expressed in standard hours per 1,000 units
T = Crews' total standard hours per shift
S = Conveyor speed in units per hour
H = Scheduled hours

Example

A conveyor line consists of eight different operations involving eight predetermined crew sizes. The conveyor speed is set at 106 units per minute and is scheduled to run 7.67 hours (allowing for two rest breaks), the production standards for each operation having previously been developed. Determine the Line Standard per 1,000 units.

Solution

Item	Operation	St'd. Hrs. Per 100 Units	No. in Crew	Total St'd Hrs. Per 8 Hr. Day.
1	A	0.0512	3.0	24.0
2	B	0.0637	4.0	32.0
3	C	0.0332	2.0	16.0
4	D	0.0352	2.0	16.0
5	E	0.0983	6.0	48.0
6	F	0.1002	6.0	48.0
7	G	0.0984	6.0	48.0
8	H	0.0164	1.0	8.0
		Totals	30.0	240.0

$$L = \frac{T}{S \times H} \times 1{,}000$$

$$L = \frac{240 \; Hrs.}{(106 \times 60) \times 7.67} \times 1{,}000$$

$$L = \frac{240 \; Hrs.}{48{,}800} \times 1{,}000$$

$$L = 4.92 \; St'd \; Hrs.$$

Line Standard is 4.92 Standard Hours / 1,000 units produced.

Problem

Determine the production performance per cent of this department if they produced 50,000 units in an 8-hour shift while consuming 240 actual paid hours.

Formula

$$P = \frac{S \times U}{H}$$

where
P = Production performance in per cent
S = Standard hours per 1,000 units
H = Actual hours worked on operations
U = Units produced

Solution

$$P = \frac{4.92 \times \dfrac{50,000}{1,000}}{240 \text{ hours}}$$

$$P = 1.025\%$$

Production performance for the department = 102.5% efficient (100.00% required).

Example

An operation in the receiving department of a factory involves unloading full cartons from trailers to the dock, opening cartons and hanging units on a moving conveyor entering plant. The empty cartons are piled nearby on pallets while a preparatory operation is performed on the recently hung units as they begin to flow through the production cycle. The conveyor line is travelling at a speed of 42 units per minute for a total of 7.67 hours per shift. A crew of four men are employed. Determine the Line Standard for this crew.

Operation	St'd Hrs Per 100 Units	No. in Crew	St'd Hrs. Per 8 Hr.
(1) Unload trailer, stack empties.	0.0419	1.0	8.0
(2) Unpack cartons, hang units on conveyor line.	0.0871	2.0	16.0
(3) Inspect, adjust and prepare units to enter production process.	0.0450	1.0	8.0
Totals		4.0	32.0

Formula

$$L = \frac{T}{S \times H} \times 1,000$$

Solution

$$L = \frac{32.0}{(42.0 \times 60) \times 7.67} \times 1,000$$

$$L = \frac{32.0}{19,000} \times 1,000$$

$$L = 1.68 \quad \text{Standard Hours /1,000 units}$$

CHART OF STANDARDS AND CREW DETERMINATIONS

(1) Examples of Formulas

Formula

$$N = P \times S$$

where
N = Number of crew required
P = Required production by crew per hour
S = Production Standard in Standard Hours per 100 units

Example

Determine the crew required for an operation performed on a conveyor running at a speed of 40 units per minute, where the production standard is 0.0852 Standard Hours per 1,000 units produced.

Solution

N = (40 × 60) × 0.0852 St'd Hrs./100 units
N = 2,400 × 0.0852 St'd Hrs./100 units
N = 24 × 0.0852
N = 2.04   operators

Crew required is two operators.

Example

Determine the crew required for an operation performed on a conveyor running at a speed of 40 units per minute, where the production standard is 0.210 Standard Hours per 100 units produced.

Solution

N = (40 × 60)   ×   0.210 St'd Hrs./100 units
N = 2,400   ×   0.210 St'd Hrs./100 units
N = 24   ×   0.210
N = 5.04 operators

Crew required is five operators.

Formula

Check the present crew complement of three operators on a conveyor line running at 60 units per minute, with a production standard of 0.0555 Standard Hours per 100 units produced.

Solution

$$N = \frac{(60 \times 60)}{100} \times 0.0555 \text{ St'd Hrs. /100 units}$$

$$N = \frac{3600}{100} \times 0.0555 \text{ St'd Hrs./100 units}$$

$$N = 36 \quad \times 0.0555 \text{ St'd Hrs./100 units}$$

$$N = 1.998 \text{ operators}$$

The crew is over-manned, requiring only two operators instead of the present crew of three.

(2) Conveyor Crews and Most Economical Line-Speed Chart

Formula (previously discussed)

$$N = P \times S$$

where

N = Number of crew required

P = Required production per hour by crew

S = Production standard in standard hours per 100 units.

UNITS PER MIN.		110	112	114	116	118	120	122	124	126	128
UNITS PER HOUR		6600	6720	6840	6960	7080	7200	7320	7440	7560	7680
OPERATION											
	STANDARD			CREWS		INDICATED					
A	.0646	4.26	4.34	4.42	4.50	4.57	4.65	4.73	4.81	4.88	4.96
	OPERATORS	5	5	5	5	5	5	5	5	5	5
B	.0848	5.60	5.70	5.80	5.90	6.00	6.11	6.21	6.31	6.41	6.51
		6	6	6	6	6	7	7	7	7	7
C	.0388	2.56	2.61	2.65	2.70	2.75	2.79	2.84	2.89	2.93	2.98
		3	3	3	3	3	3	3	3	3	3
D	.0512	3.38	3.44	3.50	3.56	3.62	3.69	3.75	3.81	3.87	3.93
		4	4	4	4	4	4	4	4	4	4
E	.0630	4.16	4.23	4.31	4.38	4.46	4.54	4.61	4.69	4.76	4.84
		5	5	5	5	5	5	5	5	5	5
F	.0737	4.86	4.95	5.04	5.13	5.22	5.31	5.39	5.48	5.57	5.66
		5	5	5	6	6	6	6	6	6	6
G	.1341	8.85	9.01	9.17	9.33	9.49	9.65	9.82	9.98	10.14	10.30
		9	9	10	10	10	10	10	10	11	11
H	.0922	6.09	6.20	6.31	6.42	6.53	6.64	6.75	6.86	6.97	7.08
		6	7	7	7	7	7	7	7	7	7
I	.0632	4.17	4.25	4.32	4.40	4.47	4.55	4.63	4.70	4.78	4.85
		5	5	5	5	5	5	5	5	5	5
J	.0725	4.79	4.87	4.96	5.05	5.13	5.22	5.31	5.39	5.48	5.57
		5	5	5	5	6	6	6	6	6	6
	TOTAL OPERATORS	53	54	55	56	57	58	58	58	59	59
	LABOR COST @ $3.00/HR	159	162	165	168	171	174	174	174	177	177
	LABOR COST Per 1,000 Units	24.1	24.1	24.1	24.1	24.1	24.1	23.7	23.3	23.4	23.0
YRLY. LABOR COST for 2,000 HRS.		318,000									354,000

EXHIBIT 4-1   Conveyor Speeds

Example

See chart (Exhibit 4–1) for operation "A" to demonstrate the manner in which the figures in the chart have been calculated.

$$N = 6600 \times 0.0646$$
$$N = 4.26 \qquad \text{(crew required is 5)}$$

Problem

Determine most economical line crews for various conveyor speeds. (This chart can be expanded in either direction.)

(3)   Analysis of Charting Results

When working with decimal figures resulting from measured work, whether it be time study, MTM, or Work Factor system, some tolerance must be permitted in these judgments. Some rounding-off of figures also is employed as this presentation (Exhibit 4–2) only represents an example in the effective use of production standards for line-balancing and use of charting for analysis. The results can also be used to advantage where excessive absenteeism dictates a change in crews and in conveyor speeds. Introduction of an incentive system would, of course, have considerable impact on these results.

ITEMS COMPARED	CONVEYOR SPEEDS		DIFFERENCES
	110	128	
(a) Number of operators	53	59	+6.0
(b) Yearly production in units	13,200,000	15,360,000	+2,160,000
(c) Yearly direct labor cost	$ 318,000	$354,000	+ $36,000
(d) Labor cost per 1,000 units	$ 24.00	$23.00	− $1.00
(e) Annual labor cost for the 16.4 % increase in production	. . . . $16.70/1,0000		
(f) Net saving annually through higher speed.	$52,000 — $36,000= $16,000		

EXHIBIT 4-2   Cost Comparison of Speed Range 110 to 128 Units /Min.

Note: There are many other cost considerations to examine, i.e., the greater absorption of burden expense through improved production, possible added expenses through higher indirect costs, additional operating problems, and related situations. All these factors will require further judgments.

Production Scheduling Performance

Formula

$$P = \frac{C}{O}$$

where

P = Scheduling performance in per cent
O = Number of orders scheduled for completion during a given period
C = Number of scheduled orders completed on time during the week

Example

Determine the scheduling performance of the machine shop for the week ending April 10th, when 250 orders were scheduled and 210 orders were actually completed in the scheduled time.

Solution

$$P = \frac{C}{O}$$
$$P = \frac{210}{250}$$
$$P = 84\%$$

Note: There are many variations to the approach given here. It can be calculated on a daily, weekly or monthly basis. Exact dates can be expected to be made or certain tolerances permitted. Adjustments can be made for unusual circumstances, etc.

How to Check on Machine Capacity (Special Purpose Machine)

Formula

$$P = S \times \frac{C}{100}$$

where
P = Standard hours produced
S = Production standard for operation
C = Capacity of machine in units per hour

Problem

Check on one operator's capability to handle a machine processing a *small* electronic element.

An operator is assigned to a job that should theoretically produce 100% of its capacity or 80 units per hour. The standard is 1.24 standard hours per 100 units.

Solution

$$P = 1.24 \times \frac{80}{100}$$

$P = 0.992$ Standard Hours produced

This checks out satisfactorily because the answer is less than one.

Formula

$$P = S \times \frac{C}{100}$$

Check on one operator's capability to handle a machine producing a *large* electronic element.

Example

An operator is assigned a job to theoretically produce 100% of its capacity of 120 units per hour with a production standard of 1.48 standard hours per 100 units.

Solution

$$P = 1.48 \times \frac{120}{100}$$

$$P = 1.78 \text{ Standard Hours produced}$$

This does *not* check out satisfactorily as 1.78 is much greater than one and would require two operators.

Note: Refer to Chapter 2 for machine utilization and operating effectiveness, including charts on one operator running more than one machine.

Determination of Machine Time

Formula

$$T = \frac{\pi}{12} \times \frac{D \times L}{S \times F}$$

where
T = Cutting time in minutes
π = 3.1416
D = Diameter in inches
L = Length of piece
S = Surface speed in feet per minute
F = Feed in thousandths of an inch per minute

Example

Determine the cutting time of a metal bar 5 inches in diameter and 9 inches long. The part has a surface speed of 80 feet per minute and the feed is 0.015 inches per minute.

Solution

$$T = \frac{3.1416}{12} \times \frac{5 \times 9}{80 \times .015}$$

$$T = 0.262 \times \frac{45}{1.20}$$

$$T = 9.83 \text{ Minutes cutting time}$$

Line Balancing

The principal objective of line balancing is to determine the manpower requirements, in conjunction with the necessary machines, to produce a given quantity of product. This element in plant operations is particularly adaptable to the quantitative, mathematical approach. Line balancing refers to the process of arranging in sequence the capacities of specific lines of machines for attendance by adequately skilled workers, to produce a given quantity of a quality product on customers' schedules with a minimum of cost. Assembly work along conveyors also lends itself well to this same treatment.

[2] Automation, or machines linked together without human beings running each machine, approaches this condition. Once the human being is introduced the problem becomes complex. Each operation time becomes a random variable, with a mean, a standard deviation, and some unknown nonnormal distribution. These operations may or may not have to be performed in some specific order.

Operations may move between operations, at some added cost of walking time. Operations may be more finely broken into smaller operations, at some additional cost for picking up and laying down each piece. Machines cannot move and must be supplied in integers. A half a machine is not feasible. Human time can be subdivided, and one man can be put on more than one operation at some increase in training cost. The time for a machine to do an operation can be reduced at the cost of some engineering time. The time for an operator to perform an operation may often be reduced by using a more highly skilled operator, a more highly motivated operator, or by applying motion economy principles.

Line balancing can get extremely complex. In fact, it is often so complex that none of the extremely sophisticated models in the professional literature is truly capable of handling real situations. For our purpose only the basic approach will be illustrated here with a simple numerical example (Exhibit 4–3).*

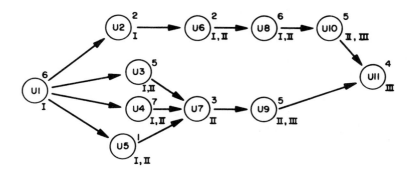

EXHIBIT 4-3

---

[2] Robert A. Olsen, *Manufacturing Management—A Quantitative Approach* (Scranton, Pa., International Textbook Co. Copyright © 1968), pp. 334–337.

* *Journal of Industrial Engineering*—July, August, 1965, p. 244.

The Roman numerals indicate zones. Zoning requires all the tasks at any one station to have at least one zone number in common. The numbers U1, U2, U3 are the operation or route sheet numbers. The other Arabic numbers are the time required for each operation, in tenths of a minute.

Exhibit 4–4 shows two possible solutions to the following line-balancing problem presented here.

The following data is supplied:

Cycle time = 1.0 minutes
Operations = only one person can work on one operation

This results in one operator being idle 30% of the time, and a second operator being idle 10% of the time.

			First Solution					*Second Solution*				
ZONES		OPERA-TION	MINS.	A	B	C	D	E	A	B	C	D	E
I		U1	0.6	0.6					0.6				
I		U2	0.2	0.2					0.2				
I	II	U3	0.5			0.5				0.5			
I	II	U4	0.7				0.7				0.7		
I	II	U5	0.1		0.1				0.1				
I	II	U6	0.2	0.2						0.2			
II		U7	0.3				0.3			0.3			
II	III	U8	0.6		0.6							0.6	
II	III	U9	0.5					0.5					0.5
II	III	U10	0.5			0.5							0.5
III		U11	0.4					0.4				0.4	
Totals			4.6	1.0	0.7	1.0	1.0	0.9	0.9	1.0	0.7	1.0	1.0

EXHIBIT 4-4

When storage is possible between operations, as with machinery operations, more than one operator can be assigned to one operation. The idle time can now be eliminated completely, as indicated in Exhibit 4–4.

Cycle time is chosen by dividing the total work to be done, 4.6 minutes, by some fixed number of operators. Here, $\frac{4.6}{5} = 0.92$ mins. per operator is chosen as the cycle time.

Instructions to operators do not have to be complicated, if they are working with you. For solution 1, Exhibit 4–5, the following might be helpful:

Operator A: Do operations U1 and U2 until a bank accumulates ahead of next operator. Do U1, U2, and U3 until bank disappears.
Operator B: Do operation U3 until a bank accumulates. Do U3 and U4 until bank disappears.
Operator C: Do operations U4, U5, U6, U7 until a bank accumulates. Do U4, U5, U6, U7, and U8 until bank disappears.
Operator D: Do operation U8 until a bank accumulates. Do U8 and U9 until bank disappears.
Operator E: Do operations U9,U10,U11 or U10 and U11, as required.

ZONES		OPERA-TION	MINS.	Solution 1 A	B	C	D	E	Solution 2 A	B	C	D	E	F	G
I		U1	0.6	.60					.60						
I		U2	0.2	.20					.06	.14					
I	II	U3	0.5	.12	.38					.50					
I	II	U4	0.7		.54	.16				.02	.66	.02			
I	II	U5	0.1			.10						.10			
I	II	U6	0.2			.20						.20			
II		U7	0.3			.30						.30			
II	III	U8	0.6			.16	.44					.04	.56		
II	III	U9	0.5				.48	.02					.10	.40	
II	III	U10	0.5					.50						.26	.24
III		U11	0.4					.40							.40
Totals			4.6	.92	.92	.92	.92	.92	.66	.66	.66	.66	.66	.66	.64

Cycle Time: $\frac{4.6}{5} = 0.92$ minutes   Cycle Time $\frac{4.6}{7} = 0.657$ mins. (Balance on 0.660 to simplify example.)

EXHIBIT 4-5

Note: Operation U4 now requires two stations.

Note: Exhibit 4–5 presents a solution to line-balancing problem.

5

Materials Handling Formulas[*]

MATERIALS HANDLING LABOR RATIO

Explanation

In industrial terminology, productivity is the ratio of the cost of incoming materials to the cost of finished products.

Productivity is of special interest to production managers because it usually represents one of the few remaining areas—internally controlled—where substantial cost reduction can be attained. Cost of incoming materials is usually set by outside suppliers; cost of production processes is generally dependent on the design and capability of the production machinery.

How can the effectiveness of material handling be measured in terms of improved production? Financial statements—balance sheet, profit-and-loss statement, and operating cost reports—provide a precise and detailed measure—eventually. But all financial statements, even operating cost reports, usually come too late to tell "how we're doing." They really state only "how we did" for a period of time that is already history. These statements do not provide day-to-day control of costs and profits. Neither do they provide all the information necessary to find improvement opportunities. Cost are often combined in a manner that hides some of them, protecting them from reduction or elimination.

Fortunately there is a more practical method of measuring all material handling costs—a method that is becoming more popular (and accurate). Its success is based on the use of seven ratios that help control costs influenced by material handling operations. Use of these seven ratios will help spotlight and control the direct costs of

* Text in this chapter, and Exhibits 5–8—5–11, from "An Introduction to Material Handling," used with permission of the Material Handling Institute, Inc. Copyright © 1966, Pittsburgh, Pa., pp. 19–26.

Exhibits 5–1—5–7 are from "Management Guide to Productivity," Yale Materials Handling Division, Yale & Towne, Inc., Eaton Yale & Towne, Inc. Copyright © 1965, Philadelphia, Pa., pp. 3, 7, 10, 13, 16, 19, and 22.

handling materials; find and understand hidden operating costs that can be reduced by better material handling practices; and explore opportunities to improve profits through utilization of newer techniques and equipment.

There is no "magic number" which can be applied to any one of the seven ratios. They merely represent a way of establishing some quantitative measure of the present ratio of material handling productivity. To use these "quantitative figures" effectively, calculations can be made for systems used in the past which, in turn, permit comparisons to be made. If figures are available, comparisons should be made with facilities having similar handling problems. Used consistently, periodically, and economically as yardsticks, the seven ratios will enable a manager to see which way handling costs are going—and by how much.

In short, the seven ratios can be used as controls to guide short and long-range profit improvement. The ratios are:

1. Material handling labor
2. Direct labor handling loss
3. Movement/operation
4. Manufacturing cycle
5. Space utilization efficiency
6. Equipment utilization
7. Aisle space potential.

1. Materials Handling Labor Ratio (See Exhibit 5-1.)

This represents the number of personnel assigned to material handling duties in proportion to the total operating force. It determines the proportion of an establishment's labor force that is directly chargeable to material handling needs.

The material handling labor ratio is given by:

Formula

$$MHL \text{ ratio} = \frac{\text{Personnel assigned to } MH \text{ duties}}{\text{Total operating personnel}}$$

The value of the material labor ratio is more universally accepted than that of any other of the ratios. It is simply a special form of the indirect labor ratio (a portion of the indirect labor ratio). Since the indirect labor ratio is so well understood and widely used, the material handling labor ratio is easily comprehended and readily accepted.

In specific terms, the material handling labor ratio can be used to:

(a) Control basic material handling expense in relation to total labor (to maintain the status quo).

(b) Determine the extent of improvement possible in moving materials into and out of the establishment, and between operations (to break through the status quo).

(c) Gauge the success of any changes in material handling in terms of labor saving.

(d) Study long-term trends in material handling operations to justify or to alter earlier decisions on change.

To calculate the material handling labor ratio accurately, first determine the

WORKSHEET No. 1—SHEET 1

For Computation of your Materials Handling Labor Ratio

Materials Handling Labor Ratio = $\dfrac{\text{Materials handling labor}}{\text{All labor}}$

Our materials handling labor consists of labor assigned solely to materials handling duties

1. Materials handling equipment operators	No. of persons	Payroll dollars per year
a. Non-powered hand trucks	1	
b. Powered industrial trucks		
(1) Walkie platform & pallet trucks	2	$ 8,600.
(2) Fork lift & platform trucks	9	42,300.
(3) Tractor trailer systems	1	
(4) Mobile cranes		
(a) Operators	1	5,440.
(b) Riggers	1	4,700.
(5) Tractor shovels, bulldozers, etc.	1	
(6)		
c. Cranes, hoists, & other overhead equipment		
(1) Operators	2	10,160.
(2) Riggers	2	9,600.
d. Conveyor loaders and unloaders	3	15,600.
e. Intraplant motor trucks	1	4,200.
f. Intraplant railroads	1	18,000.
g. Elevators	4	
h. Manual handling labor (such as palletizing)		
TOTAL	25	$118,600.

2. Workforce for activities essentially devoted to:		
a. Receiving dock	4	$ 18,000.
b. Shipping dock		
c. Raw materials storage	1	4,500.
d. Finished goods storage	5	22,000.
e. Distribution warehousing		
f. Scrap & salvage operations	1	4,200.
g.		
TOTAL	11	$ 48,700.

WORKSHEET No. 1—SHEET 2

For Computation of your Materials Handling Labor Ratio (cont'd)

3. Activities partially devoted to, or supporting, the materials handling function	No. of personnel	Annual payroll dollars	% Chargeable MH	Net annual dollars for MH
a. Tool room & supplies issue	3	$17,000.	50%	$ 8,500.
b. Maintenance of MH equipment	1	6,000.	100%	6,000.
c. Production control, particularly dispatching & expediting	3	20,000.	100%	20,000.
d. Packaging operations	4	18,400.	25%	4,600.
e. Inventory control records	4	18,000.	50%	9,000.
f. Inspection	2	11,000.	20%	2,200.
g. Traffic	—	—	—	
Total	17			$50,300.

Annual Materials Handling Labor Cost:

1. MH equipment labor   Payroll $ 118,600.
2. MH activities labor   Payroll $ 48,700.
3. Activities partially chargeable   Payroll $ 50,300.

Item A: Total MH labor $217,600.

Item B: Total annual payroll (Ordinarily computed by taking operating payroll and omitting general administrative and sales payroll) $2,400,000.

MH Labor Ratio = $\dfrac{\text{MH Labor}}{\text{Total Payroll}}$ = $\dfrac{\text{Item A}}{\text{Item B}}$ = $\dfrac{\$217,600.}{\$2,400,000.}$ = 9 %

EXHIBIT 5-1

number of people who do nothing but handle and move materials full time. Add the equivalent number of people who support this group—clerks, supervisors, and material handling maintenance men. This total is the numerator in the formula. The denominator is simply the total number of people on the operating payroll.

2.   Direct Labor Handling Loss Ratio (See Exhibit 5-2.)

This is the ratio of material handling time lost by direct labor to total direct labor time. It measures the relative amount of direct labor lost because the workers are required to handle materials when they could be producing.

The direct labor handling loss ratio is given by:

Formula

$$DLHL = \frac{MH \text{ time lost by direct labor}}{\text{Total direct labor time}}$$

This ratio is useful to check lost production time in operations with high direct labor content and to appraise the potential savings of better handling methods.

One of two methods may be used to calculate the direct labor handling loss ratio for a department or operation. Estimates based on experience can be tricky in this case; this ratio could fool an expert.

If detailed time standards data are available, handling time included in the direct labor standards can be extracted. Select a normal mix of products and operations, total the elemental handling times for these operations, then compare the sum with the total standard direct labor time. The resulting percentage is the amount of handling time buried in direct labor expense. . . . Work sampling can also be used here.

3.   Movement/Operation Ratio (See Exhibit 5-3.)

This is the ratio of the number of "moves" inherent in a process to the number of productive operations in the process. If it takes 25 productive operations to manufacture a product, and the materials and parts that have to go into the product must be moved 125 times to make the product, then this ratio is 125 to 25, or 5 to 1. Mathematically, it looks like this:

$$M/O \text{ ratio} = \frac{\text{Number of moves}}{\text{Number of productive operations}}$$

The movement/operation ratio measures the relative efficiency of the material handling plan. It shows how well the handling system works, but not necessarily the degree of mechanization of the handling plan.

Primary worth of the movement/operation ratio is its "scare value." When ratios for high-volume products are calculated, managers are frightened into trying to reduce them. This leads to better material handling methods, lower product costs, and higher profits.

To calculate this ratio, count each productive operation and inspection in one column. (List them if preferred.) In another column count each time the materials and parts are moved from one location or position to another, whether the distance be long or short. Add to this count the number of times tools and supplies are moved

Example

DIRECT LABOR HANDLING LOSS RATIO

Assume that work sampling studies (either estimated or taken by rigorous statistical methods) show the following about direct labor activities in five departments.

(1) Dept.	(2) Observations made on operator #	(3) % of time spent on handling*	(4) No. of operators in this dept.	(5) Dept. payroll per year	(6) Direct labor payroll spent on handling* —estimated. Col. 3 × Col. 5	(7) Estimated amount that could be saved by better handling practice
A	1	10	18	$120,000	$12,000	$3,000
B	6 9	12 } 15	30	$180,000	$21,600 (@ 12%)	$6,000
C	12	8	7	$48,000	$3,840	(none)
D	15 22 31	18 } 21 } 17	42	$256,000	$46,080 (@ 18%)	$25,000
E	41	6	8	$51,000	$3,060	$1,000
				$655,000	$86,580	
Total feasible savings						$35,000

Tentative conclusions:

1. Direct labor handling loss ratio is $\frac{86,580}{655,000} = 13.2\%$ (quite low).

2. Still, by reasonably critical examination of each department, it appears that materials handling improvements could readily save $35,000 per year.

3. Department D needs special attention. The handling loss seems excessive.

* *Other than work feeding which is properly part of the operator duties.*

EXHIBIT 5-2

WORKSHEET No. 2—SHEET 2

For Computation of your Movement/Operation Ratio *(cont'd)*

Productive operation	Moves or delays (x) distance		Description, or reason for activity	Handling technique used		Needs further investigating		
	No.	Ft.		Manual (√)	Mechanized (√)	Methods (√)	Layout (√)	Equipment (√)
	11	2	Put on floor	√		√		√
	X	1	await installation with other parts					
	12	2	Put on pallet	√		√		
	13	60	take truck to assembly location		√			
	X	1	await distribution	√		√		
	14	2	Put on hand truck			√		Costs this to mechanize
	15	10-30	Move to work stations	√		√	√	√
	16		broken floor	√				
	X	1	await work					
	17	8	operator moves to bench	√		√	√	√
Assembly								

WORKSHEET No. 2—SHEET 1

For Computation of your Movement/Operation Ratio

Movement/Operation Ratio = (Number of moves) / (Number of production operations)

NOTE: There are two ways to obtain these data. One, For a quick impression of your handling plan, just walk through the shop, tallying moves and operations. Then compute the ratio.

Two, For a more detailed job and information for corrective action, use the form below. Fill in data on "moves" and "operations" in sequence, and add explanatory detail as desired. Totals of Columns 1 and 2 will give data for the M/O ratio. Remaining columns provide guidance for corrective action. Add comments, queries and suggestions as needed.

Product *Components for lower drawer*   Observer *Jones*

Location *Receiving*   To *Finished Goods Storage* Date *January 19, 1961*

Productive operation	Moves or delays (x) distance		Description, or reason for activity	Handling technique used		Needs further investigating		
	No.	Ft.		Manual (√)	Mechanized (√)	Methods (√)	Layout (√)	Equipment (√)
Receiving	1	2	Cartons from R.R. rail to pallet	√		√		√
	2	20	to rec deck		√			
	X	1	await checking	√				
	3	30	to storage (incoming goods)		√			
	X	1	Identify by area/ storage plan	√		√	√	
	4	5	Unpacked/moved to storage loc.	√		√	√	
Inspect	5	2	Lot check of inspect, tally	√		√	√	
	6	3	Reorder pallet after full used	√		√	√	
	7	10	moved to storage location completed	√		√	√	
	8	2-6	Put on shelves	√		√	√	
Issues	9	2-6	Put on handcart	√		√	√	
	10	40	to assembly area pick up area	√		√		√

EXHIBIT 5-3

from one location to another to service productive operations. Divide the total number of moves by the number of operations to find the ratio.

4. Manufacturing Cycle Efficiency (See Exhibit 5-4.)

This is the ratio of actual productive time spent in making a product, to total time. In other words, it is a gauge of the time required to handle a product under ideal circumstances and under actual circumstances. This ratio measures the efficiency of a system for putting materials "through the mill" to make a product. (Note: not all delays are due to material handling.)

Formula

$$MCE \text{ ratio} = \frac{\text{Sum of all production operation times}}{\text{How long it took to make the product}}$$

Another way of calculating this ratio is to use for the denominator the total time between arrival of materials at the warehouse door, and shipment of finished product. This method also considers storage time of raw materials and finished products.

5. Space Utilization Efficiency (See Exhibit 5-5.)

This is the ratio of the *cubic* feet of space usually occupied to net usable storage. It measures the effectiveness of space utilization. This ratio should be applied regularly to warehouse and storage operations; production operations can be scrutinized less frequently because they don't change that rapidly.

This ratio can be found by:

Formula

$$SUE \text{ ratio} = \frac{\text{Space usefully occupied}}{\text{Net usable space}}$$

Space costs should be kept in line with business activity. As a long-range control, this ratio has helped many companies avoid investment in unnecessary buildings. But to get full benefit from using this ratio, include height utilization, as well as floor space, in calculations.

To monitor this ratio, it is possible to employ a crew of technicians full time to measure everything. It requires a reasonable approach. For example, total storage space (net usable) need be measured only once.

The numerator—cubic feet usually occupied— is likewise easily calculated for a stable area such as a production department. It generally won't change enough to warrant recalculation until changes of considerable magnitude have been made.

Storage areas are different—utilization changes frequently and rapidly. Estimate utilization of these areas regularly (monthly, perhaps) to spot trends, opportunities to eliminate or combine, or otherwise make more use. The trick is to estimate used space, rather than to try to measure it accurately. Miscellaneous piles, stacks, and mounds of materials can trap an analyzer into calculations more probing than the problem requires.

COMPUTATION OF MANUFACTURING CYCLE EFFICIENCY

Part 10876, pinion gear, is processed from a blank to a polished gear with groove broached in the hub for spline fit to a shaft.

Sum of all production operations times:

Machining (8 operations)	4 hr 10 min
Heat treatment	1 hr 25 min
Inspection	10 min
Rust inhibition treatment and packaging	4 min

Total required production time = 5 hr 49 min (Call it 6 hours)

Elapsed time in production system—(1 shift operation)

Part entered system	1/ 4/61	8:00 am
Part delivered to finished goods storage	1/20/61	2:30 pm

Therefore, elapsed time = $(12 \times 8 \text{ hr}) + 5\frac{1}{2} \text{ hr} = 101\frac{1}{2} \text{ hr}$

$$\text{MCE} = \frac{6}{101.5} = 6\%$$

Comment: An MCE of 6% is exceptionally high because many parts like this are put into sub-assemblies, which then await further processing and/or assembly. If the elapsed time in such a case were included we would find efficiencies of fractions of a percentage point! A common reaction is that such MCE figures are "unfair" or "unreasonable," since the parts must await other processing. This reaction is quite beside the point. The fact is the MCE is *very* low and the goal is to improve it. Our questions will fall in several areas—

1. Is the MCE for the part alone (prior to sub-assembly) reasonable?
2. Is the overall MCE excessive because it was scheduled too far in advance?
3. Is the overall MCE high because sub-assemblies were delayed through poor planning?

EXHIBIT 5-4

Example

COMPUTATION OF SPACE UTILIZATION EFFICIENCY
FOR A SMALL WAREHOUSE WITH AN OFFICE.

First, net usable cube is calculated. This is done by determining total available interior space, then deducting space occupied by obstacles.

Step 1.

Calculation of total interior space of building 40 ft wide, 100 ft long and 16 ft high:

40 x 100 x 16..64,000 cu ft

Step 2.

Calculation of space occupied by obstacles:

6 columns, 2 ft x 2 ft each...................	24 sq ft
2 washrooms, 12 ft x 10 ft each...............	240 sq ft
Office space 40 ft x 20 ft....................	800 sq ft
Shipping assembly area 20 ft x 30 ft...........	600 sq ft
Space needed to park fork truck, hand trucks, etc. 10 ft x 15 ft.........................	150 sq ft
Space needed to be kept clear in front of doors...	100 sq ft
Total...............................	1,914 sq ft

1,914 sq ft x 16 ft (height)............................30,624 cu ft

Net usable cube   33,376 cu ft

Next, cube actually in use is determined. This is done by measuring the base area of stacked material and multiplying by height. Averages should be used. (If one product of a given size is stored and the quantity in the building is known, the number of units times the net cube will quickly give the cubage in use.)

For purposes of this example, assume that material stored is calculated at 15,800 cu ft.

Therefore:

$$\frac{\text{cubic feet usefully occupied}}{\text{net usable cube}} = \frac{15,800}{33,376} = 47\% \text{ space utilization}$$

EXHIBIT 5-5

6. Equipment Utilization Ratio (See Exhibit 5-6.)

This is actual output of a production facility compared to the theoretical capacity. It measures the extent to which production facilities are utilized.

This ratio can be found by:

Formula

$$EU \text{ ratio} = \frac{\text{Actual output}}{\text{Theoretical output}}$$

If sales aren't taxing capacity, production management won't worry much about improving this ratio. But, if production is impeding sales volume, application of this ratio to selected operations will spot the bottlenecks. Better material handling may solve the problem, although the trouble may lie with one or more of several other functions, including maintenance, scheduling, materials, or operating methods.

When calculating facility, the most difficult problem is evaluating theoretical capacity. Once this has been established on a regular basis (daily, weekly, or monthly), comparison of actual with theoretical output is a simple computation.

Example

COMPUTATION OF EQUIPMENT UTILIZATION RATIO

A key stamping press forming the chassis for a piece of electronic gear is capable of one (1) cycle per minute.

A study of production records over the previous 4 weeks shows that the press produced 6,240 pieces in 9,600 running minutes (4 weeks × 40 hours per week × 60 min per hr).

Therefore, the equipment utilization is only:

$$\frac{\text{actual output}}{\text{theoretical capacity}} = \frac{6,240}{9,600} = 65\%$$

Comment. This can surely be improved. To pinpoint the trouble spots, look into the types of questions raised in the text.

EXHIBIT 5-6

7. Aisle Space Potential Ratio (See Exhibit 5-7.)

This is the ratio of current aisle floor space—to current aisle floor space—minus theoretical optimum aisle floor space. This relationship is just as tricky as it sounds.

To make sure that the apparent anomaly is evident, examine the algebraic expression:

Formula

$$ASP \text{ ratio} = \frac{(\text{Current aisle floor space}) - (\text{theoretical optimum})}{\text{Current aisle floor space}}$$

Ideal ASP ratio would be close to infinity—as the numerator approaches optimum, or zero. This will never happen. This ratio is supposed to be a measure of how much productive space can be gained by reducing space presently devoted to aisles to a minimum. Reduced aisle space means more productive space.

MATERIAL HANDLING COSTS

Explanation

Material handling costs break down into three categories:

(1) Transporting raw material to the plant, and shipping out finished products. Costs are a function of plant location in regard to raw material supply and location of market.
(2) In-plant receiving and storage; movement of materials between processing operations; and warehousing and shipping of finished product.
(3) Handling of material by the operators at machine or assembly benches.

More efficient and economical handling and transportation in any of these three phases contribute to a reduction in cost.

Checklist of Handling Costs:

In any review of material handling costs, all of the following factors should be considered:

Materials

. Purchase price of equipment.
. Cost of all materials used.
. Maintenance costs and repair-parts inventory
. Direct power costs—kilowatt-hours, fuel, etc.
. Cost of oils and lubricants required by equipment.
. Costs of auxiliary equipment.
. License fees.
. Installation costs, including all materials and labor plus alterations and rearrangement of present plant and equipment.

Example

COMPUTATION OF AISLE SPACE POTENTIAL

Measurement of the aisle space in a warehouse shows the following:

	Width	Length	Aisle area
Central aisles	12 ft	400 ft	4,800 sq ft
All others	10 ft	1,500 ft	15,000 sq ft
Total area in aisles:			19,800 sq ft

Investigation shows the 10-foot aisles are used because a 6,000 lb capacity fork truck has to operate in them to handle some heavy equipment. All other loads run under 4,000 lbs. By concentrating the heavy items in one storage area, operations by the 6,000 lb truck will not be necessary in all aisles. The layout can then be rearranged as follows:

	Width	Length	Aisle area
Central aisle for 6,000 lb truck access	12 ft	30 ft	360 sq ft
Heavy storage area aisle for 6,000 lb truck access	10 ft	150 ft	1,500 sq ft
Other central aisles	10 ft	370 ft	3,700 sq ft
All other aisles cut to 7 feet (based on purchase of straddle type, narrow aisle fork trucks)	7 ft	1,350 ft	9,450 sq ft
Total area in aisles:			15,010 sq ft

Aisle space potential is now calculated:

$$\text{Aisle potential} = \frac{\text{Current aisle floor space, minus theoretical optimum aisle floor space}}{\text{Current aisle floor space}} = \frac{19,800-15,010}{19,800} = \frac{4,790}{19,800} = 24\%$$

Thus, 24% of the aisle space could be saved. Or approximately 5,000 square feet. Assume building costs are at least $6 per sq ft. This, then, represents $30,000 in space. This space saving could be made by trading 3 current 4,000 lb trucks for 3 new narrow aisle trucks. Trucks are calculated roughly as follows:

3 new 4,000 lb narrow aisle trucks	$18,000
Less: trade-in on present trucks	4,000
Net cost of new trucks	$14,000

This means that a net equipment investment of $14,000 will release floor space worth $30,000!

But this is only part of the story. For in most instances, floor space saved means additional cubic feet of storage space. This, in turn, permits more stacking. True, it is not possible to stack in *all* the aisle space saved because of layout obstacles, multiples of pallet spaces needed, etc. Assume in this case that study shows it is possible to stack *3 high* in just *half* the space saved. Thus:

$$\frac{2,500 \text{ sq ft available}}{16 \text{ sq ft per pallet}} = \text{floor space gained for 156 pallets}$$

156 pallets x 3 high = 468 pallet loads that can be added

EXHIBIT 5-7

Salary and wages

. Salaries of departmental supervision.
. Direct labor costs (all personnel engaged in the operation of machinery or equipment).
. Clerical labor, such as time keeping and production recording.
. Training costs (cost of labor actually devoted to training or makeup guarantees while in training).
. General staff and service departments:
  a. Accounting.
  b. Purchasing.
  c. Personnel.

Financial charges

. Interest on investment.
. Insurance (fire, flood, compensation) and social security.
. Real and property taxes.
. Rate of depreciation.
. Rate of obsolescence.

Miscellaneous direct charges

. Freight on equipment and materials.
. Traveling expenses for personnel who procure and inspect equipment.

Savings

. Unamortized value of displaced equipment.
. Savings of fixed charges against displaced equipment.
. Increased production resulting from new equipment.
. Savings in labor costs provided by new equipment.
. Savings in indirect labor and departmental burden.
. Savings in time resulting from new installation.
. Savings in damage and spoilage through the use of new methods.

Cost of New Equipment

The purchase of new equipment must be justified; in other words, will the equipment pay for itself and add to company profits in the relatively near future?

Key factor in evaluating equipment cost is labor—how much direct labor is required to do a certain job with the old machine or system compared with the new. (As fringe benefits to production employees increase, managers in industry are willing to spend more and more on capital equipment that saves labor.)

In addition to labor, another important factor, especially to management, is the "time" value of money. New machinery requires capital, and it is important to show that this capital is being used in a productive fashion.

Accumulation of cost data is generally done on an annual basis and is most familiar to management on an annual cost and return basis. Therefore, methods shown in this book are based on the annual cost method of calculating payback.

Methods for Forecasting Profitability of New Equipment

In general, there are four methods of forecasting how profitable equipment will be (Exhibits 5-8—5-11):
1. Payoff period method
2. Estimating returns on investment
3. MAPI systems
4. Discounted cash-flow method.

These methods are not based on accounting procedures—they are techniques used in engineering departments (see examples). To use these methods, however, it is necessary to be familiar with several terms that are not in the general engineering vocabulary.

The annual capital cost of any piece of equipment is made up of depreciation, which is the annual shrinkage in market value of the equipment as it ages, and the interest charge on the investment that the machinery represents. Capital cost is based primarily on the following four factors: 1. First cost of the equipment. 2. Life expectancy of the equipment. 3. Salvage value at the end of its useful life. 4. Rate of return required on the investment.

FIRST COST—The first cost of new equipment is not only the purchase price, it should also include transportation costs, installation, and tryout (or debugging).

Purchase price and transportation are readily determined and are usually fixed. Due to the increasing size and complexity of modern machinery, the installation costs are more frequently becoming significant. Preparation costs often involve new power requirements, special foundations, and sometimes the moving and rearranging of entire departments.

Time required to checkout new equipment increases with a degree of complexity of the new equipment. In many cases where the machine is the first of its kind ever built by the supplier, the debugging costs approach 100 per cent of the purchase price. Failure to properly estimate debugging costs has frequently caused severe budgeting distress.

LIFE EXPECTANCY—Obsolescence is the primary consideration in the determination of this factor. With rapid strides in research and development, new improved machines are becoming available almost as frequently as new automobiles. New materials and processes are likewise reaching the market in increasing variety. These developments, coupled with changing customer demands, force each company to modify its own products to keep up. Thus, the risk of obsolescence, rather than physical life, becomes the determining factor. Neither of these are related to the tax life of the equipment under present income tax regulations.

SALVAGE VALUE—Some handling machinery is designed to perform a specific job, and because of its lack of flexibility, its value when obsolete may approach zero. Other general purpose machinery may have some salvage value regardless of its age. Estimates of the salvage value of special purpose machinery should be very conservative; salvage value of general purpose machines can be estimated by checking used equipment catalogs.

RATE OF RETURN REQUIRED—This is the price of the risk involved in purchasing equipment. Management must not underrate the risk involved in investing in automatic equipment and must compensate fully for that risk in determining the rate of return required.

Payoff-Period Method

Often called the engineers' "rule-of-thumb" method, this is the one most commonly used on capital expenditure request forms. The investment is divided by the annual earnings (net income from the investment) to give the time (in years) needed to break even.

Actually, this method does not measure over-all return on investment. It tells only how long it will take to get the money back into working capital. The actual value of the new machinery depends on how much value or useful life is left after the payoff period.

A typical payoff-period computation (below) shows a number of factors that are often considered. There is no accepted standard for the factors to be included in this kind of reckoning, which is one criticism of it. A more serious shortcoming is the assumption that a machine is likely to be the best buy if it pays for itself rapidly. This method may indicate purchase of a ramshackle machine which will pay for itself in 18 months; but which will be ready for the junkpile in 24 months. A better machine may take 24 months to pay off, but it might continue to produce earnings for many years.

The payoff method is a good risk indicator. It can serve as a coarse screen to pick out high-profit projects which are so desirable that they will not require close study, and those that show such poor promise that it isn't worthwhile to conduct a thorough study. It has a built-in safety factor against the possibility of a short economic life.

		Present Machine (5 years old)		Proposed Machine
First Cost	1a)	$20,000	1b)	$30,000
Annual Costs				
Hours of operation	2a)	2000	2b)	2000
Direct Labor	3a)	$4000	3b)	$1000
Maintenance & Repair	4a)	300	4b)	200
Power	5a)	1800	5b)	2000
Depreciation, Straight Line, 10 years	6a)	2000	6b)	3000
Interest on Undepreciated Investment, 6%	7a)	600	7b)	1710
Burden	8a)	300	8b)	260
Floor Space Cost	9a)	400	9b)	350
Property taxes and insurance	10a)	1000	10b)	3000
Totals	11a)	$10,400	11b)	$11,020
Salvage value, old machine	12a)	1020		————
Net total, new machine		————	13b)	$10,000
Output rate, units	14a)	14 per hr.	14b)	20 per hr.
Cost per unit	15a)	37.14¢	15b)	25.0¢

Total advantage ((15a-15b) x 14b x 2a) ═ $4856 per year.

Approx.* payoff period (($10,000/$4856) x 12)═25 months.
*Without adjustment of depreciation, etc., for 2nd and 3rd years.

(Note that this is biased by inclusion of depreciation.)

Estimating Returns on Investment

In common use, this method is frequently called the "accounting method" since it is closely related to many of the concepts used in conventional accounting practice. As in the payoff-period method, many variations are possible because of the different factors that may be used.

If the factors used are as crude as those for an ordinary payoff-period computation, then the only advantage of this method is that it takes into consideration the length of time the equipment will be useful. In practice, those who use the return-on-investment method tend to consider additional factors, making the computation more complete. If skillfully applied, the results can be useful, but critics object (rightfully) that the method depends on the way the estimator juggles figures. Misleading answers can result from inexpert or biased use.

Usual Accounting Procedure — 3 versions

(a) $\dfrac{\text{Initial annual earnings}}{\text{Original net investment}} = \dfrac{2695}{6000}\,(100) = 44.9\%$

(b) $\dfrac{\text{Initial annual earnings after taxes}}{\text{Original net investment}} = \dfrac{2695\,(.50)}{6000}\,(100) = 22.5\%$

(c) $\dfrac{\text{Initial annual earnings after depreciation and taxes}}{\text{Original net investment}} = \dfrac{\left(2695 - \dfrac{7500}{5}\right)(.50)}{6000}\,(100) = 10\%$

Average Book Value Procedure — 3 versions

(a) $\dfrac{\text{Average annual net earnings}}{\text{Average book value}} = \dfrac{2695}{3750}\,(100) = 71.8\%$

(b) $\dfrac{\text{Average annual net earnings after taxes}}{\text{Average book value}} = \dfrac{2695\,(.50)}{3750}\,(100) = 35.9\%$

(c) $\dfrac{\text{Average annual net earnings after depreciation and taxes}}{\text{Average book value}} = \dfrac{(2695 - 1500)(.50)}{3750}\,(100) = 15.9\%$

EXHIBIT 5-9

MAPI System

Developed by the Machinery and Allied Products Institute, this method was originally defined in about 1950, and is steadily coming into broader usage. Forecasts based on this system are known for their accuracy.

The MAPI method measures how the company would profit with and without the proposed new machine or system. It determines, by analyzing next year's figures, the timeliness of machine replacement. The method is based on an attempt to balance the alternative of replacing a machine now or replacing it later.

To use the method, it is necessary to get a set of charts and forms from the Machinery and Allied Products Institute (below). These forms provide for the systematic inclusion of more than 20 factors influencing the profit to be expected from equipment. While they require some estimating (as do all forecasts — none can be made without guessing at probable operating rates, salvage values, etc.), they enforce the use of an orderly and valid method of computation. The charts cover interest calculations, tax depreciation, and earnings patterns.

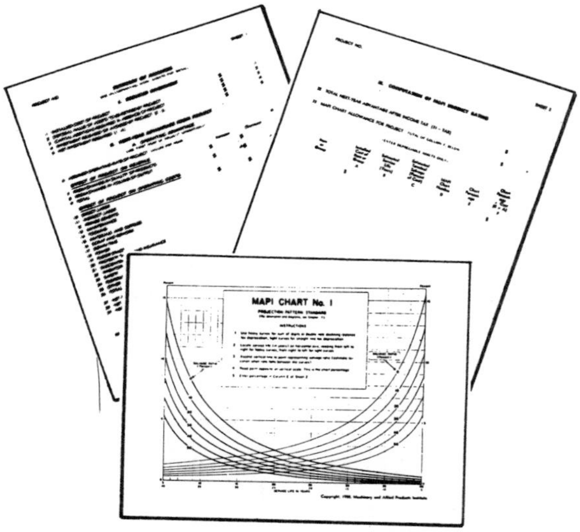

EXHIBIT 5-10

Discounted Cash Flow Method

This method is sometimes called the cash flow-back method, profitability index, or the investor's method. It treats equipment investment in the most sophisticated manner because it allows for continuous compounding of interest and present-worth factors. It is also one of the most controversial methods. The principle on which it works can be applied two ways. Both ways take into account the fact that a dollar you have in hand today is worth more than a dollar that you expect to get in some future year. Reason: the dollar you have today will earn interest; not so the dollar you expect to get in the future.

The method discounts the cash flow-back from the investment, over the years, until it equals the present worth of the investment. Essentially, you find the interest rate which discounts future earnings of a project down to a present value equal to the project cost. A tabulation is prepared covering all income and expense expected to occur over the life of the equip-

ment. The entries are extended by compound-interest rates until the present worth of that series of income equals the investment. The discount rate which yields this equality is the rate of return on investment.

You can pick an arbitrary rate — whatever amount of return your company wants. The sample calculation (below) shows a machine that will give a return of better than 18 percent (and also pay for itself with a little left over). It will provide a return of 20 percent, and just pay for itself. If you want a return of 22 percent, this equipment won't meet your standard. (These discounted amounts are taken from a regular discount table. In practice, only one of the three columns would be shown.)

Time of investment, status of project, unlike alternatives, and adding or subtracting of facilities are easily taken care of by this system. Income and expenses are noted as they occur, as is money spent before projects start and funds returned after completion.

CALCULATIONS

Year	Income before Depreciation	Depre-ciation	Cash Flow-Back after Taxes on Income	on Depre-ciation		Total after Taxes No Interest	at 12%	at 13%
1	$2695.00	$1500.00	$1347.50	$750.00	=	$2097.50	× .893 = $1873.07	× .885 = $1856.29
2	2695.00	1500.00	1347.50	750.00	=	2097.50	× .797 = 1671.71	× .783 = 1642.34
3	2695.00	1500.00	1347.50	750.00	=	2097.50	× .712 = 1493.42	× .693 = 1453.57
4	2695.00	1500.00	1347.50	750.00	=	2097.50	× .636 = 1334.01	× .613 = 1285.76
5	2695.00	1500.00	1347.50	750.00	=	2097.50	× .567 = 1189.28	× .543 = 1138.94
						$10487.50	$7561.49	$7376.90

Interpolating: $\dfrac{7561.49 - 7500}{7561.49 - 7376.90} = \dfrac{61.49}{184.59} = .33$ . . . . . . Add .33% + 12% = 12.33% Return on Investment

Pay-Out period is the cash flow-back after taxes to the point where the cumulative total equals the $7500 investment.

This occurs between the third and fourth years on the cash flow-back chart. The exact point where the money invested has returned is $1207 into the fourth year for a 3.58 YEARS Pay Out.

This method is sometimes called Cash Discount Flow. In this method, a present worth factor is determined.

$f = \dfrac{\text{Immediate cash flow (out)}}{\text{Recurring Annual Cash Flow (in)}} = \dfrac{7500}{2695} = 2.78$

Interest rate from annuity tables (5 years)   =   approximately 23.4%

This is equivalent return on investment before taxes

EXHIBIT 5-11

6

Warehousing Formulas[1]

COST OF CARRYING INVENTORIES

Following is a list of elements that should be considered in determining the cost of carrying inventories. A general discussion of each element is provided after the list.

    a. Interest
    b. Freight
    c. Labor
    d. Space
    e. Insurance and taxes
    f. Warehouse loss and damage
    g. Obsolescence

a. Interest: This is the rate of interest paid for the use of the money invested in the inventory. Depending on how the company wishes to express the cost of inventories, the interest rate used may range from the lowest rate paid for borrowed money to the highest return that can be expected if the money were invested in an alternative use, such as new plant equipment. An indicative rate would be six per cent, but it could well vary from three percent on long-term bonds to 20 percent expected return for alternative types of investment.

b. Freight: This is the transportation charge to get the inventory to the warehouse. The cost may be easily determined by checking the published freight rates or by obtaining quotations from common carriers.

c. Labor: The warehouse handling charges for receiving, shipping, and up-keep of inventories while in storage are definitely part of the total inventory costs. Warehouse administrative costs should also be included.

d. Space: The lease or depreciation cost of the building should be related to the inventory cost. An average or standard for space utilization is necessary to relate

[1] From *Modern Warehouse Management,* Creed H. Jenkins. Copyright © 1968 by McGraw-Hill, Inc. Used with permission of McGraw-Hill Book Company. pp. 20, 21, 98, 104–106, 108–111, 242, 243, 249–253.

the cost of rent to the sales unit. Included in the space charge are lease or depreciation, utilities, building maintenance, and other expenses directly associated with maintaining space.

e. Insurance and Taxes: Insurance and property taxes are inherent costs of inventory. These can be determined through the company accounting department or other departments that may be set up specifically to handle insurance and taxes.

f. Warehouse Loss and Damage: Any handling of goods will inevitably result in some loss or damage. The cost allowable for these may be negligible for certain raw materials. On the other hand, it may be one of the most significant cost factors for such items as fragile glassware and items liable to theft.

g. Obsolescence: A cost allowance should be made for inventories that may go out of style or that become less valuable because of technological changes.

An example of a cost computation for carrying a warehouse inventory is given in Exhibit 6–1.

TURNOVER

Another important cost relationship to be considered in determining an inventory plan is the turnover factor. In Exhibit 6–1, the cost of carrying product X for a year is shown to be $5.48 per unit, or 27.4 per cent of the value. This was based on an inventory which turns over once a year. If the inventory turns over twice or three times per year, the cost per unit is quite different.

COST OF CARRYING INVENTORY

PRODUCT "X"

Cost Element	Computation		Annual Cost
Interest	$20.00/unit × 6% interest		$1.20 each
Freight	$1.00/unit		1.00
Labor	$0.22/unit received	$0.22	
	0.04/unit/month upkeep		
	× 12 months	0.48	
	Total		0.70
Space	$0.08/sq ft/month × 2 sq ft/unit × 12		
	months		1.92
Insurance	$0.10/year/unit		0.10
Taxes	$10.00/$100 assessment @ 25% valuation ×		
	$20.00/unit		0.50
Loss and damage	2%/year × $20.00/unit		0.04
Obsolescence	1%/year × $20.00/unit		0.02
Total annual cost of carrying inventory			$5.48 each
Percentage	$\dfrac{\$\ 5.48}{\$20.00}$		27.4%

The cost of carrying $100,000 of product "X" for one year would be about $27,400 (27.4% × $100,000).

EXHIBIT 6-1

In order to compute the inventory carrying cost for one unit of the product, it is necessary to separate the cost elements that are directly related to the activity from the costs that are related to the inventory while it is at rest in storage. For the inventory costs shown in Exhibit 6–1, the separation of cost elements to determine unit costs would be as follows:

Warehousing Evaluation and Requirement

Elements	A Costs Related to Activity	B Costs Related to Storing
Interest	$ ...	$1.20
Freight, inbound	1.00	
Labor:		
Receiving	0.22	
Upkeep	...	0.48
Space	...	1.92
Insurance	...	0.10
Taxes	...	0.50
Loss and damage (split)	0.03	0.01
Obsolescence	...	0.02
Totals	$1.25/unit	$4.23/unit

The unit cost of carrying inventory when related to turnover is determined by using the following formula:

$$\text{Unit cost} = A + \frac{B}{\text{turnover/year}}$$

Examples of how turnover affects inventory carrying cost per unit are shown below:

Turnover per Year	Equation	Inventory Carrying Cost per Unit
1.0	$1.25 + \dfrac{\$4.23}{1.0}$	$5.48
2.0	$1.25 + \dfrac{\$4.23}{2.0}$	$3.37
4.0	$1.25 + \dfrac{\$4.23}{4.0}$	$2.31
6.0	$1.25 + \dfrac{\$4.23}{6.0}$	$1.95

EXHIBIT 6-2

To determine the total cost of warehousing materials, as opposed to inventory carrying costs shown above, simply add the unit cost of the shipping operation and the outbound freight to the unit inventory carrying costs.
For example:

Shipping operation	$0.40/unit
Freight, outbound	2.00
Inventory carrying cost	5.48
Total warehousing cost	$7.88/unit

As turnover increases, unit costs for carrying inventory decrease. As shown for product X (Exhibit 6–2), the cost decreased from $5.48 per unit (27.4 percent) for a turnover of once a year to $1.95 per unit (9.8 percent) for a 6.0–times turnover per year. Assuming that all items in a product line are supplied through the warehouse, the turnover rate becomes a very important index of inventory performance. However, under a different distribution plan turnover may be a completely meaningless or even misleading measure of performance.

HANDLING TIME STANDARDS

The basic purpose of all handling time standards is to determine how much time an activity should take to do it. To illustrate how this is done, assume that two standards cover the entire work of a warehouse. The first is 2.0 hours to receive and put into storage each inbound shipment. The second is 1.0 hours to order-pick and load each outbound shipment. If the entire activity for a month consisted of 20 inbound shipments and 460 outbound shipments, the computation to determine total standard time would be as follows:

Inbound
20 shipments at standard time of 2.0 hrs. each   . . . . . . 40 hrs/month
Outbound
460 shipments at standard time of 1.0 hrs. each   . . . . 460 hrs/month
                    Total standard time   . . . . 500 hrs/month

Now, if the warehouse actually used 600 hours to perform the work, the handling performance, or efficiency, could be expressed as 83 percent— (500 standard hours ÷ 600 actual hours).

The very simple standards shown in this example will work well for a warehouse handling only one product, but they are inadequate for warehouse operations carrying many different products. There should be separate receiving and shipping standards for each product line, with allowances for each shipment handled and with allowances for housekeeping and upkeep of inventories.

(See Chapters 2 and 3 for further explanation of work standards.)

Exhibit 6–3 is a work sheet showing the use of moderately refined standards and how they are used to compute the standard time for a month's activity at a warehouse.

SPACE UTILIZATION STANDARDS

In the process of establishing space utilization, it is important to first determine the best possible storage layout for the warehouse. The layout should be designed to provide the optimum balance between space utilization and handling efficiency. Methods should then be prescribed for storing each commodity, and the cubic measurement of each commodity should be determined. With the elements of layout, storage methods, and cubic measurements established, the necessary components are available for setting the standards.

ELEMENTS	COMMODITIES				TOTAL
	A	B	C	Z	
Receiving					
Number of inbound shipments	50	10	100	80	
Standard hours each	0.3	0.2	0.4	0.3	
Total standard hours	15	2	40	24	81
Number units received, in 1,000s	100	30	200	240	
Standard hours per 1,000 units	1.0	1.3	0.7	2.0	
Total standard hours	100	39	140	480	759
Total standard receiving time	115	41	180	504	840
Total actual receiving time	140	41	195	700	1,076
% Performance, receiving	82%	100%	92%	72%	78%
Shipping					
Number outbound shipments	150	60	400	300	
Standard hours each	0.2	0.5	0.3	0.2	
Total standard hours	30	30	120	60	240
Number units shipped, in 1,000s	120	20	150	300	
Standard hours per 1,000 units	2.0	1.5	1.0	2.0	
Total standard time	240	30	150	600	1,020
Total standard shipping hours	270	60	270	660	1,260
Total actual shipping hours	280	100	320	700	1,400
% Performance, shipping	96%	60%	84%	94%	90%
Inventory Upkeep					
Total inventory units, in 1,000s	150	80	300	450	
Standard time per 1,000 units	0.1	0.2	0.1	0.1	
Total standard time	15	16	30	45	106
Total actual time	20	20	35	50	125
% Performance, upkeep	75%	80%	86%	90%	85%
Total Time on Standards					
Standard hours	400	117	480	1,209	2,206
Actual hours	440	161	550	1,450	2,601
% Performance	91%	73%	87%	83%	85%
Recoup and Special					
Standard 85% of actual	10	- -	- -	17	27
Actual hours	12	- -	- -	20	32
Total Warehouse Time					
Standard hours including R & S	410	117	480	1,226	2,233
Actual hours	452	161	550	1,470	2,633
% Performance	91%	73%	87%	83%	85%

HANDLING TIME STANDARDS

Warehouse _____   Month _____

EXHIBIT 6-3

Following is an example of how to set space utilization standards, including the necessary elements and typical values for each.

a. General Layout Information Needed

(1) Total space available ....100,000 sq. ft.
(2) Space used for aisles   .... 30,000 sq. ft.
(3) Space used for staging areas   .... 4,000 sq. ft.
(4) Other nonusable space   .... 1,000 sq. ft.
(5) Net usuable storage space (1 less 2,3,4) .... 65,000 sq. ft.
(6) Percentage usable space of total space:
    65,000 sq. ft. ÷ 100,000 sq. ft. = 65 %
(7) Factor for unusable space, 1.0 ÷ 65% = 1.54

b. Specific Commodity and Storage Information Needed

Commodity A
(1) Dimensions: 1 ft by 1 ft by 2 ft = 2 cu ft per case.
(2) Palletized on 4 ft by 4 ft pallets, 4 tiers high, 32 cases per pallet.
(3) Pallets are stacked 4 high, without racks.
(4) Number cases per stack of pallets is 128 cases.
(5) Clearance between stacks is 2 in. per pallet.
(6) Total space per stack (4 ft 2 in. by 4 ft 2 in) = 17.4 sq. ft.
(7) Factor for inventory flexibility, 1.48.

c. Determining Space Utilization Standard

Summary of Data Required
(1) Factor for unusable space is 1.54—see a.(7).
(2) Number cases per stack of pallets is 1.28—see b.(4).
(3) Total space per stack of pallets is 17.4 sq ft—see b.(6)
(4) Factor for inventory flexibility is 1.48 (discussed later).

Formula

$$\text{Space per unit} = \frac{\text{Total space}}{\text{Units of product}}$$

$$\text{Space per unit} = \frac{17.4 \text{ sq. ft.} \times 1.48 \times 1.54}{128 \text{ cases}}$$

$$\text{Space per unit} = 0.31 \text{ sq. ft./case}$$

This converts to 310 sq. ft./1,000 cases.

The space utilization standard used in this example calls for 310 square feet of warehouse space to store each 1,000 cases of commodity A. The standard of 310 square feet per 1,000 cases includes allowances for unusable space, such as aisles and staging areas, and for inventory flexibility. The flexibility is provided to allow for the fact that all the usable space cannot be fully utilized all the time. Extra space is required to allow for pallets stored only one high for order picking and for the fact that variations in inventory levels prevent all rows of pallets from being full clear to the aisles.

The example cited is based on a commodity that is stored on pallets in bulk. Standards can also be developed in the same manner for items stored in racks. The important concept is to relate the number of units that can be stored to a specific number of square feet of floor space. Space utilization standards can be developed using cubic feet instead of square feet. The process of developing cubic space standards is essentially the same as that described in the foregoing example, except for the difference in the unit of measure. Also standards can be related to the weight of commodities rather than to their sales unit, such as for case, gallon, and foot. When there are many different sales units to deal with, it is generally best to convert all of them to pounds. This simplifies the task of working with the standards and reduces the chance of error inherent in using several different units of measure. The validity of standards is not affected by the unit of measure used, if the measure is appropriate and the arithmetic is correct.

Different space utilization standards should be developed for each different product stored in the warehouse. This may appear to be a formidable task at first, but space standards are relatively easy and quick to set compared to handling time standards. With standards for each product, the space utilization performance can be really determined by actual measurement of space used and extension of the inventory on hand by the appropriate standard. For instance, assume that in the 100, 000 square foot warehouse used in the preceding example the only commodity stored was commodity A, with the same space utilization standard of 310 square feet per 1,000 cases. If the inventory were 200,000 cases, the performance would be determined as follows:

Standard space = 200,000 cases × 310 sq. ft/1,000 cases
Standard space = 62,000 sq. ft.
Actual space available = 100,000 sq. ft.

$$\text{Performance} = \frac{\text{Standard space, 62,000 sq. ft.}}{\text{Actual space, 100,000 sq. ft.}}$$

Performance = 62.0%

Space standards can also be used to determine storage capacity. In the operation of a warehouse, it is frequently necessary to determine then umber of units that can be stored in a given amount of space. To do this is simply a matter of dividing the appropriate standard into the amount of space available. In the previous example of the 100,000 square foot warehouse and the single commodity standard of 310 square feet per 1,000 cases, the total number of cases that could be stored is determined as follows:

$$\text{Storage capacity} = \frac{\text{Actual space of 100,000 sq ft.}}{\text{Standard 310 sq ft./1,000 cases}}$$

Storage capacity = 323,000 cases

Exhibit 6–4 is a work sheet showing the use of four different commodity standards to determine the space utilization performance for each commodity and for the total warehouse. The computation and procedures illustrated by the work sheet would be the same if there were hundreds of different commodities instead of only four.

SPACE UTILIZATION STANDARDS

Warehouse _____   Month _____

ELEMENTS	A	B	C	Z	TOTAL
Inventory Number units, in 1,000s	150	80	300	450	
Standard Space St'd. square feet per 1,000 units	100	60	75	80	
Total standard square feet, in 1,000s	15,000	4,800	22,500	36,000	78,300
Actual Space Total actual square feet, in 1,000s	20,000	12,000	33,000	35,000	100,000
% Performance	75%	40%	68%	103%	78%

EXHIBIT 6-4

PERFORMANCE-CONTROL REPORTS

Exhibit 6–5 shows a performance-control report for handling time and space utilization. The report covers only four product lines and two warehouses, but the same type of report is applicable to warehousing systems involving many different product lines and many warehouses. The same kind of report is also appropriate for a single operation. Such report can provide management with valuable insight into its warehousing operations.

WAREHOUSE PERFORMANCE REPORT

Month and Year _____

HANDLING TIME
Standard versus Actual Hours

PRODUCT LINE	LOS ANGELES Warehouse			SEATTLE Warehouse			TOTAL This Month			YEAR-TO-DATE		
	St'd.	Act.	%	St'd.	Act.	%	St'd.	Act.	%	St'd.	Act.	%
A	400	500	80	100	150	67	500	650	77	1,800	2,100	86
B	300	350	86	200	200	100	500	550	91	1,400	1,500	93
C	460	600	77	150	140	107	610	740	82	1,650	1,900	87
D	150	200	75	200	220	91	350	420	83	1,000	1,400	71
Total	1,310	1,650	79	650	710	92	1,960	2,360	83			
Year-to-Date	3,720	4,650	80	2,130	2,250	95	---	---	---	5,850	6,900	85

SPACE UTILIZATION
Standard versus Actual M Sq Ft

PRODUCT LINE	LOS ANGELES Warehouse			SEATTLE Warehouse			TOTAL This Month			YEAR-TO-DATE		
	St'd.	Act.	%	St'd.	Act.	%	St'd.	Act.	%	St'd.	Act.	%
A	18M	20M	90	12M	16M	75	30M	36M	83	78M	97M	80
B	6	8	75	8	10	80	14	18	78	50	75	67
C	26	34	76	4	4	100	30	38	79	90	98	92
D	4	8	50	14	15	93	18	23	78	55	75	73
Total	54M	70M	77	38M	45M	84	92M	115M	80			
Year-to-Date	158M	210M	75	115M	135M	85	---	---	---	273M	345M	79

EXHIBIT 6-5

USE OF STANDARDS IN PLANNING

Following are examples of using standards to plan warehouse operations:

Example

(a) FORECAST—Warehouse management has just learned that it will be required to handle a new product line. After investigation, the following information and standard data are determined:

(b) ACTIVITY—*Inbound:* By rail, 1 million pounds per month for first two months to build up stock, then 500,000 pounds per month.

Outbound: By truck, average 5,000 pounds per month, 250 orders.

Inventory: 1 million pounds average.

(c) STANDARDS—*Practices:* All products are received, stored and shipped on pallets, in full-pallet loads, 2,000 pounds per pallet.

Receiving: Standard time is 0.10 hour per 1,000 pounds.

Shipping: Standard time is 0.15 hour per 1,000 pounds.

Orders: Standard time is 0.10 hour per order.

Inventory upkeep: Standard is 0.01 hour per 1,000 pounds per month.

Space: Standard space utilization is 15 square feet per 1,000 pounds.

(d) DETERMINING LABOR AND SPACE REQUIREMENTS

Function	Activity	Standard	Requirement
Receiving:			
First two months	1,000,000 lb/month	0.10 hr/1,000 lbs.	100 hr/month
Normal	500,000 lb/month	0.10 hr/1,000 lbs.	50 hr/month
Shipping:	500,000 lb/month	0.15 hr/1,000 lbs.	65 hr/month
Orders	250 orders/month	0.10 hr/order	25 hr/month
Upkeep	1,000,000 lb/month	0.01 hr/100 lbs.	10 hr/month
Space	1,000,000 lb/month	15 sq ft/1,000 lbs.	15,000 sq ft

The standard handling time must be converted to expected actual time, since the warehouse is operating at a performance level of only 90 per cent. This is done by multiplying the standard hours by the reciprocal of 90 per cent (1/0.90). It is expected that the space utilization performance will be 100 per cent. Therefore, no conversion is necessary. Making the necessary conversions and using all available information, the following summary of requirements can be stated:

(e) SUMMARY OF REQUIREMENTS—*Handling labor:* 200 standard hr $\times \dfrac{1}{0.90} = 222$ hrs., or 1.3 men will be required the first two months. After that, only 0.9 man will be needed.

Space: 15,000 sq. ft. will be required.

Equipment: Since the product is handled in full-pallet loads, the same lift-truck time will be needed as that indicated for handling labor: 1.3 men for the first two months, 0.9 man after that. No additional pallets will be needed because the product is received on pallets.

WAREHOUSING COST CONTROL

Expense Budgets

Expense budgets are the formal plans for future expenditures. The first thing to be done in preparing a warehouse budget is to establish a reliable forecast of work that will be performed during the period concerned. The next is to predict the cost of performing it.

The forecast of operations should take into consideration the following influences on warehousing expenditures:

(1) Types and quantities of products to be warehoused. This is commonly referred to as "product mix."

(2) Major repair projects for buildings and equipment.
(3) Major stock-rearrangements projects.
(4) Equipment additions and deletions.
(5) Space additions and deletions.
(6) Personnel requirements.
(7) Wage and salary adjustments.
(8) Travel and entertainment plans.

This forecast of operations should then be converted into an expense budget by assigning estimated costs to a prescribed classification of expenses. Estimates of costs should be based on the best information sources available, which includes historical data, engineered standards, and the warehouse manager's educated guesses. Standard costs, which are described later in this chapter, are one of the best means of converting operating forecasts to expense budgets. The problem is that too few warehousing systems have these standards, and if they do, the standards cover only the two major classifications—direct labor and space.

SUMMARY EXPENSE BUDGET
Warehouse Operations Division

(Forecast of Monthly Expense by Quarter)

Warehouse_____   Year_____

	AVERAGE MONTHLY TOTALS				AVERAGE MONTH FOR YEAR
	1st Qtr.	2nd Qtr.	3rd Qtr.	4th Qtr.	
Operating Forecast					
1. Shipping volume, in M units	100	175	200	150	156
2. Storage inventory, in M units	400	300	250	300	312
3. Space: office, in M ft	2	2	2	2	2
4. Space: storage, in M ft	53	53	53	53	53
5. Salary personnel	4	4	4	4	4
6. Direct labor hours	1,330	1,630	1,630	1,400	1,497
7. Lift trucks	4	4	4	4	4
Expenses Account					
1. Salaries	$ 3,500	$ 3,700	$ 3,700	$ 3,800	$ 3,675
2. Wages	5,700	6,900	6,900	6,000	6,375
3. Equipment depreciation	600	600	600	600	600
4. Equipment rental					
5. Equipment operation	200	300	300	200	250
6. Equipment repairs	200	600	200	200	300
7. Building depreciation					
8. Building rent	3,600	3,600	3,600	3,600	3,600
9. Building repairs	700	100	100	100	250
10. Packing supplies	100	180	200	160	160
11. Office supplies	50	90	100	80	80
12. Travel and lodging	200	50	50	40	85
13. Communications	250	350	350	250	300
14. Utilities	200	100	100	200	150
15. Property taxes	700	700	700	700	700
16. Miscellaneous	200	200	200	200	200
17. Total	$16,200	$17,470	$17,100	$16,130	$16,725

Approvals:_____   _____

EXHIBIT 6-6

Exhibit 6–6 shows a forecast of operations and an expense budget for a medium-size warehouse. When the budget is approved by the warehouse manager and his manager, a copy is given to accounting to use in reporting actual expenses against the budget. This performance report contains the same expense accounts as those used in the approved budget. An example of an Expense and Budget Comparison report, using budget figures which are those of the average month for the year, is shown in Exhibit 6–7. This comparison of actual and budget expenses by classification provides an overall control of warehouse expenditures. It shows by expense classification when things are getting out of line. A more thorough investigation into the detailed expenditures can be made on a project basis to isolate just what caused the trouble.

EXPENSE AND BUDGET COMPARISON
Warehouse Operations Division

Performance Report

Warehouse _____ Month __May__

Manager _____ Year _____

EXPENSE ACCOUNTS	CURRENT MONTH			YEAR-TO-DATE		
	Actual	Budget	Over*	Actual	Budget	Over*
1. Salaries	$ 3,750	$ 3,700	$ 50*	$19,050	$18,500	$ 550*
2. Wages	6,100	6,375	275	32,200	31,875	325*
3. Equipment depreciation	600	600		3,000	3,000	
4. Equipment rental						
5. Equipment operation	400	250	150*	1,500	1,250	250*
6. Equipment repairs	175	300	125	1,700	1,500	200*
7. Building depreciation						
8. Building rent	3,600	3,600		18,000	18,000	
9. Building repairs	100	250	150	900	1,250	350
10. Packing supplies	350	160	190*	750	800	50
11. Office supplies	100	80	20*	1,250	400	850*
12. Travel and lodging	20	85	65	175	425	250
13. Communications	250	300	50	1,100	1,500	400
14. Utilities	100	150	50	800	750	50*
15. Property taxes	700	700		3,500	3,500	
16. Miscellaneous	50	200	150	1,650	1,000	650*
17. Total	$16,295	$16,750	$ 455	$85,575	$83,750	$ 1,825*

Reasons for significant variances this month: _____

_____

_____

_____

_____

_____

By _____   Date _____

EXHIBIT 6-7

Developing Charge Rates for Company Warehouses

Company-operated warehouses that charge other divisions of the company for services may either adopt public warehousing rates, like those described in the preceding section, or develop their own. If public warehousing rates are used, the preceding section should serve as a basis for understanding and applying these rates. The way to develop charge rates from the warehouse's own data is discussed here. The method described enables the charge system to be integrated into the warehouse cost-control system, since both the standards and actual data used are the same for both.

There are various ways to develop charge rates. A commonly used method is described, but it is recognized that different approaches will produce equally valid rates. There are possibly as many different ways to develop charge rates as there are different types of accounting systems. The method described makes use of the following elements:

(a) Space utilization standards
(b) Handling time standards
(c) Hourly cost standards
(d) Performance measurement of actual compared to standard.

Since the space utilization and handling time standards were discussed previously, only the hourly cost standards will be explained here in depth.

For the purpose of providing a more complete picture of how charge rates are developed, without confusing the concept with too much detail, assume the following standards and performances (standard ÷ actual) have been properly developed for product X:

Element	Standard	Performance, %
Storage	7.0 sq ft./ 1,000 lbs	80
Space cost	$0.09/sq ft./month	105
Handling	$1.50 hr/1,000 lb.	90
Labor cost	$8.00/direct-labor hour	95

Using this basic information, simple but reliable storage and handling charge rates can be developed as follow:

$$\text{Storage charge rate} = \frac{7.0\,\text{sq ft}/1{,}000\,\text{lb}}{80\%} \times \frac{\$0.09/\text{month}}{105\%}$$

$$\text{Storage charge rate} = \$0.76/1{,}000\,\text{lb./month}$$

$$\text{Handling charge rate} = \frac{1.50\,\text{hr}/1{,}000\,\text{lb}}{90\%} \times \frac{\$8{,}00/\text{hour}}{95\%}$$

$$\text{Handling charge rate} = \$1.40/\,1{,}000\,\text{lb. shipped}$$

The storage charge rate is 76 cents per 1,000 pounds per month and the handling charge rate is $1.40 per 1,000 pounds shipped. These two rates will cover all the storage and normal handling costs for product X.

The one additional charge rate that is needed is for special types of work such as repair and physical inventory taking. The standard labor cost of $8.00 per hour can be used, or the rate can be refined by factoring it for performance. It would then be $8.00/95% or $8.40 per direct-labor hour.

An example of a month's activity for product X and the application of these charge rates follows:

Activity

Inventory on hand at end of month	....	600,000 lbs.
Total shipments during month	....	300,000 lbs.
Special handling	....	10.0 hours

Charges

Storage:	$0.76/1,000 lbs × 600,000 lbs	$456.00
Handling:	$1.40/1,000 × 300,000 lbs	420.00
Special Handling:	$8.40/hr. × 10 hours	84.00
	Total charge for warehousing product X.	$960.00

The assumptions that are made in developing and using this type of simplified charge system are as follows:

a. That the inventory at the end of month is the same as the average during the month.
b. That the one rate for handling is about the average of large and small, and easy-to-handle and difficult-to-handle, shipments.
c. That statically developed averages are a valid basis on which to determine charges.

Development of Direct-Hour and Space Cost Standards

Using the information in the sample Summary Expense Budget (Exhibit 6–6), the following shows how to develop direct-hour and space cost standards.

Not all costs are clearly identified with either handling or space. Some of the expense accounts as shown in the Summary Expense Budget must be allocated between the two. The allocation should be based on a detailed analysis of the accounts, but for the purposes of illustration, the allocation here will be shown by percentage. After all costs have been assigned to either handling or space, the totals are arithmetically divided by the total number of direct-labor hours, and the total number of square feet of storage space, respectively. The resultant quotients are the standard costs.

Expense Accounts	Month	%	Handling	%	Space
1. Salaries	$3,700	100	$3,700		
2. Wages	6,375	100	6,375		
3. Equipment depreciation	600	70	420	30	$180
4. Equipment rental					
5. Equipment operation	250	100	250		
6. Equipment repairs	300	70	210	30	90
7. Building depreciation					
8. Building rent	3,600	...	....	100	3,600
9. Building repairs	250	...	....	100	250
10. Packing supplies	160	100	160		
11. Office supplies	80	100	80		
12. Travel and lodging	85	100	85		
13. Communications	300	100	300		
14. Utilities	150	30	45	70	105
15. Property taxes	700	30	210	70	490
16. Miscellaneous	200	70	140	30	60
Total	$16,750		$11,975		$4,775

Total handling cost ÷ total direct hours = standard cost

$$\$11,975 \div 1,497 = \$8.00/hr.$$

Total space cost ÷ total storage space = standard cost

$$\$4,775 \div 53,000 = \$0.09/sq\ ft.$$

There are many other valid ways of developing these cost standards. Also, the standards may be subdivided into several additional standards, which may be desirable in more sophisticated cost-control systems. One common difference that will be found particularly in warehousing systems that stress cubic measurement is that space will be expressed as a standard cost per *cubic foot* rather than as a function of square feet, as shown in this sample. If space is expressed in cubic rather than square-foot measurement, the process of developing cost standards is the same as the example given. It is just a matter of using cubic feet of warehouse space throughout the storage standards, space-costs and performance-measurement systems instead of using square feet as the unit of measure.

7

Industrial Relations Formulas

CALCULATION OF LABOR TURNOVER RATES[1]

(1)   Formula and Example

Formula for Quit Rate

$$R = \frac{Q}{N} \times 100$$

where

R = Quit rate
Q = Number of employees who quit
N = Total number of employees reported during the period

Example

In one industrial sampling, 623 employees quit between January 1 and 31, while 30,062 employees worked or received pay during the week of January 11–17. The January quit rate for the industry is:

$$R = \frac{623}{30,062} \times 100$$

$$R = 2.1$$

(Current turnover rates for your particular industry can be obtained by writing to the Bureau of Labor Statistics, United States Department of Labor, Washington, D.C.)

In plants where the average work force changes, which is common in many manufacturing establishments, the turnover rate is calculated by dividing the accessions or withdrawals, whichever is *smaller,* by the average work force. This can

---

[1] Lewis R. Zeyher, *Production Manager's Desk Book* (Englewood Cliffs, N. J.: Prentice-Hall, Inc. Copyright © 1969), pp. 177–178.

be done on a monthly basis or yearly. If the turnover figure is expressed on a monthly basis, you would multiply it by 12 to get a turnover rate on a yearly basis.

Formula

$$R = \frac{W}{N} \times 100$$

where

R = Turnover ratio
W = Number of accessions or withdrawals, whichever is *smaller*.
N = Average number of employees in work force

Example

Determine *yearly* turnover rate for a plant with the following labor statistics:

Number of employees added 100
Number of employees withdrawing 80
Average work force (12 months) 400

$R = \dfrac{80}{400} \times 100$ 20%

Turnover ratio is 20 per cent.

JOB EVALUATION

Determining Slope of Wage Trend Line—using the Least Square method with simultaneous equations. (This method is also discussed in Chapter 2.)

Formula

$$T = AX + C$$

where

T = Salary (wkly)
C = Constant
A = $ Value per point
X = Total points per job

Example

See Exhibit 7-1.

Salary	Total Points	Multi- plier	$T = AX + C$	Multi- plier	$T = AX + C$
$175.00	422	1	$175 = 422A + C$	422	$73850 = 178084 + 422C$
175.00	397	1	$175 = 397A + C$	397	$69475 = 157609 + 397C$
135.00	351	1	$135 = 351A + C$	351	$47385 = 79961 + 351C$
100.00	281	1	$100 = 281A + C$	281	$28100 = 78961 + 281C$
172.00	252	1	$172 = 252A + C$	252	$43344 = 63504 + 252C$
147.00	247	1	$147 = 247A + C$	247	$36309 = 61009 + 247C$
90.00	231	1	$90 = 231A + C$	231	$20790 = 53361 + 231C$
90.00	223	1	$90 = 223A + C$	223	$20070 = 49729 + 223C$
90.00	207	1	$90 = 207A + C$	207	$18630 = 42849 + 207C$
90.00	204	1	$90 = 204A + C$	204	$18360 = 41616 + 204C$
93.00	189	1	$93 = 189A + C$	189	$17577 = 35721 + 189C$
90.00	171	1	$90 = 171A + C$	171	$15390 = 29241 + 171C$
$1447.00 = 3175A + 12C$				$409,280 = 871,645A + 3175C$	

EXHIBIT 7-1   Chart for Supervisory Salaries

Determination of Least Square Trend Line Using Simultaneous Equations

$$T = AX + C$$
$$\$1,447 = 3175A + 12C$$
$$409,280 = 871,645A + 3175C$$

Multiply by–
$\times$ 3175
$\times$ 12

$$\$4,594,225 = 10,080,625A + 38,100C$$
$$4,911,360 = 10,459,740A + 38,100C$$

Subtract

$$\$-317,135 = -379,115A$$
$$A = \$0.83651 \text{ per unit}$$

Substitute for "A" ($0.83651)

$$\$1447 = 3175 (\$0.83651) + 12C$$
$$1447 = \$2655.92 + 12C$$
$$-12C = \$1208.92$$
$$C = -\$100.74 \quad \text{Constant Value}$$

Determine Trend Points for Curve

Where:

T = Wkly Salary
C = Constant
A = $ Value per point
X = Total points per job

$T = AX + C$
$T = (\$0.83651 \times 150 \text{ points}) - \$100.74$
$T = \$125.48 - \$100.74$
$T = \$24.74 \text{ at 150 points}$

$T = (\$0.83651 \times 350 \text{ points}) - \$100.74$
$T = \$292.78 - \$100.74$
$T = \$192.04 \text{ at 350 points}$

$$T = (\$0.83651 \times 250 \text{ points}) - \$100.74$$
$$T = \$209.13 - \$100.74$$
$$T = \$108.39 \text{ at } 250 \text{ points}$$

Note: Logarithmic graph paper can be used instead of the above calculations to determine the proper slope of a line.

COMPUTING THE LABOR-TURNOVER INDEX[2]

Example

See Exhibit 7-2.

Month	Turnover per cent	Deviation	D^2	Deviations. S. D.
Jan.	4.4	0.6	0.36	1.302
Feb.	4.6	0.8	0.64	1.736
Mar.	4.8	1.0	1.00	2.170
Apr.	4.3	0.5	0.25	1.085
May	4.2	0.4	0.16	0.868
June	4.0	0.2	0.04	0.434
July	3.6	— 0.2	0.04	— 0.434
Aug.	3.4	— 0.4	0.16	— 0.868
Sept.	3.8	0.0	0.00	0.000
Oct.	4.0	0.2	0.04	0.434
Nov.	4.1	0.3	0.09	0.651
Dec.	4.0	0.2	0.04	0.434
Jan.	3.8	0.0	0.00	0.000
		Total D^2 ....2.82		

EXHIBIT 7-2   Computing the Labor-Turnover Index[2]

Average industry turnover, 3.8%

$$\text{Average } D^2 = \frac{2.82}{13} = 0.216$$

$$\text{Value of 1 } S.\ D. = \sqrt{.216} = 0.46$$

Value of 1 unit of deviation = 2.17

Explanation:

First the labor-turnover index is computed on a monthly basis and is compared with the average turnover figure for the industry, which is 3.8 per cent. The following steps are then taken to determine the standard deviations from the industry average.

(1) In column 3 (Exhibit 7–2) the deviations from the industry's average of 3.8 are shown. The 3.8 has been accepted as a satisfactory standard to achieve.

---

[2] Thomas J. Luck, *Personnel Audit and Appraisal* (New York: McGraw-Hill Book Co. Copyright © 1955), pp. 27–30.

(2) The deviations are then squared, and a total is determined for column 4. The sum is divided by the number of months used in the table, and the resulting figure of 0.216 is used to obtain the square root, which indicates the value of 1 *S.D.* In this case the value of 1 *S.D.* is 0.46 per cent.

(3) The reciprocal of 0.46 is then computed to obtain the value of 1 per cent of deviation, or in this case, 1 per cent of deviation equals 2.17 *S.D.*

(4) As the final step, the deviations in column 3 are multiplied by the value of 1 unit of deviation to obtain the deviations in standard deviation units.

COMPUTING THE PERSONNEL PERFORMANCE INDEX

Top management in general is always interested in raising the organization to new levels of performance. The top executive likes to see new records set forth for output, quality, cost reduction, and general plant efficiency. Thus, nearly every plant has certain standards, or "score cards," that are used to compare departments within the organization. One concern has developed a monthly industrial relations audit report measuring employment, turnover, absenteeism, overtime, and related items. The major officials of the company use this report to check against goals and past performances. Ten indices were selected by the before-mentioned company and are listed below.

If these ten indices are rising, it is assumed that the personnel department is not doing as effective a job as it might. The data for these factors are gathered monthly from available company records. The indices for the ten factors are then computed and combined. It is quite difficult to combine figures derived from various ratios, trends, and percentages. However, it is possible to use "standard deviations" in order to make a combination of like items. Each one of the ten factors is first measured to see how far it deviates from an accepted "norm."

Using the approach previously indicated in Exhibit 7–2, all of the other indices can be similarly calculated. It is then possible to add together figures from each of the ten indices. There is no problem of trying to add together unlike items, for each index is computed in standard deviations from an accepted norm for each index. The standard deviations for any given month can now be added together to give a composite index covering all ten factors relating to the personnel department.

For example, the figures for the month of January would be computed for each index and then added together in a manner as illustrated in Exhibit 7–3.

Each of the ten factors has been weighted in accordance with its importance in determining the value of the personnel department to the company. The weights used in Exhibit 7–3 are not meant to be absolute, but are merely examples of how weights might be used in a given company. Each company should determine its own list of factors and also decide independently upon the weights, if any, to be used. The final index figure, in this case 0.055, is posted to a graph each month. By following the trend of the graph, a quick picture of the performance of the personnel department can be obtained. If there is a sudden upswing in the index, it is advisable for management to make thorough investigations of the causes.

Computing the Personnel Department's Performance Index

January

Factors	Deviation S. D.	Weight	Weighted Factor
Labor index	0.00	2	0.00
Absenteeism	0.25	2	0.50
Participation in company programs	0.25	— 1	— 0.25
Accident rate	— 0.20	1	— 0.20
Grievance rate	0.05	2	0.10
Wage rate	0.05	1	0.05
Rework-time cost	— 0.10	1	— 0.10
Material consumption	0.05	1	0.05
Personnel load ratio	0.25	2	0.50
Overtime hours trend	— 0.10	1	— 0.10
Total			0.55
Weighted average			0.055

EXHIBIT 7-3   Computing the Personnel Department's Performance Index

THE STANDARD INJURY FREQUENCY RATE AND THE STANDARD INJURY SEVERITY RATE.

Injury Frequency Rate[3]

Formula

$$\text{Frequency rate} = \frac{\text{Number of disabling injuries} \times 1{,}000{,}000}{\text{Employee-hours of exposure}}$$

Example

A plant worked 356,000 man-hours during a year in which there were five disabling injuries with a total of 175 days charged. To determine its injury rates, substitute the appropriate figures in the formula.

Solution

$$\text{Injury frequency rate} = \frac{5 \times 1{,}000{,}000}{365{,}000}$$

Injury frequency rate = 13.70

Injury Severity Rate

Formula

$$\text{Injury severity rate} = \frac{\text{Days charged} \times 1{,}000{,}000}{\text{Employee-hours of exposure}}$$

---

[3] National Safety Council: Accident Prevention Manual for Industrial Operation, Chicago, Illinois. Copyright © 1959, pp. 9–11.

Example

A plant worked 365,000 man-hours during a year in which there were five disabling injuries with a total of 175 days charged. To determine its injury rates, substitute the appropriate figures in the formula.

Solution

$$\text{Injury severity rate} = \frac{175 \times 1,000,000}{365,000}$$

$$\text{Injury severity rate} = 479$$

Average Days Charged Per Injury

Formula

$$\text{Average days charged} = \frac{\text{Total days charged}}{\text{Total disability injuries}}$$

$$\text{Average days charged} = \frac{175}{5}$$

$$\text{Average days charged} = 35$$

or for same results use—

$$\text{Average days charged} = \frac{\text{Severity rate}}{\text{Frequency rate}}$$

$$\text{Average days charged} = \frac{479}{13.70}$$

$$\text{Average days charged} = 35$$

Dates for Compiling Rates

Injury rates should be determined as soon as possible after each period (month or year, for example) as the information becomes available. A reasonable time may be obtained for completion of reports. However, absolute accuracy in rates does not justify long delays.

(1) Annual injury frequency rates should be based on all injuries occurring within the year and reported with one month after the close of the year. Monthly injury frequency rates should be based on all injuries occurring within 20 days after the close of the month.

(2) Days charged for reported cases in which disability continues beyond the closing dates in (1) should be estimated on the basis of medical opinion as to probable ultimate disability.

(3) Cases first reported after closing dates stated in (1) need not be included in the standard injury rates for that period, or for any similar subsequent period. However, they should be included, and should replace estimates, in rates for longer periods of which that period is a part.

(4) An injury, and all days lost or charged because of it, should be charged to the date on which the injury occurred, except that for injuries such as bursitis, tenosynovitis, or silicosis, which do not arise out of specific accidents, the date of the injury should be the date when the injury is first reported.

Calculation of Employee-Hours

Employee-hours used in calculating injury rates is the total number of hours worked by all employees, including those of operating, production, maintenance, transportation, clerical, administration, sales, and other departments.

Employee-hours should be calculated from the payroll or time clock records. If this method cannot be used, they may be estimated by multiplying the total employee hours worked for the period by the number of hours worked per day.

For travelling personnel, such as salesmen, executives, and others whose working hours are not defined, an average of eight hours per day should be used in computing the hours worked.

Disabling Injuries

The standard procedure specifies that a work injury is an injury, including occupational disability, which arises out of and in the course of employment.

Definitions

To insure uniformity in the computation of injury rates and thereby provide for comparability among rates, the Standard specifies that only *disabling injuries* shall be counted in the computation of standard injury rates. In general terms, a disabling injury is one which results in death or permanent impairment or which renders the injured person unable to work for a full day on any day after the day of injury. Disabling injuries are of four classes, as follows:

(1) *Death* is any fatality resulting from a work injury, regardless of the time intervening between injury and death.

(2) *Permanent total disability* is any injury other than death which permanently and totally incapacitates an employee from following any gainful occupation, or which results in the loss or the complete loss of use of any of the following in one accident: (a) both eyes; (b) one eye and one hand, or arm, or foot, or leg; (c) any two of the following not on the same limb: hand, arm, foot, or leg.

(3) *Permanent partial disability* is any injury other than death or permanent total disability which results in the complete loss or loss of use of any member or part of a member of the body, or permanent impairment of functions of the body or part thereof, regardless of any pre-existing disability of the injured member or impaired body function.

(4) *Temporary total disability* is any injury which does not result in death or permanent impairment, but which renders the injured person unable to perform a regularly established job which is open and available to him, during the entire time interval corresponding to the hours of his regular shift or any one or more days (including Sundays, days off, or plant shutdown) subsequent to the date of the injury.

As may readily be seen, minor injuries are excluded in the calculation of the standard injury rate. The reason is that it is impossible to obtain standardization of minor cases.

Number of Lost Time Accidents Per Hours Worked

Formula

$$N = \frac{L}{T} \div 10,000$$

where

N = Number accidents per 10,000 hours worked
L = Number of lost time accidents
T = Total hours worked

Example

The maintenance department of a plant experienced 14 lost time accidents in nine months of the year. Total hours worked of 17 employees was 25,500.

Solution

$$N = \frac{14}{25,500} \div 10,000$$
$$N = 5.48$$

8

Capital Investment Decisions[1]

METHODS OF QUANTITATIVE ANALYSIS

Quantitative techniques lead to two general results:

   (1) A comparison between two projects to show which is the better. Thus, a new project may be compared with an existing project or two new projects may be compared with each other.
   (2) An index for each project which permits comparison or ranking of any number of projects.

Certain implicit assumptions underlie the various methods. Usually these are not stated but are easily recognized once they have been identified. Possible assumptions are that types of equipment being compared on the basis of cost/benefit relationships:

   (1) Perform the same function at different cost
   (2) Perform different (improved quality) functions at
     (a) The same cost
     (b) Different costs
   (3) Perform improved functions (improved productivity) yielding greater revenue at
     (a) The same costs
     (b) Different costs
   (4) Perform new functions (make a different product to yield increased profit by means of a better cost-revenue relationship)
   (5) Have different lives and patterns of costs and revenue.

Thus, methods of analysis which compare only costs or cost savings imply that all other factors remain the same.

---

[1] From *The Management of Capital Expenditures*, R. G. Murdick and D. D. Deming. Copyright © 1968, McGraw-Hill, Inc. Used by permission of McGraw-Hill Book Company. Pp. 64–72.

SOME BASIC CONCEPTS

The economics of decision making is simple in concept. The management of a company has in its possession a net aggregate of physical and cash (or cash equivalent) assets. The question which management faces is: "How can we best employ these resources so as to yield the greatest on-going return?" It does not matter what the present assets cost, or how long or short a time the company has had them, except as these matters affect future cash benefits. It conceivably might be advisable for a company to liquidate its assets and pay off the owners if the banks offer a higher return on savings than the company can earn for the owners.

The future result of a management decision is represented by a series of cash expenditures and cash revenues which, combined, are the anticipated cash benefits. This stream of cash benefits, discounted to the present time, is the measure of the value of an investment. Management should be seeking those investments which will optimize all company cash benefits in the short run, and particularly in the long run.

A second basic concept is the distinction between accounting data and the cash flows relevant to decision making. The decision maker often must rely on accounting data to compute cash-flow items. One of these accounting or "book" terms which causes considerable confusion is "depreciation" and another is "profit." The relationship between these book terms and the cash-flow items of "investment" and "cash benefits" is indicated below:

Cash benefits (net cash inflow)	Investment (cash outflow)
— depreciation (book entry)	— depreciation (book entry)
Profit (book entry)	Reduces investment (book value)

One important relationship derived from the above table is that cash benefits over the years provide for profits and capital recovery:

$$\text{Cash benefits} = \text{profit} + \text{depreciation}$$

The other more obvious conclusion is that: Investment — book value of investment = depreciation.

EXHIBIT 8-1   Symbolic Relationship Between Book and Cash Items Taking into Account Taxes

	On Books	Cash Flow
Cash benefits due to savings or increased earnings....	CB	CB
Depreciation on new machine....DN		
Less: depreciation of old machine ....DO		
Additional depreciation expense......$DN - DO = DA$		
Additional taxable income................	$CB - DA$	
Increase in income tax (50%)....	$0.5\,(CB - DA)$	$0.5\,(CB - DA)$
Net additional income after taxes......	$0.5\,(CB - DA)$	
Net annual cash benefits after taxes.........		$CB - 0.5\,(CB - DA)$
		$= 0.5\,(CB + DA)$

EXHIBIT 8-2   Numerical Example of Relationship Between Book and Cash Items Taking into Account Taxes

	ON BOOKS	*CASH FLOW*
Cash benefits due to savings or increased earnings ..	$10,000	$10,000
Depreciation on new machine $2,000		
Less: depreciation on old machine 1,000		
Additional depreciation expense .. $1,000	1,000*	
Additional taxable income	9,000	
Increase in income tax (50%) ·..............	4,500	4,500
Net additional income after taxes	$4,500*	
Net annual cash benefits after taxes		$5,500*

* $1,000 + $4,500 = $5,500

A third concept is that while past (sunk) costs and depreciation of equipment are by themselves irrelevant to the decision maker, they do affect the cash flow of taxes in the future. The relationship between depreciation (a book entry depending on past history) and taxes which represent future cash expenditures is shown first in symbolic form (Exhibit 8–1) and then in a numerical example (Exhibit 8–2).

A fourth basic concept is that $1 today is worth more than $1 offered to us at some future time. For this reason, revenues or costs which will occur in the future must be reduced or discounted to make them comparable to today's dollars. Conversely, if we loan out a dollar today to be returned at some future date, we expect a greater amount to be returned for each year it is loaned out.

At the interest rate i, a sum S will be worth A dollars at the end of one year. The relationship is:

$$A = S + iS = S (1 + i)$$

After two years,

$$A = S (1 + i) + iS (1 + i) = S (1 + i)^2$$

After n years,

$$A = S (1 + i)^n$$

If $S =$ $100, $i =$ 4%, and $n =$ 5 years, then

$$A = 100 (1.04)^5 = (100) 1.2166 = \$121.66$$

The present value of the future amount A is found by solving for S:

$$S = \frac{A}{(1 + i)^n}$$

If $A =$ $100, $i =$ 4%, and $n =$ 5 years

$$S = 100 (1.04)^{-5} = 100 (0.8219)$$
$$S = \$82.19$$

Thus $82.19 today is as good as $100 payable five years hence. Some of the cruder methods of investment analysis do not take into consideration this "time value" of money.

The fifth basic concept is also related to the time value of money. When an investment is made in equipment, the company ties up its own funds or borrows money. In either case, there is an expense associated with the use of money. If the company borrows money, it pays interest to whoever loans it. This is an expense easily identified by the accountant and his books. If the company uses its own money, it foregoes the opportunity of loaning out this money or putting it to work in some other way. The opportunity of earning this money represents a cost to the company just as real as the interest rate for borrowing. This cost does not appear in the accountant's books but must be taken into consideration in decision making. It is *implicit* cost called an "opportunity" cost.

Payback Method

The payback method is the simplest and perhaps most commonly used quantitative method for evaluating an investment. It answers the question: "How many years will it take for the cash benefits to pay for the cost of the investment?"

$$(1) \quad \text{Payback time} = \frac{\text{investment}}{\text{cash benefits per year}}$$

For example, suppose we wish to purchase a group of machines to produce a new product. The machines cost \$240,000 and the cash benefits are anticipated to be \$40,000 per year. Cash benefits are, of course, the excess of revenues over cash expenditures. Then,

$$\text{Payback time} = \frac{\$240,000}{\$\ 40,000} = 6 \text{ years}$$

If we are considering replacing a machine whose salvage value is \$5,000 with a new machine costing \$23,000, which will save us \$4,000 per year in operating expenses, the formula may be restated slightly.

$$\text{Payback time} = \frac{\text{net investment}}{\text{net cash benefits}} = \frac{\text{net investment}}{\text{savings}}$$

$$\frac{\$23,000 - 5,000}{\$4,000} = 4.5 \text{ years}$$

$$(2) \quad \text{After payback} = \frac{\text{investment}}{\text{cash benefits}} = \frac{\text{investment}}{\text{profits after taxes} + \text{depreciation}}$$

If we refer to Exhibit 8–1 and remember that *CB* represents cash benefits and *D* represents depreciation, then

$$\text{After payback time} = \frac{\text{investment}}{0.5\ (CB - D) + D} = \frac{\text{investment}}{0.5\ (CB + D)}$$

An investment of \$85,000 for plant and equipment and \$15,000 for working capital is being considered. Profits of \$10,000 per year after taxes are expected. Depreciation charges will be \$1,700 per year. What will be the after tax payback time?

$$\text{After payback time} = \frac{\text{investment}}{\text{profits after taxes} + \text{depreciation}}$$

$$\text{After payback time} = \frac{\$100,000}{\$10,000 + \$17,000} = 3.7 \text{ years}$$

Or alternately,

$$\text{After payback time} = \frac{\text{investment}}{0.5 \ (CB + D)}$$

$$\frac{\$100,000}{0.5 \ (\$37,000 + \$17,000)} = 3.7 \text{ years}$$

Return on Investment

The common accounting approach to return on investment is equivalent to the reciprocal of the payback formula.

(1) $\text{Return on investment} = \dfrac{\text{cash benefits}}{\text{investment}}$

(2) $\text{Cash-flow return after taxes} = \dfrac{\text{cash benefits}}{\text{investment}}$

$$= \frac{\text{profits after taxes} + \text{depreciation}}{\text{investment}}$$

$$= \frac{0.5 \ (CB + D)}{\text{investment}}$$

The common accounting approach often uses gross investment rather than the correct value, the net invesment, for replacement analysis. In addition, the accounting approach may use a variation which does not represent future cash flows but uses an accounting term in the numerator which reduces the apparent return.

(3) $\text{Return on Investment} = \dfrac{\text{Profits after taxes}}{\text{investment}}$

First-Year Performance

In the first-year performance method, costs are calculated for each of the alternatives for the next year only. This avoids discounting and covers the period about which most is known. Consider the case of a service company which either rents cars or uses them for company purposes such as delivery of sales.

Example:

Cost of new car – $3,000, life—5 yrs		Old car value – $ 1,500, life—3 yrs.	
Depreciation next year	$ 900	Depreciation next year	$ 700
10% interest on capital	300	10% interest on capital	150
Operating cost 8c/mile		Operating cost 11c/mile	
(10,000 miles)	800	(10,000 miles)	1,100
	$2,000		$1,950

In the above example, the quantitative analysis indicates that both alternatives are about the same from an economic viewpoint. The final decision would be based upon judgment of other factors and of the degree of uncertainty of the data employed.

Total-Life Average or Full-Life Performance

The total-life average lumps together all costs involved in owning and operating a unit of equipment over its life. This sum is divided by the estimated life of the machine to give an average annual cost. The comparison between investments is thus on a cost basis, and it is assumed that the alternatives produce the same revenue.

$$\text{Average Cost/Year} = \frac{\text{operating costs} + \text{depreciation} + \text{interest}}{\text{estimated years of life}}$$

Example:

	Old machine	*New machine*
Given:	$2,000 market value	$11,000 installed cost
	No scrap value	$1,000 scrap value 8 years of life
	2 years life remaining	$36,00 annual operating costs
	$4,500 annual operating	excluding depreciation
	costs excluding depreciation	10% interest
	10% interest	
Costs:	$ 2,000	Depreciation over remaining
	9,000 (2 years)	life $10,000
	300	Operating cost .. $28,800 (8 yrs)
	————	Interest at 10% . 5,300
	$11,300	————
		$44,100

$$\frac{\text{Average cost}}{\text{year}} = \frac{\$11,300}{2} = \$5,650$$

$$\frac{\text{Average cost}}{\text{year}} = \frac{\$44,100}{8} = \$5,512$$

Interest on old machine $= 0.10 \times \$2,000 + 0.10 \times \$1,000 = \$300$

If the new machine is depreciated on a straight-line basis from $11,000 to $1,000 over a period of eight years, the depreciation will be $1,250 per year. The interest on the new machine is computed as follows:

0.10	×	$11,000	$1,100	interest for the first year
0.10	×	9,750	975	interest for the second year
0.10	×	8,500	850	interest for the third year
0.10	×	7,250	725	interest for the fourth year
0.10	×	6,000	600	interest for the fifth year
0.10	×	4,750	475	interest for the sixth year
0.10	×	3,500	350	interest for the seventh year
0.10	×	2,250	225	interest for the eighth year

As a shortcut, the interest may be computed by multiplying the average interest payment (composed of the sum of the first and last payments divided by two) by the number of years.

$$\text{Interest} = \frac{0.10 \ (\$11,000 + \$2,250)}{2} \times 8 = \$5,300$$

Note that the last payment = salvage value + the annual depreciation × interest rate $= (\$1,000 + \$10,000/8)(0.10) = \$225$.

Average Rate of Return

The average rate-of-return method is based upon accounting terms rather than cash flow. It does not take into account the time value of money and assumes profits are fairly constant.

$$\text{Average rate of return} = \frac{\text{average annual net income after taxes}}{\text{average investment over the life of the project}}$$

$$= \frac{\text{after tax profit}}{\text{average investment over the life of the project}}$$

Example:

An investment in a plastic molding machine which costs $12,000 requires working capital of $3,000 for inventory and accounts receivable. After-tax profits are ex pected to average $2,700 per year over the six-year life of the machine. There is no salvage value at the end of life.

Average investment = $3,000 + 1/2 ($12.000) = $9,000

Average rate of return = $\frac{\$2,700}{\$9,000}$ = 0.30 or 30%

Present-Worth Method

The present-worth or present-value method is one of the better methods of investment analysis because it takes into account the time value of money. The present-worth method discounts future costs and revenues in order to compare the present value of future benefits with the present value of the investment. If the present value of the benefits does not exceed the investment, the investment should not be made.

When a new project is being considered, a profitability index may be computed. The first example below illustrates this. If two projects which have different service lives are being compared, they must be compared over the same period of time. The comparison may thus extend over multiples of the lives of each.

Thus, if the service life of one project is three years and another is four years, the comparison must be over a 12–year period with replacements occurring for each. The second example below illustrates this point in a very simple fashion.

Example:

A company is considering purchasing a No. 2 Centerless Grinder to produce a new product. The basic data on the machine are:

Initial investment $18,000
Service life .... 10 years
Disposal value at end of life $5,000
Straight -line depreciation $1,000/year
Interest rate 10%

Some refinements are neglected in the following table for simplicity of showing the general approach:

End of Year	Operating Costs	Revenue	Cash Benefits	Cash Benefits after Taxes 0.5 (CB—D)	Discount Factor	Today's Value
1	$12,000	$18,000	$ 6,000	$3,500	0.9091	$3,182
2	12,100	21,000	8,900	4,950	0.8264	4,091
3	12,200	22,000	9,800	5,400	0.7513	4,057
4	12,400	22,500	10,100	5,550	0.6830	3,791
5	12,600	21,000	8,400	4,700	0.6209	2,918
6	12,900	19,500	6,600	3,800	0.5645	2,145
7	13,300	18,000	4,700	2,850	0.5132	1,463
8	14,000	17,000	3,000	2,000	0.4665	933
9	15,000	16,000	1,000	1,000	0.4241	424
10	15,000	16,000	1,000	1,000	0.3855	386
					Present value	$23,390

Present value of the capital investment

$$= \$18,000 - \frac{\$5,000}{(1-0.10)^{10}}$$

$$= 18,000 - 5,000\ (0.3855)$$

$$= 16,072$$

Let V = present value of the cash flow

C = present value of the capital investment

(a) Profitability index $= \dfrac{V}{C} = \dfrac{\$23,390}{\$16,072} = 1.45$

(b) Return on investment $\dfrac{V - C}{C} = \dfrac{V}{C} - 1 = 0.45$

or

$\underline{45\%}$

USING PROBABILITY CONCEPTS IN MAKING DECISIONS

Managers constantly use probability concepts in making decisions. When a manager applies final judgment to deciding upon a capital expansion project, he is in effect saying that this is "very likely" to be profitable. If he were asked to place an index number between 0 and 1 on "very likely" he might say "about 0.9." Such an index number is the manager's subjective judgment of the probability that a future event will occur.

Instinctively the manager makes use of elementary probability relationships. He knows that if the successful introduction of a product requires that his company be first on the market *and* that the price prove to be attractive to customers, then the likelihood of *both* of these events occurring is less than the likelihood of only one of them occurring. This compound probability of simultaneous occurrence of two independent events is found by multiplying the individual probabilities together.

All mathematicians might not agree with the subjective concept of probability, but many applied mathematicians and businessmen have found it to be very useful in practice. Some justification for it is found in the fact that is based upon "repeated" experiences (or similar experiences) of businessmen who may be in a position to judge or "guesstimate" the number of successes which will occur out of large numbers of similar situations.

Decision-Tree Analysis of Investment[2]

The term "decision-tree" is derived from the treelike appearance of the pictorial representation of decisions and chance events. Branches of alternatives grow from the initial decisions or event. Alternative random events with their associated probability of occurrence are considered. As part of the tree, alternative *decisions* are also included.

---

[2] For further discussion of decision trees applied to capital investments, see John F. Magee, "Decision Trees for Decision Making," *Harvard Business Review*, July-August. Copyright © 1964, pp. 88–90.

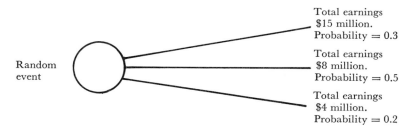

EXHIBIT 8-3   Random Event and Outcomes

Let us take a very simple case. A company makes the decision to invest $3 million in plant and equipment for a new product. The management estimates that there is a 30 per cent chance that earnings over the next eight years (the product life cycle) will be $15 million, a 50 per cent chance that earnings will be $8 million, and a 20 per cent chance that earnings will be $4 million. These are alternative outcomes (which depend upon a chance event) that various numbers of consumers will like the product and buy it. The diagram in Exhibit 8–3 with the circle representing the chance event and the branches representing the alternatives illustrates this situation. Since all possible outcomes are presumed to be shown, the sum of the probabilities must equal unity for the branches emanating from a single event.

The "expected" or weighted-average earnings which are the result of the occurrence of one of the three possible outcomes of the random event are:

$$\text{Expected earnings} = 0.3(15) + 0.5(8) + 0.2(4)$$
$$\text{Expected earnings} = \$9.3 \text{ million}$$

If the investment required is $3 million, then the expected value (in millions) of the *decision* is $9.3 - $3.0 = $6.3.

Now suppose that we have a choice of a decision among three alternatives with expected values of $6.3 million, $3.5 million, and $2.6 million. Which alternative do we choose? Obviously we select the course of action which yields the greatest return. Exhibit 8–4, with the square representing the decision point, shows the three alternatives and the cutting off of the lower payoff branches.

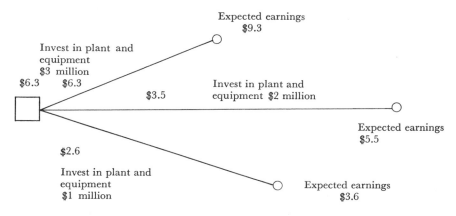

EXHIBIT 8-4   Decision with Three Alternatives

SIMULATION (MONTE CARLO) TECHNIQUE[3]

The simulation of happenings in the world by selecting outcomes of events at random is called the "Monte Carlo process." Such a simulation allows us to generate hundreds of possible occurrences and note the frequency with which various end results occur. The *relative frequency* with which a particular result appears corresponds to the probability that it will occur in real life.

With the Monte Carlo simulation, any number of random variables may be considered. To illustrate the method in the simplest way, we shall assume that we wish to determine the annual earnings (revenue-operating costs) for a proposed project. Management first estimates the probability of each set of costs which may occur as shown in Exhibit 8–5. Column (3), cumulative probability, is there computed.

EXHIBIT 8-5

(1) Possible operating cost	(2) Probability	(3) Probability of cost equal to or less than that of Column 1
$1,000	0.10	0.10
2,000	0.70	0.80
3,000	0.20	1.00

Similar computations are carried out for revenues.

Possible revenue	Probability	Probability of revenues equal to or less than that of Column 1
$2,000	0.10	0.10
3,000	0.20	0.30
4,000	0.40	0.70
5,000	0.30	1.00

The cumulative probabilities are charted in Exhibits 8–6 and 8–7. We use these charts to find pairs of cost and revenue values which combine to produce earnings. Thus,

$$\text{Earnings} = \text{Revenue} - \text{Cost}$$

In order to generate cost value, we simply select a number at random between zero and 1.00, find this value on the vertical axis of Exhibit 8–6, and then read across to the bar which is reached. For example, if we number markers from zero to 100, place them in a bowl, draw the marker numbered 28, we would read 0.28 on the vertical scale and across horizontally to the bar corresponding to $2,000. A much easier method is to use a table of random numbers (RN), an abbreviated example of which is shown in Exhibit 8–8. In practice, a much larger set of random numbers should be used. Any pattern of selection of two digits may be chosen, as long as it is selected in advance. We shall take a very simple pattern, reading the last two digits of the first column and continuing on to the second column, in order to generate costs and revenues. Twenty-five values are drawn although it is desirable to draw more. These values are shown in Exhibit 8–9.

---

[3] From *The Management of Capital Expenditures*, R. G. Murdick and D. D. Deming. Copyright © 1968, McGraw-Hill, Inc. Used by permission of McGraw-Hill Book Company. Pp. 92–95.

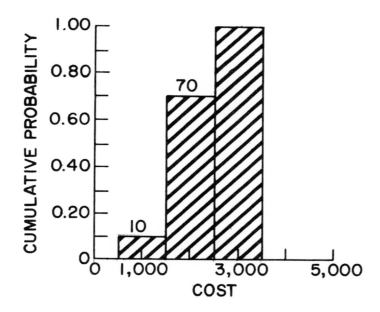

EXHIBIT 8-6   Cumulative Probability for Operating Cost

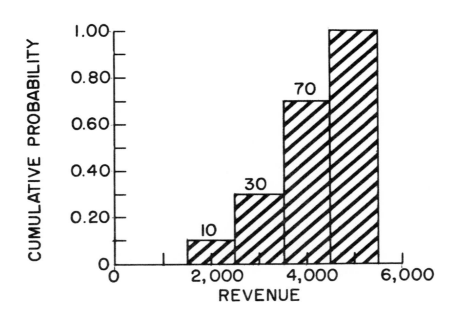

EXHIBIT 8-7   Cumulative Probability for Revenue

(1)	(2)	(3)	(4)
9225	3702	1843	4790
6329	8101	6080	5013
7954	1485	7176	1314
4452	1339	1903	5830
5951	5597	9429	0457
6533	7926	5984	7289
3193	1461	5733	0398
5271	3268	7461	7860
0627	8016	3834	1872
3459	7565	8695	9645
5113	4042	5629	6134
7005	1375	8608	9383
8908	4769	1873	9942
2619	5999	6670	6382
9007	4907	3656	1528
7083	7897	3294	6863
7304	5110	6725	4130
1933	3328	6855	3010
0264	9653	4789	1850
7868	9598	5665	4105

EXHIBIT 8-8   Table of Random Numbers

RN	Cost ($1,000)	RN	Revenue ($1,000)	Earnings ($1,000)
25	2	61	4	2
29	2	68	4	2
54	2	16	3	1
51	2	65	4	2
33	2	42	4	2
93	3	75	5	2
71	3	69	4	1
27	2	99	5	3
59	2	07	2	0
13	2	97	5	3
05	1	10	2	1
08	1	28	3	2
19	2	53	4	2
07	1	98	5	4
83	3	43	4	1
04	1	80	5	4
33	2	76	5	3
64	2	03	2	0
68	2	29	3	1
02	1	84	5	4
01	1	33	4	3
85	3	61	4	1
39	2	34	4	2
97	3	95	5	2
26	2	29	3	1

EXHIBIT 8-9   Simulation of Earnings Based on Probabilistic Values of Costs and Revenues

9

Operation Research for Solving Complex Problems[1]

INTRODUCTION:

Operations research is the approach to decision making which asks specific questions about objectives and about controllable and uncontrollable input variables, and seeks to build mathematical models to describe the systems in which these input variables and output objectives interact. Its purpose is to help management make rational decisions.

It should be kept in mind that operations research neither eliminates nor diminishes the importance of the experience of individual managers and the intuitions and instincts born of that experience. Rather, operations research encourages managers to use these often immensely valuable assets in a rational way at the appropriate points in the decision-making process as, for example, in delineating objectives, defining an appropriate value system, identifying possible courses of action, and assigning subjective probabilities to events or situations where available information is incomplete.

THE MODEL OF THE SYSTEM

. . . The purpose of the model is to predict. It is used because it, the model, is easier and less costly to manipulate than the real system in the real world, or because it is not possible to manipulate or experiment with the real-world system at all. If it is a good model, it provides useful information as to what would result from corresponding manipulations or experiments in the real world and thereby permits decisions to be made about the configuration of the controllable inputs.

---

[1] Samuel B. Richmond, *Operation Research for Management Decisions* (New York: The Ronald Press Co., Copyright © 1968), pp. 6–9, 56–58.

Let us illustrate the operations research approach by building a model to solve a problem which involves manufacturing scheduling and inventory control.

The Courbro Company, an electronics manufacturer, is producing components as a subcontractor on a large project and is required to make daily deliveries, seven days a week, to the prime contractor, who keeps no inventories but puts these components into his process each day as they are received from the subcontractor.

The Courbro Company is thus faced with a constant demand for these components. This contract is only a small part of its total business, and since it must deliver each day, it must either produce a small lot each day or produce in larger lots and keep an inventory of finished goods. The problem is to determine the optimal lot size for cost minimization. Whatever inventories Courbro carries incur capital costs and storage cost for space, watchman, spoilage, and insurance. The total of all these *holding costs* is four cents per unit per month. If the lot size is too large, Courbro will be burdened with high holding costs. On the other hand, if the lots are too small, there are lower holding costs, but the setup expense becomes high, because the plant must go through the setup and make-ready procedure for each lot, and this costs $100 each time. These may be spoken as "opposing costs."

Problem

What is the lot size that balances these costs and minimizes the total cost? The relevant facts are:

(1) Courbro must deliver 5,000 components per month in approximately equal daily shipments.
(2) The *storage* or *holding costs* are four cents per unit per month.
(3) The setup cost is $100 per run.
(4) Production costs are independent of lot size; therefore costs of materials, labor, etc., do not enter into this problem.

Formula

The model, in general conceptual form, is:

$$TC = HC + SC$$

where

$$TC = \text{Total cost}$$
$$HC = \text{Holding cost}$$
$$SC = \text{Setup cost}$$

Solution

The holding cost is $0.04 per month multiplied by the average number of units in inventory. However, if we represent the lot size by X, the average number of units in inventory is $X/2$. This can be seen in Exhibit 9–1.

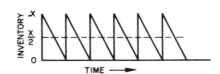

EXHIBIT 9-1   Inventory Levels of Finished Goods under Fixed Demand, with Lot Size, X and Daily Deliveries

If they make X items each run and make daily deliveries until the inventory is reduced to zero, make X items again, etc., the inventory level follows the saw-tooth shape depicted in Exhibit 9–1, and its average level over a single production cycle or over a number of production cycles is $X/2$. The monthly holding cost is thus. . . . .

$$HC = .04\left(\frac{X}{2}\right) \text{ dollars per month.}$$

We can then derive the expression for the setup cost by noting that, if the lot size is X, the total number of setups per month is 5,000 divided by X, and the monthly total setup cost, in dollars, is

$$SC = 100\left(\frac{5,000}{X}\right)$$
$$= \frac{500,000}{X}$$

Then, the model of the system becomes:

$$TC = \$0.02\ X + \frac{500,000}{X}$$

It is in the form $E = f(X_i, y_i)$, where the controllable variable is X, the manufacturing lot size, and the objective function is the total cost. Since the values of the y are fixed, we may write. . . .

$$TC = f(X)$$

and the solution requires that we find that value of X which minimizes the total cost. For arithmetic simplicity let us assume that production can take place only in lot sizes that are integral multiples of 1,000 units. Then the various possible values of X, i.e., the various possible courses of action, are evaluated in Exhibit 9–2, which is based on Equation. . . .

$$TC = \$0.02\ X + \frac{500,000}{X}$$

X (units)	$HC = \$.02\ X$	$SC = \dfrac{500,000}{X}$	TC
1,000	$ 20.00	$500.00	$520.00
2,000	40.00	250.00	290.00
3,000	60.00	166.70	226.70
4,000	80.00	125.00	205.00
5,000	100.00	100.00	200.00
6,000	120.00	83.30	203.30
7,000	140.00	71.40	211.40
8,000	160.00	62.50	222.50
9,000	180.00	55.60	235.60
10,000	200.00	50.00	250.00

EXHIBIT 9-2   Courbro Company: Evaluation of Alternate Lot Sizes

Clearly the optional lot size is 5,000 units. Exhibit 9–3, which is a graphic representation of Exhibit 9–2, indicates the behavior of costs as lot size is varied.

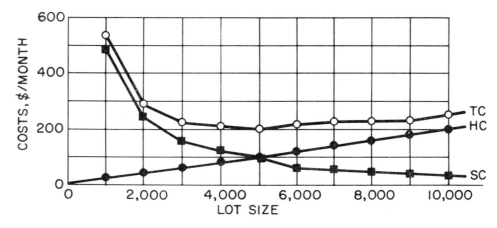

EXHIBIT 9-3

The holding cost is represented by a straight line—a *linear* function—and the setup cost is represented by a descending curved line—a *rectangular hyperbola* in this case. (The name *rectangular hyperbola* is given to the curve whose general equation is $XY =$ a constant.) The sum of these two costs, the total cost, decreases as the lot size increases to 5,000 units, and it increases thereafter. The points are connected as a visual aid; but, with the simplifying restriction requiring lot sizes, intermediate values cannot be realized. If production had not been constrained to integral multiples of 1,000 units, the lines of Exhibit 9–3 would be curved, and the optional solution could be found directly by *differential calculus* as follows:

AN OPTIONAL SOLUTION FOUND DIRECTLY BY DIFFERENTIAL CALCULUS

Formula

The derivative of a sum is the sum of the derivatives. This means that, if we are dealing with a polynomial, we may differentiate each term separately.

$$TC = \$\,0.02\,X + \frac{500,000}{X}$$

$$or\ y = \$0.02X + \frac{500,000}{X}$$

the derivative is . . . .

$$\frac{dy}{dX} = \$0.02 - \frac{500,000}{X^2}$$

The curve has a minimum point at some value of X, the lot size. The dependent variable is the cost, and the minimum-cost point corresponds to the optional lot size. Since the slope is, as seen above,

$$\frac{dy}{dx} = \$0.02 - \frac{500,000}{X^2}$$

and since the minimum-cost point is the point at which the curve has a slope of zero, we need only set the derivative equal to zero, and solve for X, the optimal lot size:

$$0 = \$0.02 - \frac{500,000}{X^2}$$

$$\$.02X^2 = 500,000$$

$$X^2 = 25,000,000$$

$$X = \pm\, 5,000$$

We *discard* the negative root, since $-5,000$ is not a feasible solution to the problem which we are solving. The positive root is the desired solution then, and the optimal lot size is 5,000 units. At that optimal lot size, the total cost is. . . . .

$$TC = \$0.02\,X + \frac{500,000}{X}$$

$$= \$0.02\,(5,000) + \frac{500,000}{5,000}$$

$$= \$100 + \$100 = \$200 \text{ per month.}$$

(This is the same answer for 5,000 units as shown in Exhibit 9–3.) Note that, at this point, the holding cost and the setup cost are the same, indicating that the optimal point turns out to be at the intersection of the two curves. This is always true for models of this type, consisting of the sum of a straight line and a rectangular hyperbola.

10

Queuing Theory Applied to Business Problems[*]

MACHINE INTERFERENCE

Mathematical Approaches

The mathematical solutions to machine interference may be generally classified as algebraic and probabilistic. The algebraic solutions imply constant times; this is extremely rare because a portion of the cycle is controlled by an operator whose work time varies from one cycle to another. The other approach, which uses probabilities, deals with completely random or partially random total cycles, with the latter best typifying the real-world setting.

1. Algebraic Analysis: Constant Time Situation

Given:

$$TET_1 \quad \text{(External Hand Time)} = 3 \text{ min}$$
$$TIT_1 \quad \text{(Internal Hand Time)} = 2 \text{ min}$$
$$TWT_1 = TET + TIT \text{ (Total Work Time)} = 5 \text{ min}$$
$$TMT_1 = \text{(Machine Time)} = 11 \text{ min}$$
$$TCT_1 = \text{(External Hand Time + Machine Time)}$$
$$= 14 \text{ min}$$
$$N = \text{Number of Machines}$$
$$MC = \text{(Machine Cost)} = \$25/\text{hr} = \$25/60$$
$$= \$.417 \text{ per min}$$
$$LC = \text{(Labor Cost)} = \$5/\text{hr} = \$5/60$$

---

* Joseph A. Panico, *Queuing Theory* (Englewood Cliffs, N. J.: Prentice-Hall, Inc., Copyright ©1969), pp. 26–35, 53–55, 58–61.

$$= \$.083 \text{ per min}$$

$$\mathcal{N} \le \frac{TCT}{TWT} = \frac{14}{5} = 2.8$$

$$\mathcal{N} = 2 \text{ (Eight-tenths machine cannot be assigned)}$$

If $\mathcal{N} = 3$, interference will be introduced into the system as the process becomes completely controlled by the operator.

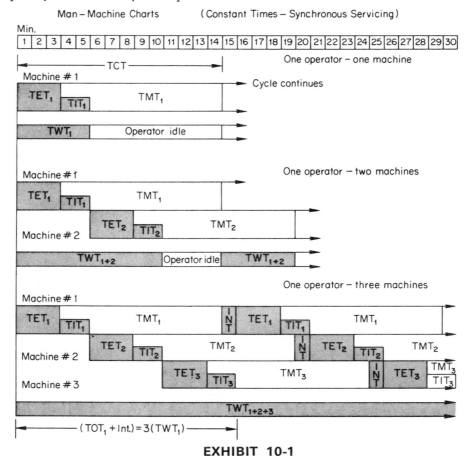

EXHIBIT 10-1

Exhibit 10–1 illustrates the various constant time situations. The first two illustrate a situation in which the cycle is partially controlled by the machine, while the third shows a situation completely controlled by the operator.

In the first illustration of one operator and one machine: Machine #1 is considered operational for the complete Total Cycle (TCT) even though idle during TET_1. Operator is also paid for idle time of $TMT_1 - TIT_1$.

$$\text{COST} = TCT \ (MC) + TCT \ (LC) = (14\tfrac{25}{60}) + 14(\tfrac{5}{60}) = \$7.00$$

The second illustration of one operator running two machines shows that:

Cost per machine remains $TCT(MC)$, while labor cost is divided between two machines.

$$\text{COST} = \frac{N\ (TCT)\ (MC) + TCT\ (LC)}{N} = \frac{2\ (14)\ \left(\frac{25}{60}\right)}{2} + \frac{14\ \left(\frac{5}{60}\right)}{2} = \$6.42$$

where

$$\text{Machine Cost} = TCT\ (MC) \text{ and Operator Cost} = \frac{TCT\ (LC)}{N}$$

When the work assignment is again increased, one operator is responsible for three machines. Notice that the machine cost for the system will remain constant until $N(TWT_1) > TCT$. When this occurs, interference has been introduced into the system since the machine must wait for the operator. The TCT is unrestricted by the machine. Interference for this situation is divided equally among the machines, which happens only in those cases involving identical operations.

$$\text{COST} = \frac{N\ (TWT_1)\ (LC) + N\ [N\ (TWT_1)\ (MC)]}{N}$$

$$= \frac{(15)\left(\frac{5}{60}\right) + (3)\ (15)\left(\frac{25}{60}\right)}{3} = \$6.67$$

Interference has increased the machine cost in the third case by one minute, which at \$25/hr is equivalent to \$.42 per cycle. Thus the impact of interference can be seen in this exceptionally simplified model. Remember that hand time has been reduced, accounting for the total cost of \$6.67.

2. Probabilistic Analysis: Random Times (Binomial solution with interference included in the down time)

This leaves for solution the situation involving completely random and partially random cycles. These are best handled through the use of probabilities. In the first solution, attention will be given the completely randomized situation.

a. Probability Tree

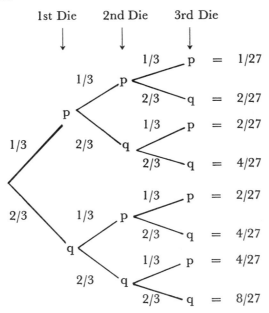

b. Stochastic Process of Independent Trials

One method frequently used is a stochastic process of independent trials which requires the use of probabilities. A traditional example is the diagram showing one throw of three dice when a two or a five is desired.

$$Pr \text{ (of a two or five)} = Pr\,(2 \lor 5)$$

thus

$$p = \text{successful event} = Pr(2 \lor 5) = \tfrac{2}{6} = \tfrac{1}{3}$$
$$q = \text{failure} \quad\quad = 1 - p = Pq(2 \lor 5) = \tfrac{4}{6} = \tfrac{3}{2}$$

All eight possibilities have been covered with respect to order as $ppp = \tfrac{1}{27}$, $pqp = \tfrac{2}{27}$, $qpp = \tfrac{2}{27}$ $pqq = \tfrac{4}{27}$ and $qqq = \tfrac{8}{27}$, etc. In this example, however, combinations instead of permutations are desired. Thus, single events pqp, ppq, and qpp are considered similar, thereby leaving the following possibilities:

Three	p's $= ppp$	$= \tfrac{1}{27}$ (all 2's or 5's)
Two	p's $= ppq, pqp, qpp = \tfrac{2}{27} + \tfrac{2}{27} + \tfrac{2}{27} = \tfrac{6}{27}$	(two 2's or 5's)
One	$p \;\; = pqq, qpq, qqp = 3(\tfrac{4}{27})$	$= \tfrac{12}{27}$ (one 2 or 5)
Zero	p's $= qqq$	$= \tfrac{8}{27}$ (zero 2's or 5's)

These could also have been predicted through the binomial expansion of $(p + q)^n$, which provides probabilistic values for all combinations. In this example it is

$$\left(\frac{1}{3} + \frac{2}{3}\right)^3 = \frac{1}{27} + \frac{6}{27} + \frac{12}{27} + \frac{8}{27}$$

Using this same principle it may be shown that $p = Pr$ (i.e., that a machine is operating) and $q = Pr$ (i.e., that a machine requires service) follow an identical expansion because of the equivalence of probabilities.

c. Consider These Possibilities: One Crew Servicing Two, Three, or Four Machines with the Following Information Given:

Machine Cycle Time = 6 hrs *(TRT)* or *(TMT)*
Machine Down Time = 3 hrs *(TDT)* Interference included[1]
Total Cycle Time = 9 hrs *(TCT)*
Machine Operational Time = 67% of Total Cycle
Machine Down Time = 33% of Total Cycle
Pr (Machine Operating) = $\tfrac{2}{3}$
Pr (Machine Down) = $\tfrac{1}{3}$

(1) Calculating machine interference for two machines

Binomial Probability Distribution $= (p + q)^n = (\tfrac{1}{3}+\tfrac{1}{2})^2$

---

[1] Interference is included in the machine down time since it could not be recorded separately during frequency studies. To find an accurate value for crew work time, subtract the calculated percent interference from the observed percent down time. Crew time is a subset of down time.

1st Term = Zero Down, Two Operating[2] = $b\left(2; 2, \frac{2}{3}\right) = \frac{4}{9}$

2nd Term = One Down, One Operating = $b\left(1; 2, \frac{2}{3}\right) = \frac{4}{9}$

3rd Term = Two Down, Zero Operating = $b\left(0; 2, \frac{2}{3}\right) = \frac{1}{9}$

Probability Tree

1st Machine	2nd Machine	Path Weight	Description
	2/3 Operating	4/9	Zero machine down
2/3 Operating	1/3 Down	2/9	One machine down
1/3 Down	2/3 Operating	2/9	One machine down
	1/3 Down	1/9	Two machines down

No. of Machines Down	Probability	Queue Size	Interpretation	Interference/ Cycle
0	$\frac{4}{9}$	0	(No crew required)	0
1	$\frac{2}{9} + \frac{2}{9} = \frac{4}{9}$	0	(Crew attending downed machine —none waiting for service)	0
2	$\frac{1}{9}$	1	(Crew attending one machine and the second must wait)	$(1)\left(\frac{1}{9}\right)$
				$\frac{1}{9}$ Total

Percent machine time lost as a result of interference:

$$\frac{\text{Interference}}{\text{Similar Machines}} = \frac{\frac{1}{9}}{2} = \frac{1}{18} = \frac{5.55\%}{\text{Machine}}$$

Percent down time per machine without interference:[3]

$$\frac{1}{3} - \frac{1}{18} = \frac{5}{18} = 27.78\% \quad \text{of total cycle}$$

or

$$33.33\% - 5.55\% = 27.78\% \quad \text{of total cycle}$$

---

[2] This is a binomial probability notation. The expression $b(2;2, \frac{2}{3}) = \frac{4}{9}$ reads: The binomial probability of two successes in two trials, when the probability of success is two in three, equals $\frac{4}{9}$.

As an example, $b(5; 10, \frac{1}{4}) = .0584$, which may be found in any book of mathematical tables that includes a table of individual terms of the binomial distribution.

The values $\frac{4}{9}$, $\frac{4}{9}$, $\frac{1}{9}$ may also be found through the binomial expansion of $(\frac{2}{3} + \frac{1}{3})^2$.

$$\left(\frac{2}{3} + \frac{1}{3}\right)^2 = \left(\frac{2}{3}\right)^2 + \frac{2}{1!}\left(\frac{2}{3}\right)^1\left(\frac{1}{3}\right)^1 + \frac{2 \cdot 1}{2!}\left(\frac{2}{3}\right)^0\left(\frac{1}{3}\right)^2 = \frac{4}{9} + \frac{4}{9} + \frac{1}{9}$$

[3] A very important fact must be noted here. Machine interference results in an increased total cycle time, but this affects only the servicing time. The machine cycle will usually remain the same irrespective of the amount of interference per machine. In the situation in which machine time is constant (e.g., numerically controlled), interference cannot extend the machine cycle. An analyst may be deceived in special situations if he is not careful, however, since machine operators may change feeds or speeds to make their operations more rhythmic. Thus, an improper analysis may occur.

This averages six hours regardless of interference. The machine may wait, but only for the crew, not for the machine time. Based on this reasoning, interference is applied only to the service time.

Percent machine time:

Production Time (% of New Cycle)

$$= (\% \; TMT) \; (\% \; TCT \; \text{minus} \; \% \; \text{Interference})^{[4,5]}$$
$$= (66.67\%) \; (100\% - 5.55\%) = 62.97\%$$

Without interference the machine would be immediately recycled, once the external work time was completed. If interference exists, however, it should be considered as extending the cycle. This prevents the crew from recycling the machine and thus reduces machine efficiency. In a complete cycle **with zero interference** one machine would produce 66.67% of nine hours. Compare this with a situation involving two machines and 5.55% interference; here each machine, instead of producing 66.67% of the total cycle, is only producing 94.45% of capacity or $(66.67\%) \; (94.45\%) = 62.97\%$.[6]

Remember that the 62.97% does not indicate a decrease in machine time. If the same formulas of TMT/TCT are used to determine machine efficiency, it will be seen that reaction (machine) time has remained the same. New $TMT = 62.97\%$ and $TCT = 94.45\%$, therefore machine efficiency equals 62.97/94.45 which remains 66.67%.

Using this information it is possible to calculate the productivity of each group of machines in a similar unit of time. The formula is:

Units produced Per Unit of Time

$$= (\text{Number of machines}) \times (\% \; \text{Production Time of One})$$

Two machines will contribute: $2(62.97\%) = 1.2594$ units in one unit time.

---

[4] $TCT = 100\%$ which includes interference.

[5] If the original total cycle time equals 100%, then the actual total cycle time is smaller, for it does not include interference. Actual total cycle time is $\% \; TCT - \%$ Interference.

[6] In the beginning of this computation, the machine was considered as producing 66.67% of the total cycle. Interference, however, is included in this figure. If there were no interference, then the machine could produce more Units of Work. So the very fact that interference exists represents lost opportunity of production.

The internal portion of the bar chart changes now that the amount of interference is known.

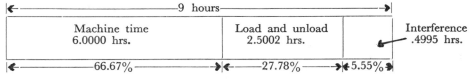

The actual cycle time is 8.5002 hrs compared to the 9 hrs of the original. Machine efficiency now equals 6/8.5002 = 70.59% of the total cycle when interference is excluded. Obviously, interference has reduced total productivity to 66.67/70.59 = 94.45% of its maximum. This is also found when 8.5002/9 = 94.45%. If the machine is only producing 94.45% of maximum, then its production time is reduced to $(66.67\%) \; (94.45\%) = 62.97\%$.

(2) Calculating machine interference for three machines

Binomial Probability Distribution $= (p + q)^n = (\frac{2}{3} + \frac{1}{3})^3$

Probability Tree

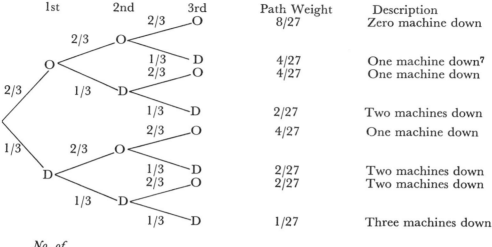

1st	2nd	3rd	Path Weight	Description
		O	8/27	Zero machine down
		D	4/27	One machine down[7]
		O	4/27	One machine down
		D	2/27	Two machines down
		O	4/27	One machine down
		D	2/27	Two machines down
		O	2/27	Two machines down
		D	1/27	Three machines down

No. of Machines Down	Probability	Queue Size	Interference/Cycle
0	$\frac{8}{27}$	0	0
1	$\frac{12}{27}$	0	0
2	$\frac{6}{27}$	1	$(1)(\frac{6}{27}) = \frac{6}{27}$
3	$\frac{1}{27}$	2	$(2)(\frac{1}{27}) = \frac{2}{27}$
			$\frac{8}{27}$ Total

Machine Time lost as a result of interference:

$$\frac{\text{Interference}}{\text{Similar Machines}} = \frac{\frac{8}{27}}{3} = \frac{8}{81} = \frac{9.88\%}{\text{Machine}}$$

Down time per machine without interference:

$$\tfrac{1}{3} - \tfrac{8}{81} \text{ or } 33.33\% - 9.88\% = \tfrac{19}{81} = 23.46\%$$

Production Time (% of New Total Cycle)

$$= 66.67\%(100\% - 9.88\%) = 60.09\%$$

Units Produced per Unit Time

$$= (\text{Number of Machines})(\% \text{ Production Time})$$
$$= (3)(60.09\%) = 1.803 \text{ Units}$$

(3) Calculating machine interference for four machines

$$(p + q)^n = \left(\frac{2}{3} + \frac{1}{3}\right)^4 = \left(\frac{2}{3}\right)^4 + \frac{4}{1!}\left(\frac{2}{3}\right)^3\left(\frac{1}{3}\right) + \frac{4 \cdot 3}{2!}\left(\frac{2}{3}\right)^2\left(\frac{1}{3}\right)^2 + \frac{4 \cdot 3 \cdot 2}{3!}\left(\frac{2}{3}\right)\left(\frac{1}{3}\right)^3$$

$$+ \frac{4 \cdot 3 \cdot 2 \cdot 1}{4!}\left(\frac{1}{3}\right)^4 = \frac{16}{81} + \frac{32}{81} + \frac{24}{81} + \frac{8}{81} + \frac{1}{81}$$

---

[7] Pr (one machine down and two are operating) $+ \frac{4}{27} + \frac{4}{27} + \frac{4}{27} + \frac{12}{27}$, which may also be expressed as $b(2; 3, \frac{2}{3})$.

No. of Machines Down	Probability	Queue Size	Interference/Cycle
0	$\frac{16}{81}$	0	0
1	$\frac{32}{81}$	0	0
2	$\frac{24}{81}$	1	$1\left(\frac{24}{81}\right) = \frac{24}{81}$
3	$\frac{8}{81}$	2	$2\left(\frac{8}{81}\right) = \frac{16}{81}$
4	$\frac{1}{81}$	3	$3\left(\frac{1}{81}\right) = \frac{3}{81}$
			Total $= \frac{43}{81}$

Machine time lost as a result of interference:

$$\frac{\text{Interference}}{\text{Similar Machine}} = \frac{\frac{43}{81}}{4} = \frac{43}{324} = \frac{13.24\%}{\text{Machine}}$$

Down time per machine without interference:

$$\tfrac{1}{3} - \tfrac{43}{324} \text{ or } 33.33\% - 13.24\% = 20.09\%$$
$$\% \text{ Production Time} = 66.67\% \ (100\% - 13.24\%) = 57.84\%$$
$$\text{Units Produced} = 4 \ (57.84\%) = 2.314 \text{ Units}$$

Exhibit 10–2 shows that two machines with a ratio of $\frac{1}{3}$ down and $\frac{2}{3}$ operating will be restricted to producing 1.26 units in any 100% unit time, when compared with other combinations of machines in the same time interval, because of interference.[8]

Cost Comparisons

COST COMPARISONS
(Using one hour as the base time for comparison purposes)
Given: Labor Cost = \$25/hr, and Machine Cost = \$65/hr

Number of Machines	% of Cycle Available for Productivity	Total* Productivity	Labor Cost (Hour)	Machine Cost (Hour)	Cost Comparisons
1	66.67%	0.667	\$25	1(\$65)	$\frac{\$25 + \ \$65}{.667} = \$134.9$
2	62.97%	1.259	\$25	2(\$65)	$\frac{\$25 + \$130}{1.259} = \$123.9$
3	60.09%	1.803	\$25	3(\$65)	$\frac{\$25 + \$195}{1.803} = \$122.0$
4	57.84%	2.314	\$25	4(\$65)	$\frac{\$25 + \$260}{2.314} = \$123.2$

*This is units produced in a unit of time. The column therefore represents a ratio of productivity for one, two, three, and four machines in any unit of time.

EXHIBIT 10-2   Cost Comparisons

This comparison shows that the lowest cost per unit is \$122.00, which ensues when one crew operates three machines. The costs here are very close, but greater differences may be expected for other ratios of man-to-machine time.

---

[8] When working with small cycle times this is usually described as pieces per hour. For example: Four machines with a 95% actual production time equals 4 (95%) = 3.80 pieces in the time period specified, i.e., 3.8 min., 3.8/hr, or 3.8/day.

SET UP AND SOLUTION OF VARIOUS QUEUING PROBLEMS

The usual procedure followed when setting up the majority of queuing problems is to: (a) make certain that data gathered are representative and do not constitute a specialized case; (b) list the data used in solving the problem; (c) express λ (the mean arrival rate) and μ (the mean service rate) *in equivalent terms,* i.e., minutes, hours, days, years, etc.; (d) determine whether input data come from a finite or infinite population; (e) include in detail only those areas necessary for decision making when the project is presented to management in its final form.[9]

1. One Line—One Server (Infinite Population)

a. General Formulas:

1. $P_0(t) = 1 - \dfrac{\lambda}{\mu} = Pr \left\{ \begin{array}{l} \text{Zero units in the system when the rate} \\ \text{of arrival } \lambda \text{ and the service } \mu \text{ are given.} \end{array} \right\}$

2. $P_n(t) = \left(\dfrac{\lambda}{\mu}\right)^n P_0(t) = \left(\dfrac{\lambda}{\mu}\right)^n \left(1 - \dfrac{\lambda}{\mu}\right) = Pr \left\{ \begin{array}{l} n \text{ units in queue plus} \\ \text{being serviced.} \end{array} \right\}$

3. $E_n = \dfrac{\lambda}{\mu - \lambda} =$ Expected number being serviced plus waiting.

4. $E_w = \dfrac{\lambda^2}{\mu(\mu - \lambda)} =$ Expected number in queue—(length of waiting line not including those being serviced).

5. $E_t = \dfrac{E_w}{\lambda} = \dfrac{\lambda}{\mu(\mu - \lambda)} =$ Expected waiting time in line.

6. $E\phi = E_t + \dfrac{1}{\mu} = \dfrac{1}{\mu - \lambda} =$ Expected waiting time in system.

7. $P(\mathcal{N} > n) = \left(\dfrac{\lambda}{\mu}\right)^{n+1} = Pr$ (The number in the waiting line plus the number being serviced is greater than n.)[10]

Restrictions: (a) First Come–First Served Discipline.

(b) $\dfrac{\lambda}{\mu} < 1$, often called the Traffic Intensity or Clearing Ratio. This restriction must hold, otherwise $P_n(t)$ is not independent of (t).

Example: A service station has pumps for both gasoline-and diesel-powered vehicles. Of its four pumps one is diesel. Two trucks cannot be serviced simultaneously although two diesel fuels are carried. Studies show that truck arrivals at this one pump follow a Poisson distribution with a mean of ten trucks per hour, while service is

---

[9] Often when a project is presented it contains too many preliminary data. This sometimes discourages management, which may cause a meaningful project to be abused.

[10] $P(\mathcal{N} > n)$ when $n = 0$ $\therefore P(\mathcal{N} > 0) = P(1 - P_0) = \dfrac{\lambda}{\mu}$;

$P(\mathcal{N} > 1) = P[1 - (P_0 + P_1)] = 1 - \left[1 - \dfrac{\lambda}{\mu} + \dfrac{\lambda}{\mu}\left(1 - \dfrac{\lambda}{\mu}\right)\right] = \left(\dfrac{\lambda}{\mu}\right)^2$; etc.

An illustration may broaden understanding. A truck operator knows from experience that his produce will perish if he must wait for more than two others to unload. Here, instead of being interested in P_2 (the probability of two trucks in the system), he is interested in $P(\mathcal{N} > 2)$ (the probability of more than two trucks in the system).

exponentially distributed with a mean of twenty per hour.[11] Find E_n, E_w, and E_t and E_ϕ.

From this information it is found that:

$\lambda = 10$ arrivals per hour[12]

$\mu = 20$ services per hour[13]

$\lambda/\mu = \frac{1}{2}$, which satisfy the restriction of $\lambda/\mu < 1$.[14]

The expected number being serviced plus waiting is:

$$E_n = \frac{\lambda}{(\mu - \lambda)} = \frac{10}{(20 - 10)} = \frac{10}{10} = 1 \text{ truck}$$

The expected number in queue—length of waiting line not including those being serviced is:

$$E_w = \frac{\lambda^2}{\mu(\mu - \lambda)} = \frac{10^2}{20(20 - 10)} = \frac{100}{20(10)} = \frac{100}{200} = \frac{1}{2} \text{ truck}$$

The expected waiting time in line is:

$$E_t = \frac{E_w}{\lambda} = \frac{\frac{1}{2}}{10} = \frac{1}{20} \text{ of an hour per truck}$$

The expected waiting time in the system is:

$$E_\phi = E_t + \frac{1}{\mu} = \frac{1}{20} + \frac{1}{20} = \frac{2}{20} = \frac{1}{10} \text{ of an hour per truck.}$$

2. One Line—One Server (Finite Population)[15]

a. General Formulas

1. $P_n = \left(\dfrac{m}{n}\right)\left(\dfrac{\lambda}{\mu}\right)^n P_0$ for $0 \leq n \leq S$[16]

---

[11] Mathematical queuing formulas are based on Poisson arrivals and exponential service. The formulas will not readily hold if input, outputs, or both do not follow these restrictions. Queuing problems having other or not easily defined distributions are best solved through simulation techniques, as described in Chapter 8.

[12] Ten elements arrive from an *infinite universe* for service. These arrivals are not equally spaced or paced—all we know is that an average of ten arrive per hour and that arrivals follow a Poisson distribution.

[13] The servicing ability is twenty per hour for one pump. This is an average which means that at varying intervals, service time may be greater or smaller than the average. The fundamental demand on service is that the distribution be exponential.

[14] λ and μ are based on averages. Thus $(\lambda/\mu) < 1$ is also based on averages. The nature of an average dictates there are times when individual occurrences of (input/output) > 1 and (input/output) $\rightarrow 0$. For example: ten elements are expected, but the nature of the distribution, with λ as its mean, indicates that the input could be greater or less than what is expected. The system can expect to service twenty, but again this will vary greatly during individual trials. Consider that we have twenty arrivals, which is a possibility, and that we can only service eight, which similarly is a possibility; arrivals then exceed capacity, and a line forms. This holds true for any time interval, unit time, and is not restricted to a one-hour period. This variable input, combined with variable output, is a description of what happens when two averages are combined. These variations allow a line to build during one time interval, and similarly allow it to be depleted during another time interval.

[15] For a comprehensive explanation see William Feller, *Introduction to Probability Theory and Its Applications* (John Wiley & Sons, Inc., New York, 1965, pp. 411–421).

[16] Any $P_n(t)$ will be designated as P_n when dealing with a finite universe, since this helps to determine which concept is being applied.

2. $P_n = \dfrac{m!}{(m-n)!}\left(\dfrac{\lambda}{\mu}\right)^n P_0$ for $S \leq n \leq m$

3. $\dfrac{P_n}{P_0} = \dfrac{m}{(m-n)!}\left(\dfrac{\lambda}{\mu}\right)^n$ [17]

4. $P_0 = \dfrac{1}{\displaystyle\sum_{n=0}^{m} \dfrac{P_n}{P_0}}$

5. $E_w = m - \dfrac{\lambda+\mu}{\lambda}(1 - P_0)$

6. $E_t = \dfrac{1}{\mu(1-P_0)}(E_w) = \dfrac{1}{\mu}\left(\dfrac{m}{1-P_0} - \dfrac{\lambda+\mu}{\lambda}\right)$

7. $E_n = E_w + (1 - P_0) - m - \dfrac{\mu}{\lambda}(1 - P_0)$

8. $E_\phi = E_t + \dfrac{1}{\mu} = \dfrac{1}{\mu}\left(\dfrac{m}{1-P_0} - \dfrac{\lambda+\mu}{\lambda} + 1\right)$

Restrictions: (a) First Come—First Served Discipline and

                (b) $\dfrac{\lambda}{\mu} < 1$.

Example: Studies show that of four machines breakdowns occur at random, the average time between breakdowns being one hour, and that a repairman averages one-eighth of an hour to service the idled machine.[18]

Thus $\lambda = 1$, $\mu = 8$, $m = 4$, and $\lambda/\mu = \frac{1}{8} = 0.125$ (which satisfy the restriction of $\lambda/\mu < 1$). With this information, solve for P_0, E_w, E_t, E_n, and E_ϕ.

The first step in the solution of this problem is to solve for P_0 and then to work this value into the formulas for E_w, E_t, E_n and E_ϕ. In this problem the universe is a small countable set, and the removal of one machine from production drastically changes the probabilities of another's being removed. Finite queuing formulas are thus used, since a strong dependency relationship has been established. If the universe had been two hundred machines, then the removal of one would not have drastically changed the probabilities of another's being removed, and infinite queuing formulas could have applied. Finite formulas are laborious, especially when large numbers are involved, and this is why analysts prefer working with infinite queuing formulas, whenever applicable.

Exhibit 10–3 was developed using finite formulas and shows the five values for all P_n.

---

[17] P_0 must be known in order to use Formula 1 or 2. From the fact that $\displaystyle\sum_{n=0}^{m} P = 1$. it follows that $\left(\dfrac{1}{P_0}\displaystyle\sum_{n=0}^{m} P_n\right)P_0 = 1$ therefore $P_0 = \dfrac{1}{\dfrac{1}{P_0}\displaystyle\sum_{n=0}^{m} P_n} = \dfrac{1}{\displaystyle\sum_{n=0}^{m}\dfrac{P_n}{P_0}}$

The individual P_n/P_0 ratios are calculated from formulas 1 and 2, as shown in formula 3. Thus, the sum of these ratios divided into one will yield the value for P_0. Once P_0 is known, all individual values for P_n are found.

[18] In finite queuing models there is a dependency relationship between arrivals, and the Poisson probability law is not applicable. Thus, instead of classifying arrivals as an average for the universe, it is necessary to classify them as an average of a unit time period.

$n =$ Number of Units in the System	Size of Queue	Ratio of Probability		P_n/P_0**
			$\dfrac{P_n}{P_0}$	P_n
0	0	$1(\frac{1}{8})^0 =$	1.00000	0.57463
1	0	$4(\frac{1}{8})^1 =$	0.50000	0.28732
2	1	$12(\frac{1}{8})^2 =$	0.18750	0.10774
3	2	$24(\frac{1}{8})^3 =$	0.04688	0.02694
4	3	$24(\frac{1}{8})^4 =$	0.00586	0.00337
		$\sum \dfrac{P_n}{P_0}* =$	1.74024	$\sum P_n = 1.00000$

* $P_0 = \dfrac{1}{\sum \dfrac{P_n}{P_0}} = \dfrac{1}{1.74024} = 0.57463$ (formula 4)

** With the value of P_0 known, each P_n follows from either formula 1 or formula 2. The notation $Pr\ (P_n|P_0)$ is read: the probability of P_n, given P_0.

EXHIBIT 10-3

Exhibit 10–3 gives the value of $P_0 = \dfrac{1}{1.74024} = 0.57463$

Thus:

$$E_w = m - \frac{\lambda + \mu}{\lambda}(1 - P_0) = 4 - 9(0.42537) = 4 - 3.82833$$

$$= 0.17167 \text{ (machines waiting for service)}$$

$$E_t = \frac{0.17167}{8(0.42537)} = \frac{0.17167}{3.40296}$$

$$= 0.05045 \text{ (hours machine waits before being serviced)}$$

$$E_n = 0.17167 + 0.42537$$

$$= 0.59704 \text{ (number of machines being serviced plus waiting)}$$

$$E_\phi = 0.05045 + 0.12500 = 0.17545 \text{ (hours machine is idle)}$$

$P_m =$ The probability that m machines are being serviced or demanding service can be found by using Erlang's Loss Formula.[19]

$$P_m = \left[1 + \frac{1}{1!}\left(\frac{\mu}{\lambda}\right)^1 + \frac{1}{2!}\left(\frac{\mu}{\lambda}\right)^2 + \ldots + \frac{1}{m!}\left(\frac{\mu}{\lambda}\right)^m\right]^{-1}$$

Individual terms of P_m can then be found as follows:

$$P_m = \left[1 + 8 + \frac{64}{2} + \frac{512}{6} + \frac{4096}{24}\right]^{-1} = [297]^{-1} = \frac{1}{297}$$

In the expansion for P_m each term represents a ratio of probability similar to that of Exhibit 10–3. The second term, $(1/1!)\ (\mu/\lambda)^1 = 8$, is four times smaller than the third term, $(1/2!)\ (\mu/\lambda)^2 = \frac{64}{2} = 32$, and their ratio, $\frac{8}{32}$, is similar to the ratio $(0.04688/0.18750)$ from Exhibit 10–3. There are five terms in the series for P_m. If each term is

---

[19] Erlang's Loss formula, named for A. K. Erlang, whom engineers regard as the first to utilize stochastic process in the solution of waiting lines.

divided by the sum of the term, the expression then becomes, through mathematical manipulation, a finite probabilistic series. Thus by using the fact that the ratio between the probabilities is similar to those found previously, it is easily verified that the first, second, third, fourth, and fifth terms equal $P_4, P_3, P_2, P_1,$ and P_0, respectively.

Thus,

$$P_m = P_4 = \frac{1}{297} = 0.00337, \; P_{m-1} = P_3 = \frac{8}{297} = 0.02694,$$

$$P_{m-2} = P_2 = \frac{32}{297} = 0.10774, \; P_{m-3} = P_1 = \frac{85.33333}{297} = 0.28732,$$

$$\text{and } P_{m-m} - P_0 = \frac{170.66667}{297} = 0.57463.$$

Looking back at the expression for $P_m = [1 + 8 + 32 + 85.33333 + 170.66667]^{-1}$, it is seen that each term of the series represents the ordered numerators for the various P_n.

b. Cost Analysis

With machine costs of $ 15.00/hr, operator costs of $ 3.00/hr, and repairman costs of $ 4.00/hr, the breakdown costs are:

(Machine Idle Time Costs)

$+$ (Lost Productivity of Machine Operator) $+$ (Service Costs)

$=$ (Hours Machine Idle) (Machine Costs)

$+ \dfrac{\text{(Hours Machine Idle) (Operator Costs)}}{\text{Number of Machines}}$

$+$ (Service Time) (Repairman Costs)

$= E_\phi \, (\$15.00) + \frac{1}{4}(E_\phi) \, (\$3.00) + (E_\phi - E_t) \, (\$4.00)$

$= (0.17545) \, (\$15.00) + \frac{1}{4}(0.17545) \, (\$3.00) + (0.125) \, (\$4.00)$

$= \$3.26334$ (average cost per breakdown).

11

Inventory Formulas[1]

ESTIMATING INVENTORY—GROSS PROFIT METHOD

Formula:

$$I_2 = I_1 + P - [S(1 - R)]$$

where

I_2 = Ending Inventory
I_1 = Beginning Inventory
P = Purchases
S = Sales
R = Estimated Gross Profit Percentage on Sales

Example:

It is desired to prepare interim financial statements for the quarter ended March 31, 19__. However, it is not considered practicable to take a physical inventory and the company does not maintain a perpetual inventory. The following data are obtained from the firm's books and other records, sales for the period $100,000, purchases for the period $60,000. Inventory: Jan. 1, 19__ $20,000. For the prior year the profit and loss statement showed a gross profit ratio of 38%. However, management has indicated that the costs have risen, but the selling price has remained more or less constant and they believe that a 35% gross profit ratio is probably being maintained. Compute the ending inventory.

Solution:

$$I_2 = I_1 + P - [S(1 - R)]$$
$$I_2 = 20,000 + 60,000 - [100,000(1 - .35)]$$
$$I_2 = 20,000 + 60,000 - [100,000(.65)]$$
$$I_2 = 20,000 + 60,000 - 65,000$$
$$I_2 = 80,000 - 65,000$$
$$I_2 = \$15,000$$

---

[1] From the book *Accountant's Handbook of Formulas and Tables*, by Lipkin, Feinstein & Derrick. Copyright © 1963 by Prentice-Hall, Inc., Englewood Cliffs, N. J., pp. 73–78.

ESTIMATING INVENTORY—RETAIL INVENTORY METHOD

Formula:

$$I_{r2} = (I_{r1} + P_r + M_u) - (S + M_j + E + W_r)$$

$$I_{c2} = I_{r2} \times \frac{I_{c1} + P_c + T}{I_{r1} + P_r + M_u}$$

where

I_{c2} = Ending Inventory at Cost (Approximate Lower of Cost or Market)

I_{r2} = Ending Inventory at Retail

I_{r1} = Beginning Inventory at Retail

P_r = Purchases at Retail

M_u = Net Additional Markups (Additional Markup Minus Markup Cancellation)

S = Sales

M_j = Net Markdowns (Markdown Minus Markdown Cancellation)

E = Employee Discounts (If not recorded on Books)

W_r = Worthless Inventory at Retail—Due to Breakage, spoilage etc. or possibly known theft

I_{c1} = Beginning Inventory at Cost

P_c = Purchases at Cost

T = Transportation—In

Example:

The Downtown Department Store uses the retail inventory method of accounting. At the end of the first quarter of operations it is necessary to evaluate the inventories of the various departments in order to prepare financial statements. An actual physical inventory is taken only at the end of the fiscal year. From the data obtained from the books and other records estimate the inventory for the Junior Miss Department at the end of the first quarter.

Information obtained from the General Ledger:

	Retail	*Dr.*	*Cr.*
Sales			$36,625
Purchases	$80,000	$52,000	
Transportation In		3,666	
Inventory	30,000	20,000	

The Buyer's records on markups and markdowns is summarized as follows:

Additional Markups	$4000	
Markup Cancellations	500	$3500
Markdowns	$3500	
Markdown Cancellations	2000	$1500

10 suits damaged—shop worn and completely unsaleable or returnable to manufacturer price at retail $20 each or $200.

Employee Discounts totaled $175.

Solution:

$$I_{r_2} = (I_{r_1} + P_r + M_u) - (S + M_j + E + W_r)$$

$$I_{c_2} = I_{r_2} \times \frac{I_{c_1} + P_c + T}{I_{r_1} + P_r + M_u}$$

$$I_{r_2} = (30{,}000 + 80{,}000 + 3{,}500) - (36{,}625 + 1{,}500 + 175 + 200)$$

$$I_{r_2} = \$113{,}500 - 38{,}500$$

$$I_{r_2} = \$75{,}000$$

$$I_{c_2} = 75{,}000 \times \frac{20{,}000 + 52{,}000 + 3{,}666}{30{,}000 + 80{,}000 + 3{,}500}$$

$$I_{c_2} = 75{,}000 \times \frac{75{,}666}{113{,}500}$$

$$I_{c_2} = 75{,}000 \times .66 \ 2/3$$

$$I_{c_2} = \$50{,}000$$

MAXIMUM INVENTORY LIMIT

Formula:

$$I_2 = O + P$$

where

I_2 = Maximum Inventory

O = Standard Order

P = Order Point

Example:

From the following information determine what the maximum amount of inventory of part X532 could be. When the inventory reaches 4050 units an order is automatically placed for 5000 units.

Solution:

$$I_2 = O + P$$
$$I_2 = 5{,}000 + 4{,}050$$
$$I_2 = 9{,}050$$

MINIMUM INVENTORY LIMIT

Formula:

$$I_1 = D \times U$$

where

I_1 = Minimum Inventory
D = Number of Days Inventory Desired on Hand
U = Maximum Daily Inventory Usage

Example:

Determine the minimum inventory from the following information. The inventory control manager has ascertained that the stock for item X532 should never fall below an 8 day supply. The company uses this part in several products and for the past 2 years the number of parts placed into production for the various products averaged 150 units per day. However, it is possible if all product lines are producing at capacity that 225 units per day would be required. What should be established as the minimum inventory?

Solution:

$$I_1 = D \times U$$
$$I_1 = 8 \times 225$$
$$I_1 = 1800$$

DETERMINING THE ORDER POINT

Formula:

$$P. = (T \times U) + I_1$$

where

$P.$ = Order Point
T = Number of Days Required to Fill Order
U = Maximum Usage
I_1 = Minimum Inventory

Example:

From the following information determine the order point for item X532. The purchasing agent has informed you that on the average it takes 10 working days from the time the order is initiated from inventory control until it is received and placed in stock by the company. The maximum possible usage 225 units per day, and the minimum inventory is 1800 units.

Solution:

$$P. = (T \times U) + I_1$$
$$P. = (10 \times 225) + I_1$$
$$P. = 2250 + 1800$$
$$P. = 4050$$

OPTIMUM ORDER SIZE

Formula:

$$Q = \sqrt{\frac{2(CN)}{UI + A}}$$

where

Q = Optimum Order Size
C = Cost of Placing an Order
N = Number of Units Consumed in One Year
U = Unit Price of Material
I = Assumed Rate of Interest
A = Annual Carrying Cost Per Unit

Example:

A purchasing agent desires to know the optimum order size for material #X53. A study of the inventory records for the past several years indicated the average annual consumption per year is 2,500 units. Cost studies showed that the average cost to prepare a purchase order and handle the invoices and other paper work amounted to $20 per order. The standard price of the material obtained from the cost accounting department is $10 per unit. A review of the financial pages of a leading business paper revealed that the current market for a relatively "risk free" investment is yielding 4%. Statistical studies prepared for the warehouse operations revealed that, on the average, annual carrying costs amounted to 5% of the inventory cost. Determine the optimum order size.

Solution:

$$Q = \sqrt{\frac{2(CN)}{UI + A}}$$

$$Q = \sqrt{\frac{2\,(20 \times 2{,}500)}{10 \times .04 + .50}}$$

$$Q = \sqrt{\frac{100{,}000}{.90}} = \sqrt{90{,}000}$$

$$Q = 300 \text{ units}$$

Note: A equals $10 times 5% or .50.

12

Cost and Production Formulas[1]

EQUIVALENT UNITS OF PRODUCTION

Formula:

$$E.U. = T - \%_1(W/P_1) + \%_2(W/P_2)$$

where

$E.U.$ = Equivalent units of production
T = Number of units transferred to next department
$\%_1$ = Percentage of completion of beginning work in process
$\%_2$ = Percentage of completion of ending work in process
W/P_1 = Number of units in beginning work in process
W/P_2 = Number of units in ending work in process

Example:

From the following information compute the equivalent number of units for department No. 10 of the General Manufacturing Company. The production records of the department showed that 200,000 units were actually transferred to Dept. No. 11. The physical inventory at the end of the year revealed that there were 15,000 units of production still in process. The department foreman estimated that these units were on the average 80% completed. Reference to prior records revealed that at the beginning of the accounting period there were 45,000 units in process which were estimated to be on the average 20% completed.

Solution:

$$E.U. = T - \%_1(W/P_1) + \%_2(W/P_2)$$
$$E.U. = 200,000 - .20(45,000) + .80(15,000)$$
$$E.U. = 200,000 - 9,000 + 12,000$$
$$E.U. = 203,000$$

---

[1] From the book *Accountant's Handbook of Formulas and Tables,* by Lipkin, Feinstein & Derrick. Copyright © 1963 by Prentice-Hall, Inc., Englewood Cliffs, N. J., pp. 99–109.

BREAKEVEN POINT

Formula:

$$B/E = \frac{F/E}{1 - \dfrac{V}{S}}$$

where

B/E = Breakeven point
F/E = Total fixed expenses
V = Total variable expenses
S = Normal, budget, or capacity sales volume

Example:

The following data were assembled from the company's flexible budget.

Budget sales volume $1,800,000

Item	Amount	Fixed	Variable
Materials	$ 400,000		$ 400,000
Labor	175,000		175,000
Manufacturing			
Expenses	600,000	$100,000	500,000
Selling Expenses	100,000	30,000	70,000
General Expenses	300,000	245,000	55,000
Totals	$1,575,000	$375,000	$1,200,000

Determine the estimated breakeven point.

Solution:

$$B/E = \frac{F/E}{1 - \dfrac{V}{S}}$$

$$B/E = \frac{375,000}{1 - \dfrac{1,200,000}{1,800,000}}$$

$$B/E = \frac{375,000}{1 - .6667} = \frac{375,000}{.3333}$$

$$B/E = \$1,125,000$$

NORMAL BURDEN RATE
(Job Order Cost)

Formula:

$$N.B.R. = \frac{M/E}{B}$$

where

$N.B.R.$ = Normal burden rate
M/E = Estimated manufacturing expenses at normal capacity
B = Base at normal capacity (direct labor cost, direct labor hours, machine hours, etc.)

Example:

Determine the normal burden rate from the following data gathered from the books and records of the General Manufacturing Company for department number 3.

Estimates of the departmental capacity:

Engineering Estimate	100,000 units	Theoretical
Foreman's Estimate	80,000 units	Practical
Controller's Estimate	60,000 units	Normal

The controller's staff made the following estimate of manufacturing expenses based on normal capacity:

Fixed Expenses	$ 80,000
Variable Expenses	40,000
Total	$120,000

Solution:

$$N.B.R. = \frac{M/E}{B}$$

$$N.B.R. = \frac{120,000}{60,000}$$

$$N.B.R. = \$2.00$$

VOLUME VARIATION
(Job Order Cost)

Formula:

$$V_v = A\left(\frac{V/E}{B}\right) + F/E - A\,(N.B.R.)$$

where

V_v = Volume variance
A = Attained capacity
V/E = Variable expenses at normal capacity
B = Base at normal capacity
F/E = Fixed expenses at normal capacity
$N.B.R.$ = Normal burden rate

Example:

From the following data determine the volume variation for dept No. 3 of the General Manufacturing Co..

Manufacturing expense per general ledger: $115,000
Normal burden rate: $2.00 per direct labor hour.
Normal capacity: 60,000 direct labor hours
Attained capacity per production records: 50,000 hours
Budgeted expenses at normal capacity:
         Fixed expenses $80,000
         Variable expenses   40,000

Solution:

$$V_v = A\left(\frac{V/E}{B}\right) + F/E - A\,(N.B.R.)$$

$$V_v = 50{,}000\left(\frac{40{,}000}{60{,}000}\right) + 80{,}000 - 50{,}000\,(2.00)$$

$$V_v = 33{,}333.33 + 80{,}000 - 100{,}000$$

$$V_v = \$13{,}333.33$$

BUDGET VARIATION
(Job Order Cost)

Formula:

$$V_b = M/E - A\left(\frac{V/E}{B}\right) - F/E$$

where

V_b = Budget variation
M/E = Actual manufacturing expenses for the period
A = Attained capacity
V/E = Variable expenses at normal capacity
B = Base at normal capacity
F/E = Fixed expenses at normal capacity

Example:

See example for volume variation, page 176.

Solution:

$$V_b = M/E - A \left(\frac{V/E}{B} \right) - F/E$$

$$V_b = 115,000 - 50,000 \left(\frac{40,000}{60,000} \right) - 80,000$$

$$V_b = 115,000 - 33,333.33 - 80,000$$

$$V_b = \$1,666.67$$

Proof of Variations:

Actual Manufacturing Expenses	$115,000.00
Less: Applied Manufacturing	
Expenses (2.00 × 50,000)	100,000.00
Total Variation	15,000.00
Represented By:	
Volume Variation	13,333.33
Budget Variation	1,666.67
Total Variation	$ 15,000.00

QUANTITY VARIATION
(Standard Cost)

Formula:

$$V_q = P_s (Q_a - Q_s)$$

where

V_q = Quantity Variation
P_s = Standard Price
Q_a = Actual Quantity
Q_s = Standard Quantity

Example:

The bill of materials for assembly No. 56 shows that 2 units of part No. 137 are required for each assembly. The standard price for part No. 137 is 30¢ each. The production records indicate that 3,000 assemblies were produced during the period. The material requisition register shows that 6,100 units of part No. 137 were consumed.

Determine the quantity variation.

Solution:

$$V_q = P_s \, (Q_a - Q_s)$$
$$V_q = .30 \, (6{,}100 - 6{,}000)$$
$$V_q = .30 \times 100$$
$$V_q = \$30$$

PRICE VARIATION
(Standard Cost)

Formula:

$$V_p = Q_a \, (P_a - P_s)$$

where

V_p = Price Variation
Q_a = Actual Quantity
P_a = Actual Price
P_s = Standard Price

Example:

The General Manufacturing Company has established that the standard price for part No. 137 should be 30¢ per unit. The accounting department received an invoice for 5,000 units at 33¢ per unit.

Determine the price variation.

Solution:

$$V_p = Q_a \, (P_a - P_s)$$
$$V_p = 5{,}000 \, (.33 - .30)$$
$$V_p = 5{,}000 \times .03$$
$$V_p = \$150$$

LABOR WAGE RATE VARIATION
(Standard Cost)

Formula:

$$V_w = T_a \, (w_a - w_s)$$

where

V_w = Wage Rate Variation
T_a = Actual Time
W_a = Actual Wage Rate
W_s = Standard Wage Rate

Example:

From the following data taken from the records of Dept. No. 15 of the General Manufacturing Company determine the labor wage rate variation.

Data

	Standard	Actual
Wage Rate	$2.00 per hour	$2.10 per hour
Direct Labor hours	250	260

Solution:

$$V_w = T_a (W_a - W_s)$$
$$V_w = 260 \, (2.10 - 2.00)$$
$$V_w = 260 \times .10$$
$$V_w = \$26.00$$

LABOR EFFICIENCY VARIATION
(Standard Cost)

Formula:

$$V_E = W_s (T_a - T_s)$$

where

V_E = Labor Efficiency Variation
W_s = Standard Wage Rate
T_a = Actual Time
T_s = Standard Time

Example:

See example for labor wage rate variation, page 178.

Solution:

$$V_E = W_s (T_a - T_s)$$
$$V_E = 2.00 \, (260 - 250)$$
$$V_E = 2.00 \times 10$$
$$V_E = \$20.00$$

VOLUME VARIATION
(Standard Cost—3 Factor Analysis)

Formula:

$$V_v = F/E + A \left(\frac{V/E}{B} - S.B.R. \right)$$

where

V_v = Volume variation

F/E = Fixed expenses at normal capacity

A = Attained capacity

V/E = Variable expenses at normal capacity

B = Base at normal capacity

$S.B.R.$ = Standard burden rate

Example:

See example for budget variation below.

Solution:

$$V_v = F/E + A\left(\frac{V/E}{B} - S.B.R.\right)$$

$$V_v = 30,000 + 170,000\left(\frac{70,000}{200,000} - .50\right)$$

$$V_v = 30,000 + 59,500 - 85,000$$

$$V_v = 89,500 - 85,000$$

$$V_v = \$4,500$$

BUDGET VARIATION
(Standard Cost—3 Factor Analysis)

Formula:

$$V_b = M/E_a - A\left(\frac{V/E}{B}\right) - F/E$$

where

V_b = Budget variation

M/E_a = Actual manufacturing expenses

A = Attained capacity

V/E = Variable expenses at normal capacity

F/E = Fixed expenses at normal capacity

B = Base at normal capacity

C = Percentage of normal capacity attained

Example:

From the following data pertaining to manufacturing expenses under a standard cost system, analyze the variation by the following factors: budget, volume, efficiency.

Data

	Standard	*Actual*
Capacity	Normal 100%	80%
Fixed Expenses	$ 30,000	$ 32,000
Variable Expenses	70,000	56,000
Total	$100,000	$ 88,000
Direct Labor Hours (base)	200,000	170,000
Standard Overhead Rate	$.50	...

Solution:

$$V_b = M/E_a - A\left(\frac{V/E}{B}\right) - F/E$$

$$V_b = 88,000 - 170,000\left(\frac{70,000}{200,000}\right) - 30,000$$

$$V_b = 88,000 - 59,500 - 30,000$$

$$V_b = -\$1,500$$

EFFICIENCY VARIATION
(Standard Cost—3 Factor Analysis)

Formula:

$$V_e = S.B.R. \ [A - C(B)]$$

where

V_e = Efficiency variation
$S.B.R.$ = Standard Burden Rate
A = Attained capacity
C = Percentage of normal capacity attained
B = Base at normal capacity

Example:

See example for budget variation, page 180.

Solution:

$$V_e = S.B.R. \ [A - C(B)]$$
$$V_e = .50 \ [170,000 - .80 \ (200,000)]$$
$$V_e = .50 \ (170,000 - 160,000)$$
$$V_e = .50 \ (10,000)$$
$$V_e = \$5,000$$

CONTROLLABLE VARIATION
(Standard Cost—2 Factor Analysis)

Formula:

$$V_c = M/E_a - F/E - C(V/E)$$

where

V_c = Controllable variation

M/E_a = Actual manufacturing expenses

F/E = Fixed expenses at normal capacity

C = Percentage of normal capacity attained

V/E = Variable expenses at normal capacity

Example:

See example for budget variation, page 180.

Solution:

$$V_c = M/E_a - F/E - C(V/E)$$
$$V_c = 88,000 - 30,000 - .80\,(70,000)$$
$$V_c = 88,000 - 30,000 - 56,000$$
$$V_c = \$2,000$$

VOLUME VARIATION
(Standard Cost—2 Factor Analysis)

Formula:

$$V_v = F/E + C[V/E - S.B.R.\,(B)]$$

where

V_v = Volume variation

F/E = Fixed expenses at normal capacity

C = Percentage of normal capacity attained

V/E = Variable expenses at normal capacity

$S.B.R.$ = Standard Burden Rate

B = Base at normal capacity

Example:

See example for budget variation, page 180.

Solution:

$$V_v = F/E + C[V/E - S.B.R.(B)]$$
$$V_v = 30,000 + .80[70,000 - .50(200,000)]$$
$$V_v = 30,000 + .80(70,000 - 100,000)$$
$$V_v = 30,000 + .80(-30,000)$$
$$V_v = 30,000 - 24,000$$
$$V_v = \$6,000$$

13

Statistics: Sampling[1]

STANDARD ERROR OF THE MEAN. Variates. Sample small relative to universe. Standard deviation of universe known.

Formula:

$$\sigma_{\overline{X}} = \frac{\sigma}{\sqrt{n}}$$

where

$\sigma_{\overline{X}}$ = standard error of mean
σ = standard deviation of the universe
n = number of items in sample

Example:

A very large number of samples have been taken where each sample was composed of 9 observations on the number of errors made by a worker during 9 different days. The arithmetic mean of the 9 observations was then found for each of the large number of samples. How variable were these means if it was known that the standard deviation for all errors of all workers like those in the samples taken is equal to 3?

Solution:

$$\sigma_{\overline{X}} = \frac{\sigma}{\sqrt{n}}$$

$$\sigma_{\overline{X}} = \frac{3}{\sqrt{9}} = 1$$

Assuming a normal sampling distribution of means, one would expect the workers with least errors to average about 6 less than the average for the workers with most errors. (If the distribution is normal, six times the standard error of the mean, in this case 1, would include 99.7 per cent of the means).

Also review Chapter 2, "Work Measurement: Time Study Standards and Work Sampling."

---

[1] From the book *Accountant's Handbook of Formulas and Tables*, by Lipkin, Feinstein & Derrick. Copyright © 1963 by Prentice-Hall, Inc., Englewood Cliffs, N. J., pp. 38–47.

STANDARD ERROR OF A PROPORTION. Universe large relative to sample.

Formula:

$$\sigma_P = \sqrt{\frac{P(1-P)}{n}}$$

where

σ_P = standard error of a proportion

P = proportion of "successes" in universe (success is a favorable event happening)

$1 - P$ = proportion of "failures" in universe (failure is unfavorable event happening)

n = number in the sample

Example:

Can a sample of 100 having a 60 per cent preference of office workers for a 30 minute lunch hour come from a universe of all workers like these where there is actually an even division between preference for a 30 minute and a one-hour lunch period? (Assume that the 1 per cent level of significance is desired).

Solution:

$$\sigma_P = \sqrt{\frac{50(1-50)}{100}} = \sqrt{\frac{2500}{100}} = 5$$

$$= 5 \text{ per cent}$$

$$Z = \frac{60\% - 50\%}{5\%} = 2$$

Yes, the sample could come from such a universe of equal division in preferences. Z indicates that the sample proportion (60) differs from the universe proportion (50) by only 2 standard errors. If one is making his decision at the 1 per cent level of significance, this difference would have to be at least as great as 2.33 standard errors. Since this difference is only 2 standard errors, it is therefore not regarded as significant.

STANDARD ERROR OF THE MEAN. Variates. Sample size small relative to size of universe. Standard deviation of universe not known.

Formula:

$$\sigma_{\overline{x}} = \frac{s}{\sqrt{n-1}}$$

where

$\sigma_{\overline{x}}$ = standard error of mean
s = standard deviation for items in sample
n = number of items in sample

Example:

Book inventories have been compared to physical inventories and the size of the discrepancies noted. There are too many figures to use all of them. A random sample of 101 is selected. The standard deviation for these 101 is then computed and found to be $2,000. What is the standard error for a sampling distribution of means for all samples like the one taken? (Thus a measure of the variation among the discrepancies could be obtained).

Solution:

$$\sigma_{\overline{x}} = \frac{s}{\sqrt{n-1}} = \frac{2,000}{\sqrt{101-1}} = 200$$

Therefore the standard error of the mean discrepancy is $200.

STANDARD ERROR OF MEAN. Variates. Sample size large relative to size of universe.

Formula:

$$\sigma_{\overline{x}} = \frac{\sigma}{\sqrt{n}} \sqrt{\frac{N-n}{N-1}}$$

where

$\sigma_{\overline{x}}$ = standard error of the mean
σ = standard deviation of the universe
n = number in sample
N = number in universe

Example:

What is the standard error for the sampling distribution of means for samples of petty cash amounts if there are a total of 101 such accounts and a random sample of 36 of the 101 have been studied? It is known that the standard deviation of the 101 accounts is $12.

Solution:

$$\sigma_{\bar{X}} = \frac{\$12}{\sqrt{36}} \sqrt{\frac{101-36}{101-1}} = \$2.00 \sqrt{\frac{65}{100}} = \$2(.8) = \$1.60$$

Therefore the standard error of the mean is $1.60.

STANDARD ERROR OF PROPORTION. Sample size large relative to size of universe.

Formula:

$$\sigma_P = \sqrt{\frac{P(1-P)}{n}} \sqrt{\frac{N-n}{N-1}}$$

where

σ_P = standard error of a proportion

P = proportion of "successes" in universe. (A success is the occurrence of a favorable event)

$1 - P$ = proportion of "failures" in universe. (A failure is the occurrence of an unfavorable event)

n = number of items in the sample

N = number of items in the universe

Example:

Can a sample of 100 having a 60 per cent preference of office workers for a 30 minute lunch hour come from a universe of 1000 where the workers are evenly divided between preferences for a 30 minute and a 1-hour lunch period, if one is willing to be off in his decision 1 time in 100?

Solution:

$$\sigma_P = \sqrt{\frac{50 \times 50}{100}} \sqrt{\frac{1000-100}{1000-1}} = (5)\left(\frac{30}{31.6}\right) = 4.74$$

$$= 4.74\%.$$

$$Z = \frac{60\% - 50\%}{4.74\%} = \frac{10}{4.74} = 2.10$$

Yes, the sample could come from the universe where preferences were equally divided. The Z value is 2.10. This value would have to equal or exceed 2.33 in order to make the decision in this case that the 60 is significantly different from the 50.

STANDARD ERROR OF DIFFERENCES BETWEEN ARITHMETIC MEANS
Variates. Assuming independence.

Formula:

$$\sigma_{\overline{X}_1 - \overline{X}_2} = \sqrt{\sigma_{\overline{X}_1}^2 + \sigma_{\overline{X}_2}^2}$$

where

$\sigma_{\overline{X}_1 - \overline{X}_2}$ = standard error of difference between arithmetic means

$\sigma_{\overline{X}_1}^2$ = standard error squared of mean 1. (See pages 185, 187 for method of computation of this standard error)

$\sigma_{\overline{X}_2}^2$ = standard error squared for mean 2. (See pages 185, 187 for method of computation of this standard error)

Example:

If the average working capital requirement per week this year, based on a sample of 16 representative weeks, was $200,000, and last year based on the sample of the same number of weeks, it was $195,000, was this a significant difference this year from last based on the 95 per cent confidence level? The standard deviation for all weeks is known to be $10,000.

Solution:

$$\sigma_{\overline{X}_1}^2 = \left(\frac{10,000}{\sqrt{16}} \right)^2 = \left(\frac{10,000}{4} \right)^2 = (2,500)^2 = 6,250,000$$

$$= \$6,250,000$$

$$\sigma_{\overline{X}_2}^2 = \$6,250,000 \text{ also}$$

$$\sigma_{\overline{X}_1 - \overline{X}_2} = \sqrt{6,250,000 + 6,250,000} = \sqrt{12,500,000} = 3,535$$

$$= \$3,535$$

$$Z = \frac{\$200,000 - \$195,000}{\$3,535} = \frac{5,000}{3,535} = 1.41$$

Therefore the difference could be attributed to sampling error and is not significant. Z equals 1.41. It would have to be 1.96 or greater to show significance at the 95% level, where the decision was to be made.

STANDARD ERROR OF DIFFERENCE BETWEEN PROPORTIONS.
Assuming independence.

Formula:

$$\sigma_{P_1-P_2} = \sqrt{\sigma^2_{P_1} + \sigma^2_{P_2}}$$

where

$\sigma_{P_1-P_2}$ = standard error of difference between proportions
$\sigma^2_{P_1}$ = standard error squared of proportion 1. (See pp. 186, 188 for method of computation).
$\sigma^2_{P_2}$ = standard error squared of proportion 2. (See pp. 186, 188 for method of computation).

Example:

The proportion of 100 overdue accounts receivable, which were overdue for as much or more than 3 months, selected at random from this year's list, was 5%. The comparable proportion from last year's list was 3%. Is this a significant increase over last year at the 5% level of significance?

Solution:

$$\sigma^2_{P_1} = \frac{P(1-P)}{n} = \frac{(5)(95)}{100} = 4.75$$

$$= 4.75\%$$

$$\sigma^2_{P_2} = \frac{P(1-P)}{n} = \frac{(3)(97)}{100} = 2.91$$

$$= 2.91\%$$

$$\sigma_{P_1-P_2} = \sqrt{4.75 + 2.91} = \sqrt{7.66} = 2.77$$

$$= 2.77\%$$

$$Z = \frac{5\% - 3\%}{2.77\%} = \frac{2}{2.77} = .72$$

It is therefore not a significant difference at the 5% level of significance. The difference could be attributed to errors of sampling. (The Z value of .72 is less than 1.96 which is necessary at the 95% confidence level.)

STANDARD ERROR OF MEDIAN

Formula:

$$\sigma_{Md.} = 1.25\, \sigma_{\bar{X}}$$

where

$$\sigma_{Md.} = \text{standard error of median}$$
$$\sigma_{\overline{X}} = \text{standard error of arithmetic mean}$$

Example:

A random sample of 121 persons having expense accounts was selected. The vouchers belonging to each of these 121 persons were then classified under the appropriate name to whom each belonged. The median amounts of the expenses and the arithmetic mean amounts were then computed for each of the 121 individuals. The standard error of the arithmetic mean amounts was then computed and found to be $2.20. What was the standard error of the median amounts?

Solution:

$$\begin{aligned}
\sigma_{Md.} &= 1.25\,\sigma_{\overline{X}} \\
&= 1.25\,(\$2.20) \\
&= \$2.75
\end{aligned}$$

Therefore the standard error of the median amounts in this case is $2.75.

SAMPLE SIZE. Variables. Sample small in proportion to size of universe. Standard deviation of universe can be determined.

Formula:

$$n = \frac{Z^2 \sigma^2}{E^2}$$

where

n = number of items in sample
Z^2 = square of confidence level in standard error units
σ^2 = square of standard deviation of universe
E^2 = square of maximum difference between true mean and sample mean
    —the allowance for sampling error.

Example:

A random sample is to be taken from 4000 statements to verify the amounts on the statements. The allowable sampling error is $5. The sample size is to be large enough that this error is not exceeded more than 1 time in 100. What size should the sample be?

Solution:

$Z^2 = (2.58)^2$

$\sigma^2 = (31.34)^2$

> (This was obtained from the knowledge that the total range in the statements was from \$120 to \$276.70 or \$156.70. The standard deviation could be estimated to be 1/5 of \$156.70 or \$31.34)

$E^2 = (\$5)^2$

$$n = \frac{(2.58)^2 \, (31.34)^2}{(5)^2} = \frac{6.6564 \times 982.20}{25} = \frac{6538}{25}$$

$$= 262$$

Therefore the required sample size would be 262.

SAMPLE SIZE. Proportions. Sample small in relation to size of universe.

Formula:

$$n = \frac{Z^2 P (1 - P)}{E^2}$$

where

n = number of items in sample

Z^2 = square of confidence level in standard error units

P = proportion of "successes" in universe. (A success is the occurrence of a favorable event)

$1 - P$ = proportion of "failures" in universe. (A failure is the occurrence of an unfavorable event)

E^2 = square of maximum difference between true proportion and sample proportion—the allowance for sampling error.

Example:

From 10,000 items, a random sample is to be taken for verification purposes. It is believed that 90% of the totals on the forms to be studied are correct. The sample size is to be large enough that an observed difference in the sample proportion from the true proportion would not vary more than 3% five times out of 100.

Solution:

$Z^2 = (1.96)^2$

$P = 90$

$$1 - P = 10$$

$$E^2 = (3)^2$$

$$n = \frac{(1.96)^2 \, (.90) \, (.10)}{(.03)^2} = \frac{(3.84) \, (.09)}{.0009} = \frac{.3456}{.0009}$$

$$= 384, \text{ sample size}$$

SAMPLE SIZE. Variables. Finite Universe.

Formula:

$$n = \frac{N}{\dfrac{(N-1)E^2}{Z^2 \sigma^2} + 1}$$

where

n = number of items in sample
N = number of items in universe
Z^2 = square of confidence level in standard error units
σ^2 = square of standard deviation of universe
E^2 = square of maximum allowance for sampling error

Example:

A sample is to be taken from a universe of 2,000 items for verification. The universe standard deviation was estimated from a pilot study to be $50. The error is not to exceed $10, two times out of 100. How many should be selected for the sample to be verified?

Solution:

$$n = ?$$

$$N = 2,000$$

$$Z^2 = (2.33)^2$$

$$\sigma^2 = (50)^2$$

$$E^2 = (10)^2$$

$$n = \frac{2,000}{\dfrac{(1999)(100)}{(2.33)^2 (50)^2} + 1} = \frac{2000}{\dfrac{199900}{13572} + 1} = \frac{2,000}{15.7288}$$

$$= 127, \text{ sample size}$$

SAMPLE SIZE. Proportions. Finite Universe.

Formula:

$$n = \frac{N}{\dfrac{(N-1)E^2}{Z^2 P(1-P)} + 1}$$

where

n = number in sample
N = number in universe
E^2 = square of maximum allowance for sampling error
Z^2 = square of confidence level in standard error units
P = proportion of successes in universe. (A success is the occurrence of a favorable event)
$1 - P$ = proportion of failures in universe. (A failure is the occurrence of an unfavorable event)

Example:

A sample is to be taken from a total of 2,000 items for verification. It is estimated that the universe proportion of error is 10 per cent. An allowance of .05 is to be made for sampling error. Z is to be 2.33. What size random sample should be selected?

Solution:

$$N = 2,000$$
$$E^2 = (.05)^2$$
$$Z^2 = (2.33)^2$$
$$P = .90 \qquad 1 - P = .10$$

Therefore
$$n = \frac{2,000}{\dfrac{(2,000-1)(.05)^2}{(2.33)^2(.90)(.10)} + 1} = \frac{2,000}{\dfrac{4.9975}{.4886} + 1} = \frac{2,000}{11.228}$$

$$= 178, \text{ sample size}$$

14

Statistics: Correlation[1]

REGRESSION.

Linear. One independent variable.

Formula:

$$Y' = b_{11} + b_{12} X$$

where

Y' = any value of the dependent variable on the regression line
b_{11} = Y intercept of regression line (value of Y' where $X = O$)
b_{12} = slope of regression line (ratio of the change in Y for a given change in X)
X = any value of the independent variable

Example:

(a) What is the average relationship between total cost and units produced on a given job?

(b) What is the fixed cost, the average variable cost?

Data

Job Lot Number	Number Units Produced X	Total Cost (Dollars) Y	XY	X²	Y²
1	5	9	45	25	81
2	4	8	32	16	64
3	3	6	18	9	36
4	8	12	96	64	144
5	9	14	126	81	196
6	6	12	72	36	144
7	10	14	140	100	196
8	3	8	24	9	64
9	2	7	14	4	49

[1] From the book *Accountant's Handbook of Formulas and Tables*, by Lipkin, Feinstein & Derrick. Copyright © 1963 by Prentice-Hall, Inc., Englewood Cliffs, N. J., pp. 48–60.

Data (continued)

Job Lot Number	Number Units Produced X	Total Cost (Dollars) Y	XY	X^2	Y^2
10	2	6	12	4	36
11	3	10	30	9	100
12	6	10	60	36	100
13	9	15	135	81	225
14	7	14	98	49	196
15	5	11	55	25	121
16	7	13	91	49	169
17	6	11	66	36	121
18	6	13	78	36	169
19	7	10	70	49	100
20	5	10	50	25	100
21	5	12	60	25	144
22	6	14	84	36	196
23	4	9	36	16	81
24	8	14	112	64	196
25	9	13	117	81	169
	145	275	1721	965	3197
	ΣX	ΣY	ΣXY	ΣX^2	ΣY^2

Solution (Least Squares Method):

I $\Sigma Y = b_{11} N + b_{12} \Sigma X$ (with summations over all

II $\Sigma XY = b_{11} \Sigma X + b_{12} \Sigma X^2$ values of variables indicated)

I $275 = 25\, b_{11} + 145\, b_{12}$

II $1721 = 145\, b_{11} + 965\, b_{12}$

I (5.8) $1595 = 145\, b_{11} + 841\, b_{12}$

II-I (5.8) $126 = \phantom{145\, b_{11} +} 124\, b_{12}$

$$b_{12} = \frac{126}{124} = 1.02$$

Substituting 1.02 for b_{12} in equation I:

$$b_{11} = \frac{275 - 145\,(1.02)}{25} = 5.11$$

(a) Therefore the average line of relationship is:

$$Y' = 5.11 + 1.02\, X$$

(b) Average fixed cost: $5.11

Average variable cost: for each additional unit produced, total cost increased $1.02.

STANDARD ERROR OF ESTIMATE.
Linear Regression. One Independent variable.

Formula:

$$S_{yx} = \sqrt{\frac{\Sigma Y^2 - b_{11} \Sigma Y - b_{12} \Sigma XY}{N - 2}}$$

where

S_{yx} = standard error of estimate

ΣY^2 = sum of all squared values of dependent variable

b_{11} = Y intercept of regression line (Y value where X = O)

ΣY = sum of all values of dependent variable

b_{12} = slope of regression line

ΣXY = sum of all individual products formed by multiplying each X value by its corresponding Y value

N = total number of paired observations

Example:

Using the data given on pages 195,196 and assuming a normal distribution of Y values for each X, what would be the expected total cost of 20 units? What statement could be made concerning the reliability of this result?

Solution:

Expected total cost for 20 units:

$$Y' = \$5.11 + \$1.02\,(20) = \$5.11 + \$20.40 = \$25.51$$

Statement concerning reliability of the $25.51:

$$S_{yx} = \sqrt{\frac{3197 - 5.11\,(275) - 1.02(1721)}{23}} = \sqrt{\frac{36.33}{23}} = \sqrt{1.5796} = 1.26$$

If the statement is made that the cost of producing 20 units will average $25.51, but that the cost may be anywhere in the range of $25.51 ± $1.26 (between $24.25 and $26.77), that statement would hold in approximately 2/3 (68%) of the cases like this, but would be expected to be outside this range of $24.25 to $26.77 in the other 1/3 of the cases like this.

$25.51 \pm (1.96)(\$1.26)$ would give a 95% confidence interval, and $25.51 \pm (2.58)(\$1.26)$ would give a 99% confidence interval, when the above stated assumptions hold.

COEFFICIENT OF DETERMINATION.
Linear regression. One independent variable.

Formula:

$$r^2 = 1 - \frac{S^2_{yx}}{S^2_y}$$

where

r^2 = coefficient of determination
S^2_{yx} = standard error of estimate squared (see page 197)
S^2_y = standard deviation of Y variable squared

Example:

What is the per cent of variation in cost of producing which is associated with the number of pieces produced? (See data on pages 195, 196 and computation of $S_{yx} = 1.26$ on page 197.)

Solution:

$$r^2 = 1 - \frac{(1.26)^2}{S^2_y} \quad \text{where} \quad S^2_y = \frac{\Sigma Y^2}{N} - \frac{(\Sigma Y)^2}{N^2} = \frac{3197}{25} - \frac{(275)^2}{(25)^2}$$

$$= 127.88 - 121 = 6.88$$

$$= 1 - \frac{(1.26)^2}{6.88} = 1 - \frac{1.59}{6.88} = 1 - .23 = .77$$

Therefore 77 per cent of the variance in cost of producing is associated with the variation in the number of pieces produced. This leaves 23 per cent of the variation in cost which is *not* related to variation in number of units produced. This 23 per cent might be due to any of several other factors, one of which might be variation in materials quality.

COEFFICIENT OF CORRELATION.
Linear Regression. One independent variable.

A. Where coefficient of determination, r^2, is known:

Formula:

$$r = \pm\sqrt{r^2}$$

where

r = coefficient of correlation
r^2 = coefficient of determination

Example:

What is the coefficient of correlation for the data given on pages 195,196 where r^2, or coefficient of determination, is .77?

Solution:

$$r = \pm\sqrt{.77} = \pm .87$$

Note: The sign given to r depends on the sign of b_{12}, which for this example is plus. (See page 196.) Therefore the result would be plus .87.

B. Where coefficient of determination, r^2, is not known:

Formula:

$$r = \frac{N\Sigma XY - (\Sigma X)(\Sigma Y)}{\sqrt{[N\Sigma X^2 - (\Sigma X)^2][N\Sigma Y^2 - (\Sigma Y)^2]}}$$

where

r = coefficient of correlation

$N\Sigma XY$ = Multiply each X by each corresponding Y. Add these products. Then multiply this sum by N, the total number of paired observations.

$(\Sigma X)(\Sigma Y)$ = Add all X values. Add all Y values. Multiply sum of X values by sum of Y values.

$N\Sigma X^2$ = square each X value. Add squared X values. Multiply this sum by N, the total number of paired observations.

$(\Sigma X)^2$ = Add all X values. Square this sum.

$N\Sigma Y^2$ = Square each Y value. Add squares. Multiply sum by N.

$(\Sigma Y)^2$ = Add all Y values. Square sum.

Example:

What is the coefficient of correlation for the data given on page 196 where:

$$N = 25 \qquad \Sigma X = 145$$
$$\Sigma XY = 1721 \qquad \Sigma Y = 275$$
$$\Sigma X^2 = 965 \qquad \Sigma Y^2 = 3197$$

Solution:

$$r = \frac{(25)(1721) - (145)(275)}{\sqrt{[(25)(965) - (145)^2][(25)(3197) - (275)^2]}}$$

$$= \frac{43025 - 39875}{\sqrt{(24125 - 21025)(79925 - 75625)}}$$

$$= \frac{3150}{\sqrt{(3100)(4300)}} = \frac{3150}{3651} = .86$$

Note: the difference in the .86 obtained here and the .87 obtained in A above is due to rounding.

REGRESSION.
Linear relationships. More than one independent variable.

Formula:

$$X_1 = b_{11} + b_{12} X_2 + b_{13} X_3$$

where

X_1 = any value of dependent variable on the net regression line (same as Y' used on page 196).

b_{11} = point of intersection of the regression lines of the planes

b_{12} = rate of change in X_1 as X_2 changes

X_2 = any value of first independent variable

b_{13} = rate of change in X_1 as X_3 changes

X_3 = any value of second independent variable

Example:

(a) What is the average relationship between total cost (X_1), units produced (X_2), and thickness of materials (X_3)?

(b) What total cost would be expected if 10 units were produced and the material thickness were 5 thousands of an inch?

(c) What is the change in cost if pieces produced changed from 10 to 11, while material thickness did not change?

(d) What is the change in cost if materials thickness changed from 4 to 5 thousands of an inch, while pieces produced did not change?

Data

Job Lot Number	Total Cost ($) X_1	Number Units Produced X_2	Material Thickness (Thous. inch) X_3	X_1^2	X_2^2	X_3^2	$X_1 X_2$	$X_1 X_3$	$X_2 X_3$
1	9	5	4	81	25	16	45	36	20
2	8	4	3	64	16	9	32	24	12
3	6	3	2	36	9	4	18	12	6
4	12	8	5	144	64	25	96	60	40
5	12	6	4	144	36	16	72	48	24
6	14	9	6	196	81	36	126	84	54
7	14	10	6	196	100	36	140	84	60

Data (continued)

Job Lot Number	Total Cost ($)	Number Units Pro-duced	Material Thickness (Thous. inch)						
	X_1	X_2	X_3	X_1^2	X_2^2	X_3^2	X_1X_2	X_1X_3	X_2X_3
8	8	3	4	64	9	16	24	32	12
9	7	2	3	49	4	9	14	21	6
10	6	2	2	36	4	4	12	12	4
11	10	3	6	100	9	36	30	60	18
12	10	6	5	100	36	25	60	50	30
13	15	9	9	225	81	81	135	135	81
14	14	7	7	196	49	49	98	98	49
15	11	5	6	121	25	36	55	66	30
16	13	7	7	169	49	49	91	91	49
17	11	6	5	121	36	25	66	55	30
18	13	6	8	169	36	64	78	104	48
19	10	7	6	100	49	36	70	60	42
20	10	5	5	100	25	25	50	50	25
21	12	5	7	144	25	49	60	84	35
22	14	6	9	196	36	81	84	126	54
23	9	4	4	81	16	16	36	36	16
24	14	8	9	196	64	81	112	126	72
25	13	9	8	169	81	64	117	104	72
	275	145	140	3197	965	888	1721	1658	889
	ΣX_1	ΣX_2	ΣX_3	ΣX_1^2	ΣX_2^2	ΣX_3^2	ΣX_1X_2	ΣX_1X_3	ΣX_2X_3

Solution: (Least Squares method)

$$\text{I} \qquad \Sigma X_1 = b_{11}N + b_{12}\Sigma X_2 + b_{13}\Sigma X_3$$

$$\text{II} \qquad \Sigma X_1X_2 = b_{11}\Sigma X_2 + b_{12}\Sigma X_2^2 + b_{13}\Sigma X_2X_3$$

$$\text{III} \qquad \Sigma X_1X_3 = b_{11}\Sigma X_3 + b_{12}\Sigma X_2X_3 + b_{13}\Sigma X_3^2$$

$$\text{I} \qquad 275 = 25b_{11} + 145b_{12} + 140b_{13}$$

$$\text{II} \qquad 1721 = 145b_{11} + 965b_{12} + 889b_{13}$$

$$\text{III} \qquad 1658 = 140b_{11} + 889b_{12} + 888b_{13}$$

$$\text{II} \qquad 1721 = 145b_{11} + 965b_{12} + 889b_{13}$$

$$\text{I}' = (\text{I})(5.8) \qquad \underline{1595 = 145b_{11} + 841b_{12} + 812b_{13}}$$

$$\text{IV} = \text{II} - \text{I}' \qquad 126 = \qquad\quad 124b_{12} + 77b_{13}$$

$$\begin{aligned}
\text{III} && 1658 &= 140b_{11} + 889b_{12} + 888b_{13} \\
\text{I}'' = \text{I}(5.6) && 1540 &= 140b_{11} + 812b_{12} + 784b_{13} \\
\text{V} = \text{III} - \text{I}(5.6) \quad 118 &= && 77b_{12} + 104b_{13} \\
\text{IV} && 126 &= 124b_{12} + 77b_{13} \\
\text{V}' = \text{V}(1.61) && 190 &= 124b_{12} + 167b_{13} \\
\text{V}' - \text{IV} \quad 64 &= && 90b_{13} && b_{13} = .71
\end{aligned}$$

Substituting in V: $\quad b_{12} = \dfrac{118 - 104b_{13}}{77} = \dfrac{118 - (104)(.71)}{77} \qquad b_{12} = .57$

Substituting in I: $\quad b_{11} = \dfrac{275 - 145b_{12} - 140b_{13}}{25}$

$$= \frac{275 - 145(.57) - 140(.71)}{25} = \frac{93}{25} \qquad b_{11} = 3.72$$

(a) Therefore the average relationship is: $X_1 = 3.72 + .57\,X_2 + .71\,X_3$

(b) If 10 units were produced and material thickness were 5 thousands of an inch, a total cost of \$12.97 would be expected:

$$X_1 = 3.72 + .57(10) + .71(5) = 12.97$$

(c) If pieces produced is increased one piece, say from 10 to 11, while materials thickness does not change, the expected change in total cost would be \$0.57.

(d) If materials thickness changed one thousands of an inch, say from 4 to 5 thousands of an inch, while number of pieces produced remained constant, the total cost would be expected to change \$0.71 (the value of b_{13}).

MULTIPLE COEFFICIENT OF DETERMINATION.

Linear regression. More than one independent variable.

Formula:

$$R^2_{1.23} = 1 - \frac{S^2_{1.23}}{S_1^2}$$

where

$R^2_{1.23}$ = the coefficient of determination for variable 1 (the dependent variable) with variables 2 and 3 (the two independent variables)

$S^2_{1.23}$ = standard error of estimate squared for variable 1 as the dependent and variables 2 and 3 as independent

S^2_1 = standard deviation squared for variable 1 (dependent variable)

Example:

What is the per cent of variation in cost of producing dependent upon the number of pieces produced and the thickness of the materials used? (See data on page 201. See $S^2_1 = 6.88$ computed on page 198 and there indicated as $S_y{}^2$). (See page 201 for values of data needed for $S^2_{1.23}$).

Solution:

$$R^2_{1.23} = 1 - \frac{S^2_{1.23}}{S^2_1} = 1 - \frac{S^2_{1.23}}{6.88}$$

where

$$S^2_{1.23} = \frac{\Sigma X_1{}^2 - b_{11}\,\Sigma X_1 - b_{12}\,\Sigma X_1 X_2 - b_{13}\,\Sigma X_1 X_3}{n - 3}$$

$$= \frac{3197 - (3.72)\,(275) - (.57)\,(1721) - (.71)\,(1658)}{22} = .72$$

$$R^2_{1.23} = 1 - \frac{.72}{6.88} = 1 - .11 = .89$$

Therefore 89 per cent of the variation in cost of producing is dependent upon number of pieces produced and thickness of material. By referring to page 198, it can be noted that the coefficient of determination is increased from 77 per cent to 89 per cent by adding a second independent variable.

PARTIAL CORRELATION.

Linear relationship. Two independent variables.

Formula:

$$r_{12.3} = \frac{r_{12} - (r_{13})\,(r_{23})}{\sqrt{(1 - r^2_{13})\,(1 - r^2_{23})}}$$

where

$r_{12.3}$ = the partial correlation coefficient, i.e. the relationship between variables 1 and 2 with the effects of variable 3 taken out.

r_{12} = zero order correlation coefficient for variables 1 and 2

r_{13} = zero order correlation coefficient for variables 1 and 3

r_{23} = zero order correlation coefficient for variables 2 and 3

r^2_{13} = the zero order correlation coefficient squared for variables 1 and 3

r^2_{23} = the zero order correlation coefficient squared for variables 2 and 3

Example:

It can be seen by reference to page 199 that the correlation between total cost of producing and number of pieces produced is .86. This is a gross relationship. How much would this relationship be reduced if the effect on either one or both of the given variables of the thickness of material used were to be taken out?

Solution:

$r_{12} = .86$ (see page 199 for computations).

$$r_{13} = \frac{(25)(1658) - (275)(140)}{\sqrt{[(25)(3197) - (275)^2][(25)(888) - (140)^2]}}$$

$$= \frac{41450 - 38500}{\sqrt{(79925 - 75625)(22200 - 19600)}} = \frac{2950}{\sqrt{11,180,000}} = .88$$

$$r_{23} = \frac{25(889) - (145)(140)}{\sqrt{[25(965) - (145)^2][25(888) - (140)^2]}}$$

$$= \frac{22225 - 20300}{\sqrt{(24125 - 21025)(22200 - 19600)}} = \frac{1925}{\sqrt{8,060,000}} = .68$$

Therefore:

$$r_{12.3} = \frac{.86 - (.88)(.68)}{\sqrt{[1 - (.88)^2][1 - (.68)^2]}} = \frac{.86 - .60}{\sqrt{(1 - .77)(1 - .46)}}$$

$$= \frac{.26}{\sqrt{(.23)(.54)}} = \frac{.26}{\sqrt{.1242}} = \frac{.26}{.35} = .74$$

Partialling out the effects of material thickness reduced the gross relationship between total cost and number of units produced from .86 to .74.

RANK ORDER COEFFICIENT OF CORRELATION

Formula:

$$\rho = 1 - \frac{6\Sigma d^2}{N(N^2 - 1)}$$

where

ρ = rank order correlation coefficient

Σd^2 = sum of squared differences between each pair of corresponding ranks

N = total number of pairs of ranks

Example:

For 20 jobs, it was possible to know their relative positions with respect to time spent and cost, but not possible to measure either variable precisely enough to have confidence in its exact size. The time spent and costs were therefore ranked as to their relative position among the 20 jobs as follows:

Job number	Rank in amount of time spent on job	Rank in cost of job	Difference in ranks	Difference squared
1	3	1	2	4
2	7	9	2	4
3	13	15	2	4
4	17	16	1	1
5	20	18	2	4
6	1	2	1	1
7	8	8	0	0
8	2	4	2	4
9	14	12	2	4
10	10	7	3	9
11	16	19	3	9
12	4	5	1	1
13	19	20	1	1
14	11	10	1	1
15	5	6	1	1
16	18	14	4	16
17	15	13	2	4
18	12	17	5	25
19	6	3	3	9
20	9	11	2	4

$$106 = \Sigma d^2$$

$$\rho = 1 - \frac{6(106)}{20(20^2 - 1)} = 1 - \frac{636}{7980} = 1 - .08 = .92$$

The rank order correlation coefficient is .92.

RELIABILITY OF CORRELATION COEFFICIENTS.
Linear Relationships

Formula:

$$\sigma_r = \frac{1 - r^2}{\sqrt{N - 2}}$$

where

σ_r = standard error of the correlation coefficient, r.

r^2 = coefficient of correlation, squared

N = number of pairs of observations used to compute r

Example:

A correlation coefficient of .40 was found from a sample of 51 clerical workers when the total number of minutes late reporting for work each morning for a month was matched against the total job production for that month for each of the 51. How reliable was this .40? Could the .40 have arisen due to sampling error while in reality there was no real basis for believing that there was a relationship between "reporting on time for work in the morning" and "production."

Solution:

The hypothesis to be tested is that there is no correlation in this type of universe.

Therefore:
$$\sigma_r = \frac{1 - 0}{\sqrt{51 - 2}} = \frac{1}{7} = .14$$

If one wishes to maintain his confidence at the 95 per cent level, he will have to reject the above hypothesis and say there does seem to exist a relationship between the two variables since .40 is outside the limits of $0 \pm 1.96 \, \sigma_r$, outside the range of $\pm.2744$.

RELIABILITY OF CORRELATION COEFFICIENTS

Formula:

$$Z = \frac{1}{2} \log_e \frac{1 + r}{1 - r} = 1.15129 \log_{10} \frac{1 + r}{1 - r}$$

where

Z = a transformation of r and a statistic whose sampling distribution is approximately normal

r = zero order coefficient of correlation between two variables

Formula:

$$\sigma_z = \frac{1}{\sqrt{N - m - 1}}$$

where

σ_z = standard error of Z

N = number of pairs of observations used to compute r

m = number of variables used to compute r

Example:

A correlation coefficient of .40 was found from a sample of 52 clerical workers when the total number of minutes late reporting for work each morning for a month was matched against the total job production for that month for each of the 52. What is the probable range within which the true value of the universe relationship probably falls if one accepts a 95 per cent confidence interval?

Solution:

$$Z = 1.15129 \ \log_{10} \frac{1 + .40}{1 - .40} = 1.15129 \ \log_{10} \frac{1.40}{.60} = (1.15129)(0.367915)$$

$$= .42$$

$$\sigma_z = \frac{1}{\sqrt{52 - 2 - 1}} = \frac{1}{7} = .14$$

Therefore, a 95 per cent confidence interval = (1.96)(.14) = .27 on each side of the mean.

$$Z \pm 1.96 \ \sigma_z = .42 \pm .27 = .15 \text{ to } .69$$

The true correlation would therefore have a Z value range of .15 to .69.

15

Depreciation Formulas[1]

DEPRECIATION—STRAIGHT LINE

Formula:

$$D_j = \frac{C - S}{L}$$

where

D_j = Depreciation for particular year
C = Cost
S = Salvage Value
L = Estimated Life of asset

Example:

An asset cost $70,000 and has an estimated salvage value of $10,000. It is estimated that the life of the asset will be 15 years. What is the annual charge for depreciation?

Solution:

$$D_j = \frac{C - S}{L}$$

$$D_j = \frac{70,000 - 10,000}{15}$$

$$D_j = \frac{60,000}{15}$$

$$D_j = \$4,000$$

DEPRECIATION—UNITS OF PRODUCTION

Formula:

$$D_j = (C - S)\, \frac{U_x}{U_n}$$

---

[1] From the book *Accountant's Handbook of Formulas and Tables,* by Lipkin, Feinstein & Derrick. Copyright © 1963 by Prentice-Hall, Inc., Englewood Cliffs, N. J., pp. 79–86.

where

D_j = Depreciation for Particular Year
C = Cost
S = Salvage Value
U_x = Units Produced in Particular Year
U_n = Estimated Number of Units That Asset Will Produce During its
    Life.

Example:

The XYZ Company purchased a machine that cost $100,000. From past experience and future projections it is estimated that the asset will have a salvage value of $20,000 at the end of its useful life. Engineering estimates indicate that with adequate maintenance the machine will probably produce 500,000 units during its efficient life. During the 8th year of its life the machine produced 30,000 units. What is the annual charge for depreciation by the units of production method of depreciation?

Solution:

$$D_j = (C - S)\,\frac{U_x}{U_n}$$

$$D_8 = (\$100{,}000 - 20{,}000)\,\frac{30{,}000}{500{,}000}$$

$$D_8 = 80{,}000 \times .06$$

$$D_8 = \$4{,}800$$

Note: Units may be interpreted to be units, labor hours, machine hours, etc.

DEPRECIATION—SUM OF THE YEARS' DIGITS METHOD

Formula:

$$D_j = \frac{Y - D}{\Sigma Y} \times (C - S)$$

where

D_j = Depreciation for particular year
Y = Estimated life of asset
D = Number of years of prior depreciation $(D = 1 - j)$
ΣY = Sum of the years
C = Cost of the asset
S = Estimated salvage value

Example:

An asset cost $100,000 and has an estimated life of 20 years. The expected salvage value is $10,000. What is the amount of depreciation for the 14th year?

Solution:

$$D_j = \frac{Y - D}{\Sigma Y} \times (C - S)$$

$$D_{14} = \frac{20 - 13}{210} \times (100,000 - 10,000)$$

$$D_{14} = \frac{7}{210} \times 90,000$$

$$D_{14} = \$3,000$$

Note: If the ΣY is not included in the tables, it may be calculated by the following formula:

$$\Sigma Y = \frac{Y + 1}{2} \times Y$$

$$\Sigma Y = \frac{20 + 1}{2} \times 20$$

$$\Sigma Y = 210$$

DEPRECIATION—DECLINING BALANCE (REAL)

Formulas:

1. $R = 1 - \sqrt[L]{\dfrac{S}{C}}$

2. $D_j = R(C - A)$

where

R = Constant Rate
L = Estimated Life of Asset
D_j = Depreciation for Particular Year
C = Cost of Asset
S = Salvage Value
A = Accumulated Depreciation balance at beginning of year

Example:

What is the amount of depreciation for the third year, if an asset cost $20,000, has a salvage value of $5,000 and an estimated life of 10 years?

The accumulated depreciation per books on the first day of the fiscal year is $4,884.60

Solution:

1. $R = 1 - \sqrt[L]{\dfrac{S}{C}}$

   $R = 1 - \sqrt[10]{\dfrac{5,000}{20,000}}$

   $R = 1 - \sqrt[10]{.25}$ (See Note for Solving.)

   $R = 1 - .8705$

   $R = .1295$ or 12.95%

2. $D_j = R(C - A)$

   $D_3 = .1295(20,000.00 - 4,844.60)$

   $D_3 = .1295(15,155.40)$

   $D_3 = 1,962.62$

Note: Solution for $\sqrt[10]{.25}$.

> Log $.25 = 9.397940 - 10$
> Divide by 10 $= .939794 - 1$
> Antilog Closest $= 8,705$
> Place decimal point $.8705$
>
> $\therefore \sqrt[10]{.25} = .8705$

DEPRECIATION—200% DECLINING BALANCE

Formula:

$$D_j = 2\left[\frac{(C - S) - A}{L}\right]$$

where

D_j = Depreciation for Particular year
C = Cost of Asset
S = Salvage Value of Asset
A = Balance in Reserve at beginning of year
L = Estimated Life of Asset

Example:

What is the amount of depreciation in year 2 of an asset that cost $11,000, has a 10 year life and an estimated salvage value of $1,000, and the balance in the reserve account at the beginning of the fiscal year is $2,000?

Solution:

$$D_j \text{ or } D_2 = 2\left[\frac{(C - S) - A}{L}\right]$$

$$D_2 = 2\left[\frac{(\$11,000 - \$1,000) - \$2,000}{10}\right]$$

$$D_2 = 2\left[\frac{8,000}{10}\right]$$

$$D_2 = 2\,(800)$$

$$D_2 = \$1,600$$

DEPRECIATION—COMPOSITE RATE

Formulas:

$$1. \quad R = \frac{\dfrac{(C - S)_1}{L_1} + \dfrac{(C - S)_2}{L_2} + \dfrac{(C - S)_3}{L_3} + \ldots + \dfrac{(C - S)_n}{L_n}}{\Sigma C}$$

$$2. \quad D_j = \Sigma C_x \times R$$

where

R = Composite Rate of Depreciation
C = Cost
S = Salvage Value
ΣC = Sum of the Cost
ΣC_x = Sum of the Asset Costs at end of Particular accounting period
D_j = Depreciation for a particular year
$L_1 - L_2 - L_3$ = Estimated Life of Assets

Example:

At the end of the year 19A a composite rate of depreciation was computed for office equipment from the following facts:

Item	Cost	Salvage	Estimated Life
Typewriters	300	50	5 years
Calculator	600	100	10 years
Adding Machine	400	40	9 years
Desks & Chairs	1,500	0	15 years
Total	2,800		

What is the composite rate of Depreciation?

If at the end of the fiscal year 19C the total assets in the group amounted to $3,500, what amount should be charged to depreciation expense for that year?

Solution:

1. $R = \dfrac{\dfrac{(C-S)_1}{L_1} + \dfrac{(C-S)_2}{L_2} + \dfrac{(C-S)_3}{L_3} + \dfrac{(C-S)_4}{L_4}}{\Sigma C}$

$R = \dfrac{\dfrac{300-50}{5} + \dfrac{600-100}{10} + \dfrac{400-40}{9} + \dfrac{1,500-0}{15}}{2,800}$

$R = \dfrac{50 + 50 + 40 + 100}{2,800}$

$R = \dfrac{240}{2,800}$

$R = .0857$ or 8.57%

2. $D_j = \Sigma C_x \times R$

$D_j = 3,500 \times .0857$

$D_j = \$299.95$

DETERMINING ACCUMULATED DEPRECIATION BALANCE DECLINING BALANCE METHOD OF DEPRECIATION (REAL)

Formula:

$$A_n = C\,[1 - (1 - R)^n]$$

where

A_n = Accumulated Depreciation at the end of a series of accounting periods

C = Cost of Asset

R = Constant rate of depreciation

n = The number of years

Example:

What should be the balance in the reserve account at the end of year two, if an asset cost $20,000 and has a scrap value of $5,000, and the constant rate is 12.95%.

Solution:

$$A_n = C\left[1 - (1 - R)^n\right]$$
$$A_2 = 20,000\left[1 - (1 - .1295)^2\right]$$
$$A_2 = 20,000\left[1 - (.8705)^2\right]$$
$$A_2 = 20,000\left[1 - .75,777,025\right]$$
$$A_2 = 20,000 \times .24,222,975$$
$$A_2 = 4844.60$$

DETERMINING ACCUMULATED DEPRECIATION BALANCE DECLINING BALANCE METHOD OF DEPRECIATION (200% METHOD)

Formula:

$$A_n = C - S\left[1 - \left(1 - \frac{2}{L}\right)^n\right]$$

where

A_n = Accumulated depreciation at the end of a series of accounting periods

C = Cost of Asset

S = Salvage value (if used)

L = Estimated life of the asset

n = Number of years of life expired

Example:

An asset cost $11,000, has a salvage value of $1,000, and an estimated life of 10 years. If 200% declining balance method is used, what is the balance in the accumulated depreciation account at the end of the 2nd year?

Solution:

$$A_n = C - S\left[1 - \left(1 - \frac{2}{L}\right)^n\right]$$
$$A_2 = 11,000 - 1,000\left[1 - \left(1 - \frac{2}{10}\right)^2\right]$$
$$A_2 = 10,000\left[1 - (1 - .20)^2\right]$$
$$A_2 = 10,000\left[1 - .80^2\right]$$
$$A_2 = 10,000\left[1 - .64\right]$$
$$A_2 = 10,000 \times .36$$
$$A_2 = \$3,600$$

DETERMINING ACCUMULATED DEPRECIATION BALANCE
SUM OF THE YEARS' DIGIT METHOD OF DEPRECIATION

Formula:

$$A_n = C - S \times \left[1 - \frac{R\,(R+1)}{2\,\Sigma Y} \right]$$

where

A_n = Accumulated depreciation at the end of a series of accounting periods

C = Cost of the asset

S = Salvage value of the asset

Y = Estimated life of the asset

ΣY = Sum of the years

R = Remaining life of the asset ($R = Y - n$)

Example:

An asset cost $7,000 and has an estimated useful life of 8 years; the salvage value is estimated to be $2,000. What should be the balance in the accumulated depreciation account at the end of the 5th year?

Solution:

$$A_n = C - S \times \left[1 - \frac{R\,(R+1)}{2\,\Sigma Y} \right]$$

$$A_5 = 7,000 - 2,000 \times \left[1 - \frac{3\,(3+1)}{2 \times 36} \right]$$

$$A_5 = 5,000 \times \left[1 - \frac{12}{72} \right]$$

$$A_5 = 5,000 \times .8333$$

$$A_5 = \$4,166.50$$

16

Miscellaneous Formulas[1]

RETURN ON INVESTMENT—NETWORTH

Formula:

$$R.I._{n/w} = \frac{N/I}{\dfrac{N/W_1 + N/W_2}{2}}$$

where

$R.I._{n/w}$ = Return on Investment Based on Networth
N/I = Net Income for the period
N/W_1 = Networth at beginning of period
N/W_2 = Networth at end of period

Example:

From the data tabulated below compute the return on investment as measured by the networth or total capital per books.

	Current Year	Prior Year
Assets	$300,000	$200,000
Liabilities	120,000	60,000
Networth	$180,000	$140,000
Sales	$400,000	$350,000
Expenses	360,000	330,000
Net Income	$ 40,000	$ 20,000

---

[1] From the book *Accountant's Handbook of Formulas and Tables,* by Lipkin, Feinstein & Derrick. Copyright © 1963 by Prentice-Hall, Inc., Englewood Cliffs, N. J., pp. 116–117, 129–130, 133–134.

Solution:

$$R.I._{n/w} = \dfrac{N/I}{\dfrac{N/W_1 + N/W_2}{2}}$$

$$R.I._{n/w} = \dfrac{40,000}{\dfrac{140,000 + 180,000}{2}} = \dfrac{40,000}{\dfrac{320,000}{2}}$$

$$R.I._{n/w} = \dfrac{40,000}{160,000}$$

$$R.I._{n/w} = .25 \text{ or } 25\%$$

RETURN ON INVESTMENT—TOTAL ASSETS

Formula:

$$R.I._a = \dfrac{N/I}{\dfrac{A_1 + A_2}{2}}$$

where

$R.I._a$ = Return on Investment Based on Total Assets
N/I = Net Income
A_1 = Total Assets at Beginning of Period
A_2 = Total Assets at End of Period

Example:

See example for return on investment—networth, page 217.

Solution:

$$R.I._a = \dfrac{N/I}{\dfrac{A_1 + A_2}{2}}$$

$$R.I._a = \dfrac{40,000}{\dfrac{200,000 + 300,000}{2}}$$

$$R.I._a = \dfrac{40,000}{\dfrac{500,000}{2}} = \dfrac{40,000}{250,000}$$

$$R.I._a = .16 \text{ or } 16\%$$

DETERMINING GROSS WAGES WHEN NET WAGES IS GIVEN

Formula:

$$G = N + R_w[G - D(A)] + R_f(G)$$

where

G = Gross Wages

N = Net Wages

R_w = Rate of Withholding by Percentage Method

A = Amount of Exemption per Dependent

D = Number of Dependents Claimed

R_f = F.I.C.A. Tax Rate Imposed on Employee

Example:

Smith an employee agrees to work for Jones & Co. Smith wants $75 per week take home pay regardless of deductions required by law for withholding and F.I.C.A. taxes. Smith claimed 3 dependents on his W-4 form. The current rate of F.I.C.A. tax is 3% on employer and 3% on employee. The current withholding regulations allow a $13.00 exemption for each dependent on a weekly wage basis and an 18% tax rate for computing the amount to be withheld by the percentage method. What is the gross amount of wages that should be recorded in the payroll records for Smith?

Solution:

$$G = N + R_w[G - D(A)] + R_f(G)$$
$$G = 75 + .18[G - 3(13)] + .03(G)$$
$$G = 75 + .18(G - 39) + .03G$$
$$G = 75 + .18G - 7.02 + .03G$$
$$G = 67.98 + .21G$$
$$G - .21G = 67.98$$
$$.79G = 67.98$$
$$G = \$86.05$$

CO-INSURANCE

Formula:

$$R = \frac{C}{\% \, (F.V.)} \times L$$

where

R = Recovery from insurance company
$\%$ = Co-insurance requirement
$F.V.$ = Fair value or sound value of property at date of disaster
L = Loss due to peril insured against
C = The coverage or face value of the policy

Example:

The XYZ Company had a fire insurance policy on its building for a face value of $100,000. The policy contained an 80% co-insurance requirement. On February 17, 19__ a fire occurred which caused damage in the amount of $60,000. Several days later insurance appraisers estimated the fair value of the building before the fire to be $150,000. What amount is recoverable from the insurance company?

Solution:

$$R = \frac{C}{\% \, (F.V.)} \times L$$

$$R = \frac{100,000}{.80 \, (150,000)} \times 60,000$$

$$R = \frac{100,000}{120,000} \times 60,000$$

$$R = 50,000$$

Note (1): Insurer will never pay more than the face value of the policy.

Note (2): Fair value for insurance purposes is sometimes called sound value. Sound value is the replacement cost minus accrued depreciation. For instance, if in the above example the replacement cost is $300,000 for the building new and it would have a 50 year life and the old building is 25 years old, the sound value would be (300,000 − 150,000) = $150,000.

APPENDICES

Appendix A: Make or Buy[1]

EXPLANATION:

Management decision-making has become more exacting and specific in its application during recent years.

Make or buy decisions cover many departmental functions in operations involving substantial costs. The overall objective of this brief exposition is to provide you with guidelines to assist in analyzing, formulating, and concluding make or buy decisions. It is intended to make available to departmental executives specific criteria to be applied in order to choose alternatives of make or buy.

Certain limitations have been considered in this presentation. There cannot be specific formulae to universally resolve every conceivable type of make or buy problem. However, there are general guidelines of a quantitative and qualitative nature which can be generally applied to make or buy decisions.

The general categories warranting vital consideration are:

(A) THE CONTEMPLATED GROWTH PATTERN OF THE COMPANY

1. Initiating the Choice

A fundamental error in formulating make or buy policy is making a decision which is contrary to the long range growth pattern planned for the business. Every company should establish, early in its formative stages, a tentative blueprint of the nature and extent of its manufacturing functions as applied to specific components making up its end products.

2. Stages of Growth

Company growth pattern is usually developed in several stages concurrent with the expansion of operations resulting from increased sales volume. Frequently, manufacturers will purchase component parts up to an estimated point, where the ex-

---

[1] Extracted from the book *Make or Buy* by Harry Gross. Copyright © 1966. Used by permission of Prentice-Hall, Inc., Englewood Cliffs, New Jersey, Pp. xi, 3–16, 57, 60–63, 66–71, 137, 139, 141.

panded sales volume of usage of such components warrants fabrication at more economical costs, instead of buying.

3. Purchasing the Manufacturing Company

Companies with extremely dynamic sales organizations may function strictly as jobbers, by purchasing the completed finished product with the anticipation of buying entire manufacturing companies intact at the opportune time. Under such circumstances, manufacturing is usually not contemplated from inception.

4. Extreme Complexity and Responsibility in Manufacturing

There are some industries, such as drugs and electronics, where the responsibility of technical production is so extreme that the product must be manufactured almost in its entirety to better avoid possible defects. In these circumstances it is anticipated that all manufacturing will be done on the plant premises regardless of possible excessive costs.

(B) ANALYSIS OF SALES VOLUME

1. Potential and Consistency of Sales

A major pitfall that can be readily overlooked is not adequately considering the impact of sales volume on make or buy decisions. There should be a direct relationship between the sales trends of the company's products and the feasibility and profitability of manufacturing or purchasing such products, as indicated in Exhibit A–1.

2. Improper Decisions Related to Sales Volume

An obvious error that can readily be averted is deciding to manufacture where the costs, as related to sales volume, would be in excess of the costs of purchasing. It is important, in this type of analysis (see Exhibit A–1), to consider all special elements of costs, such as facilities, research, development inspection, and engineering as applied to manufacturing. These costs should be currently amortized over the known sales volume. In connection with the purchasing of specific products, such related costs as freight, inventory, storage, and obsolescence should be computed and applied.

3. Authentic Sources of Sales Data

Reliable sales data is essential for any accurate volume forecast. Various government agencies such as the Department of Commerce, Census, Federal Reserve, Security and Exchange Commission, and Small Business Administration provide extensive and regular literature containing invaluable statistics. Various banks, industrial bulletins, and private business periodicals supply data related to sales volume by products and companies.

4. Special Areas Providing Additional Sales

It is important to determine several significant sources for developing increased

ABC COMPANY (A FURNITURE MANUFACTURER)

ANALYSIS OF PRESENT AND PROJECTED OPERATIONS FOR CHAIR UPHOLSTERING

	Present Operation	Projected Operation	Variations in Projected Operation	Projections of Chair Upholstering Costs					
				25,000 Units		50,000 Units		75,000 Units	
				Amounts	Unit Costs	Amounts	Unit Costs	Amounts	Unit Costs
SALES	$4,000,000.00	$4,000,000.00	$ —						
Less:									
Purchases – Upholstery (75,000 Units @ $3.50)	262,500.00	—	(262,500.00)						
Material Costs	1,000,000.00	1,150,000.00	150,000.00	$ 55,000.00	$2.20	$105,000.00	$2.10	$150,000.00	$2.00
Direct Labor	1,000,000.00	1,082,500.00	82,500.00	37,500.00	1.50	65,000.00	1.30	82,500.00	1.10
Factory Overhead	600,000.00	622,500.00	22,500.00	15,000.00	.60	20,000.00	.40	22,500.00	.30
(% of Direct Labor)	(60.0%)	(57.5%)							
TOTAL PRIME COSTS	2,862,500.00	2,855,000.00	(7,500.00)	107,500.00	4.30	190,000.00	3.80	255,000.00	3.40
GROSS PROFIT	1,137,500.00	1,145,000.00	7,500.00						
Selling and Administrative	858,750.00	858,750.00	—						
(% of Prime Costs)	(30.0%)	(30.1%)							
NET PROFIT (BEFORE TAXES)	$ 278,750.00	$ 286,250.00	$ 7,500.00						

EXHIBIT A-1

sales volume by a company. In some industries, such as electronics, defense agencies are a substantial customer for many products. Foreign export provides capacity for expanded sales volume within the limitations of tariff regulations. Manufacture of an improved product, coupled with aggressive sales promotion, will likewise help boost sales volume for a company.

(C) AVAILABILITY AND QUALITY OF MATERIALS

1. Procuring Reliable Sources of Supply

Extreme productive complications and excessive costs may readily result when insufficient investigation is made of the reliability of the suppliers of materials. A survey, made well in advance of the letting of purchase contracts, should include several possible vendors, especially when such materials involve products not previously made. Wherever practicable, purchases should be arranged with more than one supplier.

2. Checking the Quality of Materials Supplied

Of equal importance is the establishment of high quality standards for purchased materials or component parts. Great production loss could result from the rejection of materials, particularly when these items are to be fabricated into additional components leading to an end product. Comprehensive specification requirements should be spelled out in detail, and rigid inspection procedures must be established. All these standards pertaining to quality controls should be specifically included as *referred clauses* in the purchase contracts.

3. Other Vital Factors Regarding Material Purchases

Not to be overlooked is the financial stability of the vendor. In many instances, materials are purchased from a vendor who is infringing on the patent rights of others. Protection can be achieved through specific warranties, or through investigation of possible patent violations. Freight and shipping facilities should be analyzed with regard to costs as well as the expediting of material shipments. The economy and efficiency of purchasing more complete component units from one supplier, as contrasted with diversified fabrication by several suppliers, should be considered. These principles also apply to raw materials.

(D) EXTENT AND PROFICIENCY OF LABOR

1. Survey of Labor Market

A company should never attempt its own product manufacturing program without evaluating the availability of the employees required for the work to be performed. This would apply to an even greater extent in cases where the manufacturing is being initiated without the benefit of extensive prior experience. This analytical study would include consideration of the maximum number of workers for each specific category of work as required under the long range expanded production program.

2. The Technical Supervisors

Even more vital than the rank and file workers are the abilities of technical supervisory personnel in each department. A skeleton staff of key employees, with provision for planned expansion for more trained personnel, is guided by these supervisors.

3. Budget of Labor Costs

A labor study can be of little value, or misleading, if it does not include an exacting estimate of the cost of applicable labor. The basis for this budget can be provided by labor statistics (for rates) procured from appropriate unions and the Department of Labor, as well as industry associations. Computations of the performance of similar operations must be coordinated with rates of pay to properly evaluate these labor costs.

(E) ADEQUACY OF PROPER FACILITIES

1. Complete Facilities Necessary for Manufacturing

An absolute must for completing the make cycle in the manufacturing process is the availability of the facilities required for production. Substantial capital and financing should be kept in reserve, ready to purchase or rent machinery to carry on the manufacturing function on a well-balanced, efficient basis. In its early stages, it may not be wise for a business to acquire facilities at the most economical cost, and risk not fulfilling its production requirements.

2. Relationship of Sales, Production, and Facilities.

The first step in a well-coordinated facility procurement is to establish a long term sales forecast on a realistic basis. This projection is then coordinated with the most economical volume production runs which can be fabricated by a specific machine. The manufacturing efficiency of the facility would be best determined by experience data accumulated by the facility vendor.

3. When and How to Acquire the Facility

Once it is definitely determined that it is advantageous to procure a specific machine at the proper time, bids should be requested of several vendors and lessors for detailed operational features and costs. If it is anticipated that the machine may soon become obsolete, or that there are limited finances available, the most suitable course would be to lease. Should the choice be to purchase, arrangements can be made for the most reasonable cost of long-term financing, if required, with the vendor or the local bank. An alternate method of purchase in frequent use is the lease with option to purchase.

(F) RELIABILITY OF TECHNICAL AND ADMINISTRATIVE PERSONNEL

1. The Management Team

A catastrophic mistake of many businesses is failure to build a competent organization in each department around the key personnel. In a manufacturing company this group would be represented by sales, production, purchasing, engineering, accounting, and fiscal managers. This team plays a vital role in coordinating the policy and responsibility of the make or buy program, in addition to their regular department functions.

2. Recruiting the Staff

The most essential employee in any organization in connection with the make or buy function, if any one can be earmarked, is the sales manager. This individual plays an equally important role whether the company stresses purchasing or manufacturing its products, due to the basic importance of forecasting and maintaining consistent sales volume.

(G) PROPER COMPARISON OF COSTS UNDER "MAKE OR BUY"

1. Lower Costs as Basis for Decision

While it does not constitute the one and only basis in itself, the lower cost applicable to either make or buy will constitute the greatest weight to influence either alternate choice. This conclusion stems from the basic economic profit objectives of any business operation. Reference is made to Exhibit A–1 for a simplified crystalized summation of the pertinent cost data, which is the integral factor justifying the ultimate make or buy decision.

2. Inclusion of All Cost Elements

While the customary orthodox cost elements as derived from routine cost records provide limited complexities in compilation and interpretation, misleading conclusions can be reached by erroneously omitting or miscalculating marginal or unusual items of cost. In the manufacturing process a major item frequently overlooked or poorly estimated is the setting-up costs to be incurred prior to and in conjunction with the production assembly line. Items of cost such as engineering, research, and development are subject to wide fluctuations in connection with their applicability to the end product cost. Another expense to be considered is the amortization of machinery and equipment, specifically based on the ultimate quantity to be produced. With reference to the purchasing of materials, items frequently improperly considered and evaluated are shipping costs, excessive inventory costs for carrying charges, and obsolescence.

3. Consistent Comparability of Costs

Analysis made on an inconsistent basis of comparison, compilation, or summation can lead to fake conclusions in cost analysis. Such inconsistencies may result in including or excluding cost elements in several similar studies of either the manufacturing or the purchasing process. An example of this would be the possible omission of an item of cost such as freight-in which could be included in the statistical compilation of *making* the product, and excluded in the summation of *buying* the item. Any of these inconsistencies would obviously result in misleading comparative analysis. Erroneous cost data may be compiled by including historical data in one instance and comparing it with estimated costs in another.

(H) PROVISIONS FOR FINANCING

1. Sufficient Funds Available for Specific Purposes

Either program, make or buy, may collapse due to inadequate financial planning for carrying it out to completion. While funds need not be on hand initially for all purposes, planned commitments should have been arranged to meet specific requirements as they occur. This can be accomplished through the use of a cash flow budget for the period to be projected to anticipate the needs.

All substantial expenditures, such as for machinery and equipment, should be particularly noted and provided for at the required time in this type of presentation. As in other areas of make or buy, great care should be exercised to properly consider such special items of expense as setting up costs, research, and development. One specific item which is a drain on available cash over extended periods in the buying function is the purchasing of imported items.

2. How, When and Where to Procure Funds

While there is no dogmatic rule to follow, it generally is preferable to attempt to arrange long term financing for long term needs and short term borrowing for short term requirements. Items such as machinery and equipment are most adequately financed through time payments with the vendor or the bank over a period extended as close to its maximum productive use as possible. Long term financing can sometimes be arranged through Government agencies such as the Small Business Administration, State Development companies, and, in unusual growth situations, a stock underwriting may be most appropriate.

(I) BACKGROUND AND EXPERIENCE IN THE FIELD.

(1) Experience Essential in New Business

One of the greatest handicaps to be overcome in starting and establishing a new business occurs in cases where top management lacks extensive technical background and experience in the specific operation. In most instances, however, the principals of the business have been in similar enterprises, or have held administrative positions in that industry. There is no theoretical substitute for this background, whether applied to make policy decisions or to the buy alternate function.

2. Familiarity as an Aid in Acquisitions

In recent years, many companies have made serious economic mistakes in acquiring companies in diverse industries foreign to their regular operations. Many of these decisions were prompted by the necessity for diversification due to anti-trust requirements or for corporate capital growth. These costly blunders were made most extensively by attempting to make a product which should have been bought. In such situations, the acquisitions probably would not have been made if management had had extensive experience in the field. In some cases these non-profitable acquisitions could have been recognized by experienced management as businesses with declining sales volume or costly productive operations, possibly brought about by obsolete machinery. Where experience of top management is lacking in an acquisition, key personnel of the acquired company can be retrained to alleviate some of the complications inherent in such a project. Many of these problems are made less costly where the acquisition is made in an allied field; this makes it more practical to coordinate the manufacturing process through the utilization of similar dıoduction facilities, technical personnel, and more economical material purchases.

3. Knowledge in Adding a New Product

Similar principles regarding background and experience, which apply to make or buy for a new business and acquisitions, pertain to adding a new product. Certainly no attempt should be made in this direction without adequate knowledge in various aspects of the contemplated new product, particularly as to where it is to be manufactured. Repercussions can be drastic in operations where supervision and inspection are controlled by government regulatory agencies, as in the drug and dairy industries.

FIXED AND VARIABLE COSTS DEFINED

Fixed Costs

These costs are independent and not influenced by the volume of items produced. As long as this item of expense is incurred, it is at a specific level. Typical of this type of expense would be the fixed salary of a production manager hired to make a specific product. His salary cost will be incurred at the same amount, independent of the volume of production his department produces. Unless alternative uses for this employee's time can be found, his weekly cost is directly attributable to the particular product or products for which he was hired. Similarly, special setting up costs incurred for a particular use is fixed expense. If, as a result of the decision to make or buy the product, specific employees' salaries or other costs are acquired for that purpose only, most of these incurred costs will be of the fixed type. The meaningful characteristics of these costs are that they will not change or vary to any degree, even though they are utilized in the manufacture or purchase of varying quantities of the product.

Exhibit A–2 graphically illustrates the nature of fixed costs. The manufacturer in this case incurred a total cost of $ 875 as setting-up costs for a power grinder. This cost was incurred once and normally would not be incurred again. The cost for grinder setting-up is independent of the number of units produced over the life period of this machine.

GRAPHICAL ILLUSTRATION OF FIXED COSTS

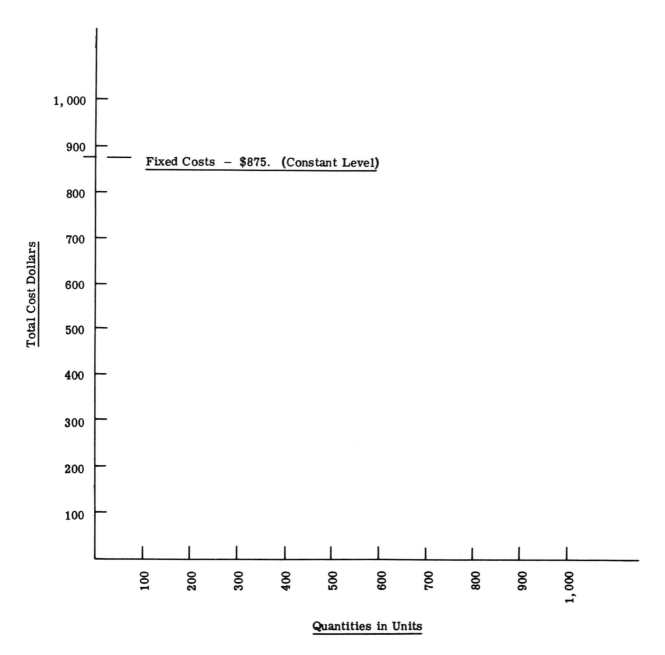

Fixed Costs − $875. (Constant Level)

Total Cost Dollars

Quantities in Units

EXHIBIT A-2

EXHIBIT A-3   VARIABLE COST LINES

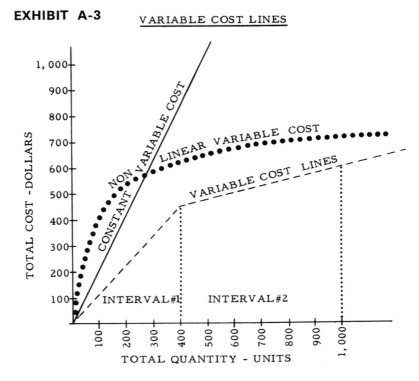

EXHIBIT A-4

SEMI-VARIABLE COST LINES

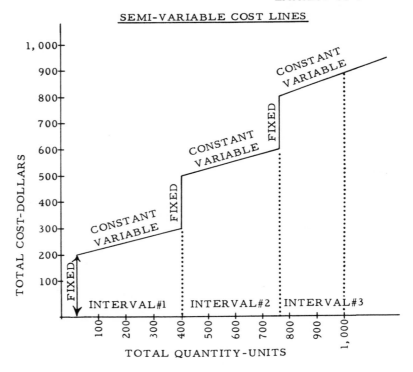

Variable Costs

These variable costs vary in proportion to the quantity made or bought. Variable costs do not always vary directly in proportion with quantities produced. Economies of scale sometimes cause variable costs to vary non-linearly with decreased labor costs from increased production, or through discount savings of large scale purchases. Thus, within a quantity interval the per unit cost is constant, while over a larger quantity interval the unit cost would usually decrease. This would be an important consideration in the make or buy decision where additional potential quantity requirements might change.

Semi-Variable Costs

These costs are those which have combined characteristics of fixed and variable costs over several quantity intervals of production units. Semi-variable costs can be more commonly viewed as occurring in "chunks," or step group increase in costs. Such costs remain constant over short intervals. When one quantity interval is passed the total costs increase in steps. Thus, the unit costs over that interval remain constant, while over several intervals the costs per unit change. The unit costs over several intervals may be approximated by a straight line following the trend of the total semi-variable cost curves.

DESCRIPTION OF EXHIBITS A–3 AND A–4 (VARIABLE COST LINES)

These several cost lines graphically illustrate three different types of cost relationships:

(1) Constant Variable Cost

In this situation, total costs increase uniformly with increasing quantities of units. Another descriptive way of presenting this would be that each quantity increment carries with it a uniformly proportionate increment.

(2) Non-Linear Variable Cost

In this instance, there is no direct proportionate relationship between quantity of units and cost dollars.

(3) Variable Cost Lines

These cost lines vary with quantities. However, the proportionate relationship between quantities of units, and unit cost, is different and varies over different intervals.

Semi-Variable Costs Line

These costs combine the characteristics of fixed and variable costs. Over each interval, costs increase uniformly in proportion with that quantity. Once a "threshold" is reached, the costs increase with a jump to another level. For example, a machine will be purchased to perform an operation. The depreciation expense of the equipment is a fixed cost. For production use up to the capacity of the equipment, there will be an additional variable cost. Where more capacity is required, another machine must be acquired. With this additional machine, the firm will incur the applicable fixed depreciation cost associated with this new equipment.

RELEVANT COSTS

These costs are termed relevant because of their relationship to the particular make or buy decision. Relevant costs are those costs which would be affected by the decision. An example would be where a cost such as a new warehouse would be incurred for the storage of raw materials for the manufacture of a product, which cost would be avoided if the same item were purchased. All pertinent periodic warehouse expenses, such as depreciation and maintenance, are also warehouse attributable costs relevant to the particular make or buy decision. If, on the other hand, a warehouse is currently in existence with excess capacity which can't be used for alternative uses other than in the make or buy activity, the warehouse costs would not be relevant. Since the warehouse expenses are not affected by the make or buy decision, they are not relevant costs.

A COST STUDY HISTORY (EXHIBIT A–5)

This is an example of the detailed cost analysis and cost comparative procedures as made by the GHI Company (a machine shop), involving the make or buy decision for a necessary part. This manufacturer needed a bearing assembly for locating and drilling holes and spotting counterbores. The firm could acquire this part through one of the following three alternatives:

(1) Make the part by using the firm's present facilities which were operating at their full load (less normal downtime).
(2) Make the part by purchasing and using a special numerically controlled N/C machine. This N/C machine could be used to its capacity for other operations and products.
(3) Buy the bearing assembly part directly from a reliable supplier.

The initial step was to list all the possible departmental costs to be analyzed for applicability to each of the three alternatives. A worksheet utilized known as the "Make or Buy Cost Accounts Analysis," (Exhibit A–5), includes relevant cost accounts as well as all pertinent details for analysis. This product group incurred many costs in the course of operations. It was necessary to cull out these costs not affected by the make or buy decision, as well as making certain to include all those items of cost which would be influenced. Similarly, within those items of cost affected by the decision, there might be some portion of cost not relevant to the analysis for the decision. These would be the fixed portions of expense of some of the semivariable cost items already on hand, such as the cost of labor. If there would be an alternative use for the men's time, the cost of another worker would be viewed as completely variable, and not as a step function with fixed cost over the worker's applicable quantity interval of production. In the same way, the analysis would note that the cost of the N/C machine installation is a fixed cost, the per unit cost of which is directly dependent on the quantity produced.

GHI COMPANY (A MACHINE SHOP)

MAKE OR BUY COST ACCOUNTS' ANALYSIS FOR BEARING ASSEMBLY PART

Cost Accounts	Department Responsible	Type of Cost	Applicable Costs to Decision				Basis for Computation	Comments
			Make With Facilities	Make With N/C Machine	Buy Part From Supplier	Costs Not Applicable		
Direct Labor	Production	Semi-Variable	X	X			Estimate of Time @ Rate	2 Shifts, 40 hr. wk.
Direct Material to Make	Purchasing	Variable	X	X			Usage & Quotation	2% Discount
Direct Purchase to Buy	Purchasing	Variable			X		Direct Quotation	No Discounts
Machine Cost – Variable	Production	Variable	X	X			Estimate of Hourly Use	Other Use
Machine Cost – Fixed	Production	Fixed				X	–	N/C Machines No Other Machine Use
Tooling Fixture Installation	Engineering	Fixed	X				Estimate of Costs	Modification Costs
Tooling Fixture Modification	Engineering	Fixed	X				Estimate of Costs	Plant Engineering
Special N/C Machine Installation	Purchasing	Fixed		X			Direct Quotation	Maximum Time Limit
Special N/C Engineering	Engineering	Semi-Variable	X	X			Direct Quotation	Outside Consultant
Setting Up Costs	Production	Fixed	X	X			Estimate of Costs	Overtime Required
In-Process Storage	Engineering	Semi-Variable	X	X			Estimate of Costs	Special Handling
Freight In	Shipping	Variable	X	X	X		Rate and Weight	Less Than Minimum
Freight Out	Shipping	Variable			X			Same Costs for all
Insurance on Equipment	Purchasing	Fixed				X		No Reduction
Insurance on Inventory	Purchasing	Variable	X	X	X		Rate and Coverage	Policy Extension
Supervision	Production	Semi-Variable	X	X	X		Estimate of Time	Proportional Allocation
Indirect Overhead (Storage)	Purchasing	Semi-Variable	X	X	X		Estimate of Costs	Additional Space
OTHER FIXED EXPENSES								
Supervision	Production	Fixed Portion				X		Part Non-Allocable
Storage	Purchasing	Fixed Portion				X		Part Non-Allocable
Factory Overhead Expenses	Production	Fixed				X		Non-Allocable
Sales Expenses	Sales	Fixed				X		Non-Allocable
Shipping Expenses	Shipping	Fixed				X		Non-Allocable
General and Administrative	Controller	Fixed				X		Non-Allocable

EXHIBIT A-5

Following are specific detailed descriptions of the cost accounts indicated in Exhibit A–5, particularly concerning their relevancy to the make or buy analysis:

Direct Labor—Semi-Variable

The cost of labor is relevant to the analysis and is semi-variable. Labor costs are added and subtracted in chunks. These chunks can either be hours or full shifts. In this case the company can cost its labor on the basis of the overtime cost of machine operators. However, if the firm were presently operating at 100% capacity the costs for labor would have to be those for full shifts.

Direct Material and Purchase—Variable

The costs of material purchased are in direct relationship to the quantity of materials purchased. Materials or parts are purchased for a specific purpose, there-fore, the applicable costs for these materials or parts are directly attributable to the make or buy decision.

Machine Costs

There are two types of machine cost. Assuming a previous decision to keep these machines on hand, such fixed machine costs as fixed period maintenance are inde-pendent of the make or buy decision. However, if there is alternative use for this machine, the variable cost of using this machine is relevant to the make or buy analysis. Depreciation is computed on the basis of proportionate hours that the machine is in use.

N/C Machine Cost

Since this firm has alternative uses for this machine, the applicable costs are handled in the same way as all other variable machine costs. If, however, the only use for this special equipment were for the particular bearing assembly, all costs would be fixed and included in the economic analysis.

Tooling Fixture Costs

These costs are directly attributable to make or buy decision. Under the make decision, a fixture would have to be built or an existing fixture would be modified, both of which would be includable.

N/C Installation Costs

These costs are handled in the same manner as conventional equipment, and are directly attributable to the make decision with this special machine.

Setting-Up Costs

These costs are directly attributable to both make alternatives, and are therefore directly attributable to the make decision.

In-Process Storage

There is a different in-process storage requirement for the alternative courses of action in making or buying. Therefore, the variable portion of the semi-variable costs of in-process storage of the various alternatives is relevant.

Freight Costs

Freight *in* is variable, but is relevant to the decision. The same costs for freight *out* are incurred under all alternatives, but it is independent of the make or buy decision and is therefore not relevant to the analysis.

Insurance Costs

Insurance costs on equipment are irrelevant to the make or buy analysis, and therefore should be deleted. Inventory insurance costs are directly related to the quantity and type of inventory on hand as a result of the make or buy decision. Thus these alternative inventory costs, as applicable to different inventory amounts, are included in the analysis. It should be noted that since the buy decision requires a larger quantity of inventory its related cost is greater.

Indirect Expenses—Variable Portion:

The proportionate expenses of supervision and storage are relevant to the make or buy analysis, and are includable.

Indirect Expenses—Fixed Portion:

These costs are either fixed or non-controllable. In either event, their incurrence is independent of any of the make or buy decisions. They are, therefore, not relevant costs, and should not be included in the analysis.

When this comprehensive analysis of detailed costs is completed, the analyst is ready for the final stage of presentation of "Make or Buy Comparative Cost Summary," as indicated by Exhibit A–6. The objective is to project the data as compiled in the "Make or Buy Cost Accounts Analysis." Then by applying applicable computations to each cost element, the complete summary of each and all costs are worked out for each of the make and buy alternatives. Thus, by comparing relative costs, a final make or buy decision can be made on an intelligent basis. In this exhibit, the variable costs are constant. If variable costs were subject to change in succeedingly greater quantity intervals, these quantity cost breaks would be indicated in a new column. Such a presentation would lead management to the question of projected quantity requirements. After these costs were collected and segregated, the summary total costs indicated that to buy for $ 825 purchase price and $ 911 total costs was the most economical decision. It was decided to choose this alternative of buying the bearing assembly since there were no special major advantages in making which might outweigh the cost-saving features of buying this part. It is important to note that the make or buy costs which have been discussed in Exhibit A–6 as total and unit costs, are only costs relevant to the make or buy decision. The use of these costs as presented is a guide for the make or buy analysis. Management accounting would

GHI COMPANY (A MACHINE SHOP)

MAKE OR BUY COMPARATIVE COST SUMMARY

FOR BEARING ASSEMBLY PART

QUANTITY - 125 UNITS

Description of Costs	Make With Present Facilities	Make With Special N/C Machine	Buy Part From Supplier
Direct Labor	$ 180	$120	$ —
Direct Material to Make	250	250	—
Direct Purchase to Buy	—	—	825
Variable Machine Cost (Note 2 Machines)	375	315	—
Tooling Fixture Installation	145	—	—
Tooling Fixture Modification	60	—	—
Special N/C Machine Installation	—	110	—
Special N/C Engineering	—	60	—
Setting Up Costs	122	90	—
In Process Storage	12	12	40
Freight In	5	5	31
Insurance on Inventory	2	2	7
Supervision	20	5	5
Indirect Overhead (Storage)	26	15	3
Total Costs	1, 197	984	911
Unit Costs (125 Units)	9. 57	7. 87	7. 29

EXHIBIT A-6

later adjust for other direct and indirect expenses not relevant for make or buy analysis, but which might be applicable for costs in the budget and income statement.

Follow-Up Procedures

Management must consistently take steps to examine and check these costs to determine that they are truly accurate descriptions of the firm's past experience. Also, determination should be made that these costs are consistent with what the firm can expect to experience in the future. After the analysis has been provided to management, the decision makers must be sure that these costs are accurate descriptions of what expenses can be anticipated. The manager should get reasonable assurance from all affected responsible personnel that the data used in the analysis will be consistent with anticipated actual costs. After the decision is made, it is important in actual operations that all department personnel responsible for costing and implementing the make or buy decision adhere to the projected costs, and to the factors upon which the budget was based. This can be accomplished by periodic checks, and reporting of operational results as compared with the original make or buy analysis data. A step often overlooked by a firm involved in the make or buy decision is that of assuming that the firm's costs are in line with those of competitors. A manufacturer would require this information to alert him as to whether there are areas in his operations which are detracting from his ability to price his product competitively. Methods of obtaining information of this type are: (a) through trade associations, (b) governmental bulletins, and (c) data from suppliers.

REACHING THE DECISION TO "MAKE" THE PRODUCT

Management's Basic Objectives

A basic concept of the make or buy decision is that since it is a management decision vitally affecting many company areas, it therefore should not be contrary to management's broad ultimate goals for the business. If a primary objective of a company was specified, it would probably be the net operating profit on sales. However, this profit rate can only be properly measured in relation to the return on investment. Thus, a furniture company having an investment capital of $ 100,000.00, doing $ 1,000,000.00 in sales, and making a net profit of 10% or $ 100,000.00 (before taxes), would have a very good composite return on investment of 100%.

$$\text{Return on Investment} = \frac{\text{Profit}}{\text{Sales}} \times \frac{\text{Sales}}{\text{Capital}}$$

$$\text{Return on Investment} = \frac{\$100,000}{\$1,000,000} \times \frac{\$1,000,000}{\$100,000}$$

$$\text{Return on Investment} = 100\%$$

In many cases, a return on investment of 50% may be satisfactory, which would be the result in the example presented if the capital were $ 200,000.

$$\text{Return on Investment} = \frac{\$100,000}{\$1,000,000} \times \frac{\$1,000,000}{\$200,000}$$

$$= 50\%$$

As indicated by this ratio, sound make or buy decisions should attempt, by key strategic moves, to improve the integral elements making up this vital formula, which will contribute to the ultimate maximum return on investment. To put it another way, each operational process cycle involved in the make or buy decision represents a partial segment of the firm's overall business activity.

Steps to Accomplish Objectives

With reference to the make decision, industrial management will strive to improve the firm's profit investment performance standards. This is achieved through one or several of the following methods, as taken from a production investment tally sheet of an auto parts manufacturer involved in a recent management survey:

(1) Reduction of Unit Costs:
    a. Direct labor
        1. Improved efficiency
        2. Repeat orders
        3. Larger volume
        4. Better automation techniques
        5. Minimum idle time
        6. Lower salary rates
    b. Material
        1. Scrap recoupment
        2. Reduced spoilage
        3. Lower prices
        4. Cheaper freight rates
        5. Improved handling
        6. Better quality
    c. Overhead
        1. Reduced supervisory help
        2. Improved inspection
        3. Automated office procedures
(2) Minimizing Investment Requirements:
    a. Lower inventories carried
        1. Raw materials
        2. Work in process
        3. Finished goods
        4. Parts standardization
        5. Shipments expedited
    b. Extended credit terms
    c. Lower investments in facilities
        1. Less factory space
        2. Standard equipment used
        3. Improved maintenance
(3) Accelerated Turnover:
    a. Faster production cycle
    b. Better handling
    c. Efficient inspection
    d. Cash collections
(4) Increased Profits:

    a. Higher profit margins
    b. Lower sales discounts
    c. Economical borrowings
    d. Sales of profitable items

These can be considered to be the vital objectives for achieving a higher composite investment return, through improved production planning and control. Such objectives would not in any way detract from the firm's basic policy commitments regarding the fulfillment of its making requirements for quality, price, and shipping schedules. A large manufacturer of surgical instruments lost one of its best customers by lowering expensive inspection standards and suffering a high ratio of rejects. Make or buy decisions, more specifically the make alternative, can and often does affect many of these indicated production objectives. Therefore, any production operation which will reduce unit costs must be weighed in relation to its corresponding effect on investment cost. Conversely, investment costs should be appraised with regard to their potential effect on reducing unit costs. The proper application of this composite return on investment formula will serve to direct balanced attention to the pertinent factors involved in the make or buy decision.

TYPES OF COST DATA REQUIRED

There are three basic strategies, all of which are employed on a composite basis in order to resolve the make decision. The initial (Stage 1) analysis involves the proper classification of costs, whether variable or fixed. Fixed costs are further broken down into two categories, cash and non-cash. Cash costs consist of such expenses as electricity, taxes or machine rentals, whereas a non-cash type of expense would be depreciation. In many cases, a cash cost may be substantially reduced if the product is purchased from a supplier, such as water used in metal plating by a machine parts manufacturer. Sometimes, even non-cash costs are decreased where a facility is bought, thus reducing fixed depreciation charges. Essentially, for the make decision, a detailed screening of all fully variable and partially variable costs is made for all integral elements of the product, including (a) material, (b) labor, (c) tooling, (d) factory overhead, and (e) other overhead and intangible costs. Fixed costs would not be included in this analysis.

COMPILATION OF COST DATA

The next step (Stage 2) involves the formal presentation of the detailed elements of cost, both for make or for buy. This would serve the purpose of forecasting and measuring unit costs on an accurate basis in order to arrive at the most advantageous economic alternative. Exhibit A–7 illustrates the technical mechanics utilized in presenting details in the "Analysis of Unit Manufacturing Costs" for PQR Electronic Company, here treated as a typical case history. In this situation, the company was considering the alternatives of either making or buying 10,000 voltage meters. Either possibility was reasonably practical for the company, since no other major influencing factor was evident. Final determination was to be based on the composite comparison of profitability as measured by unit costs combined with return on investment for each of the alternatives. There was sufficient general purpose tooling available, except

PQR COMPANY (AN ELECTRONICS MANUFACTURER)

ANALYSIS OF UNIT MANUFACTURING COSTS

FOR MAKE OR BUY DECISION

10,000 VOLTAGE METERS

	Unit Costs to "Make"	Unit Costs to "Buy"
a.) Material:		
Material or Purchase	$ 1.60	$18.50
Freight	.15	.75
Waste and Scrap	.10	1.50
Total Material	1.85	20.75
b.) Direct Labor:		
Standard - Production	3.75	—
- Setup	.50	—
Rework, etc.	1.25	—
Total Direct Labor	5.50	—
c.) Material Overhead:		
Fixed	.10	—
Variable	.15	.75
Total Material Overhead	.25	.75
d.) Factory Overhead:		
Fixed	2.40	—
Variable	7.50	—
Total Factory Overhead	9.90	—
e.) Tooling Costs:		
Fixed	2.50	—
Variable	1.20	—
Total Tooling Costs	3.70	—
f.) General Overhead:		
Fixed	1.50	—
Variable	2.50	.50
Total General Overhead	4.00	.50
Total Unit Costs	$25.20	$22.00
Total Variable Costs	18.70	22.00
Total Fixed Costs	6.50	—

EXHIBIT A-7

for a special tool, and current manpower was readily available to accommodate the production schedule over a twelve-month period. Following is a breakdown for the various cost elements, indicating the specific departments responsible for rendering necessary data and the source of procuring essential cost details:

COST ELEMENTS	DEPARTMENTS SPECIFICALLY RESPONSIBLE
a. Material. Sources: vendors' quotations of prices, specifications of usage and waste.	Purchasing, Production, Engineering
b. Direct Labor. Sources: projected salary rates, similar operating standards, time for production, set-up, and rework.	Engineering, Production
c. Material Overhead. Sources: historical financial statements, projected budgets.	Accounting, Production
d. Factory Overhead. Sources: historical financial statements, projected budgets.	Accounting, Production
e. Tooling Costs. Sources: engineering specifications, vendors' quotations.	Purchasing, Engineering
f. General Overhead. Sources: historical financial statements, projected budgets.	Accounting

In addition to the specific departments responsible for rendering necessary cost data, other departments which were consulted for general pertinent details were: (a) marketing on sales forecasting, (b) cost accounting in cost computing, and (c) accounting for financing. The final decision was resolved by the plant vice president. As presented in the "Analysis of Unit Manufacturing Costs," the total for all cost elements indicated $25.20 to make as compared with a total of $22.00 to buy. However, unit fixed costs totaling $6.50 would be incurred under either alternative of making or buying the product. It was therefore proper procedure to consider only the net viable unit costs of $18.70 as the true comparable unit cost applicable to measuring the alternatives in this case. Therefore, the true basis for comparison of unit costs indicated that the most advantageous unit cost involved in this decision was the cost of $18.70 to make. After resolving that the preferred alternative based on comparable unit costs was to make the product, it was then necessary to measure the extent of the advantages for either alternatives by the final strategy of the "Analysis of Average Investment."

MEASURING INVESTMENT COSTS

Management was then required to develop the final strategy (Stage 3) of measuring the investment required for each alternative of making or buying the product. This technique was also compiled on a projected estimated basis for the purpose of computing the investment (on an average basis) necessary to process the order under either alternative of making or buying the product. Since management did not have unlimited funds and facilities, its objective was to invest its funds and facilities in the most profitable projects. It should be pointed out that the investment in working capital (besides facilities), in this case, was a major investment cost in the decision to make the product.

Exhibit A–8 (Analysis of Average Investment)

This exhibit lists the various elements which represent the breakdown of the factors comprising investment requirements in this case. Also presented is the listing of each individual method used as a basis for projecting the estimated amount of investment required for each element involved. Analysis indicated that $31,870 was the average total investment to make, as compared with the equivalent average total investment to buy of $17,650. To arrive at a true net investment on a comparable working basis, a further adjustment was applied for the deduction of items representing funds' sources. Such items in operations reduce the gross amounts entailed as investment for effectively completing the processing of the order. These funds' sources deducted from gross investment required were $2,510 for make and $13,380 to buy. Final investment costs on a net basis were $29,360 to make and $4,270 to buy. Although the results of utilizing this method of measuring net investment costs appeared to indicate that the alternative to buy the product was preferred, this by itself was not conclusive. In order to determine overall comprehensive and composite advantages conclusively, as indicated by each method, further analyses of findings were made to make sure that no adverse factors were involved.

REACHING THE DECISION TO "MAKE"

An examination of the findings was in order, as indicated by Exhibit A–7 in conjunction with Exhibit A–8, in order to confirm the apparent course of action to make the product. The net return (based on variable costs) of $93,000 represented a net return percentage of 33.2% based on $280,000 sales. Since the net return (on a cash basis of variable costs) was $33,000 in excess of net return for buy, and also greater than the 10% average net profit of the company, the cost basis to make justified the decision. In relationship to the return on investment presented in Exhibit A–8, the net return of $93,000 amounted to a net return of investment of $29,360, or 3.5% for make. The conclusive appraisal is to consider that the net investment required to manufacture the product amounted to $25,090 more than if it were bought. Since the net return amounted to $33,000 more by making the product, there was a 132% return attributed to the investment. Also, the firm had sufficient working capital to invest in the manufacturing operation. Although there would have been approximately a 1,400% net return on investment by buying the product, the preferred economic decision was to choose the make alternative, which resulted in a net profit of $33,000 as well as a satisfactory return on investment. There were no average influencing factors in this case which were sufficiently severe to negate the sound economic decision to proceed in making the product. It was the long range policy of the company to expand rather than contract in its manufacturing operations, since facilities and personnel were adequate for desired performance. Similarly, conditions outside the company, such as competition in the industry or relationships

PQR COMPANY (AN ELECTRONICS MANUFACTURER)

ANALYSIS OF AVERAGE INVESTMENT

FOR MAKE OR BUY DECISION

	Amounts For "Make"	Amounts For "Buy"	Projection Basis
Total Volume:			
Number of Units	10,000	10,000	Based on Marketing Forecast
Material Costs	$ 18,500	$207,500	
Investment Required:			
Cash	650	3,400	15% of Accounts Receivable and Accrued Items
Inventories			
Receiving	660	6,920	Average 12 days required
In Process	16,830	–	Average 60 days value
Finished Goods	6,230	7,330	12 Days Total Costs
Tooling Costs	7,500	–	Average Cost
Total Required	31,870	17,650	
Less: Funds Sources			
Accounts Payable	1,030	12,970	20 Days Terms for Material
Accrued Payroll	1,480	410	Labor Costs, 1 week
Total Sources	2,510	13,380	
Net Investment (On Average Basis)	29,360	4,270	

EXHIBIT A-8

VITAL STATISTICS		
Total Costs (Variable)	187,000	220,000
Total Selling Price @ $28.	280,000	280,000
NET RETURN (On Sales)	93,000	60,000

PQR COMPANY (AN ELECTRONICS MANUFACTURER)
SUMMARY COMPARISON OF ALTERNATIVES FOR MAKE OR BUY

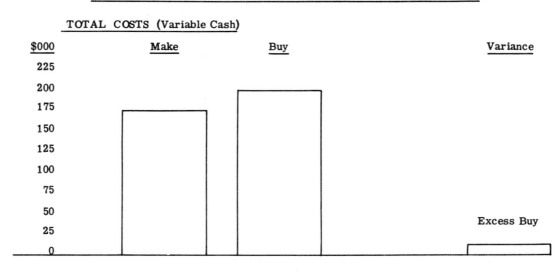

TOTAL COSTS (Variable Cash)

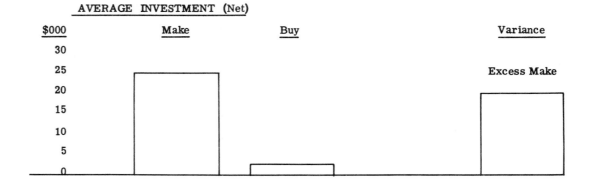

AVERAGE INVESTMENT (Net)

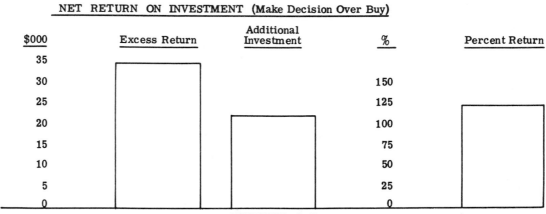

NET RETURN ON INVESTMENT (Make Decision Over Buy)

EXHIBIT A-9

with suppliers, were not serious enough to cause the company to refrain from proceeding with the make process. If the costs for specialized facilities to make the product had been considerably higher for this situation, then the ultimate decision might have been in favor of buying. A graphic presentation which summarized the final results of the detailed analysis made is presented in Exhibit A–9. Key data includes the following comparative basis results involved in the appraisal:

- a. Total Costs (variable cash)
  1. Costs to make
  2. Costs to buy
  3. Excess of costs to buy
- b. Average Investment
  1. Investment to make
  2. Investment to buy
  3. Excess investment to make
- c. Net Return on Investment (make decision over buy)
  1. Excess profit to make
  2. Additional investment required
  3. Per cent return for additional investment

During the course of operations, periodic checks were made of the costs and investment elements applicable to the alternative chosen, in this case making the product. Although there were slight variations between actual results at various stages and early projections, the final results were very close to the original estimates.

Appendix B: Warehousing

EXHIBIT B-1   Pallet Terminology and Sizes[1]

1. SCOPE AND PURPOSE

1.1 This standard defines the terminology associated with the use of pallets in the unit load method of handling of materials and products.

1.2 This standard provides a series of pallet sizes; nine rectangular and three square.

1.3 The principal objective was the attainment of the minimum number of pallet sizes for non-captive use which would provide the maximum of efficiency, economy, interchangeability and flexibility when used in United States rail, truck, maritime and air transport services. The selection was made with the objective of providing sizes suitable for a wide range of captive use.

1.4 The standard applies to pallets, irrespective of the material, (wood, metal, paper, etc.) used in their construction. References to wood or wood pallets in this standard shall include plywood and plywood pallets, respectively.

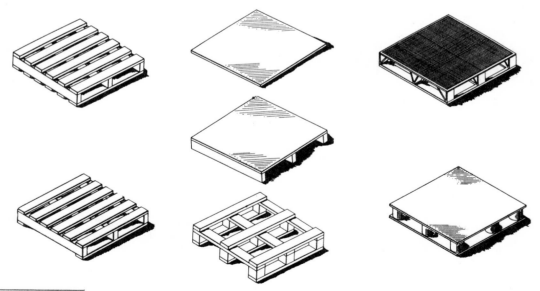

[1] Extracted from American National Standard Pallet Terminology and Sizes (ANSI MH.1-1965), with the permission of the publisher, The American Society of Mechanical Engineers, 345 E. 47th St., New York, N.Y. 10017.

2. DEFINITIONS

For the purpose of this American Standard, the following definitions shall apply:

2.1 Pallet

A horizontal platform device used as a base for assembling, storing and handling materials in a unit load. (Ref: ISO/TC 51-(1) "Load board" and (2) "Pallet").

2.1.1 *Captive Pallet*. A pallet whose use cycle remains within a single enterprise (private, corporate, or government).

2.1.2 *Non-Captive Pallet*. A pallet whose use cycle extends through one or more enterprises (private, corporate or government) and may include a common carrier service.

2.2 Pallet Measurement

2.2.1 *Pallet Size*. The dimensional length and width of the top surface of a pallet.

2.2.2 *Pallet Length*. The dimension which shall be stated first in designating a pallet size (See Note A and 2.4)

NOTE A: Based on common practice among pallet manufacturers and users, the first dimension is defined as follows. This convention is permitted in this standard.

2.2.2.1 ONE-WAY, TWO-WAY AND PARTIAL FOUR-WAY PALLETS By the length of the stringer.

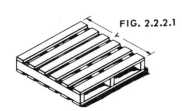

FIG. 2.2.2.1

2.2.2.2 FOUR-WAY AND ALL-WAY PALLETS HAVING TWO OR MORE DECKBOARDS AT THE TOP DECK By the length of the stringer board.

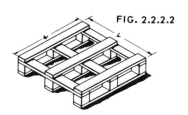

FIG. 2.2.2.2

2.2.2.3 FOUR-WAY AND ALL-WAY PALLETS HAVING A SINGLE DECKBOARD AND A SOLID FACE, TOP DECK

2.2.2.3.1 SINGLE FACED OR DOUBLE FACED REVERSIBLE } Longest dimension of the pallet.

FIG. 2.2.2.3.1

2.2.2.3.2 DOUBLE FACED NON-REVERSIBLE } Over-all pallet dimension perpendicular to the length of the bottom deckboards.

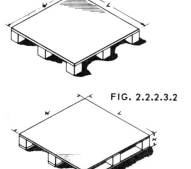

FIG. 2.2.2.3.2

2.2.3 *Pallet Width*. The horizontal dimension at right angles to the pallet length.

2.2.4 *Pallet Height*. The total vertical dimension between the outer surfaces of the top and bottom bearing surfaces.

2-3 Pallet Classifications

2.3.1 *Expendable Pallet.* A pallet, inexpensive and of simple configuration, intended to be discarded after one cycle of use.

2.3.2 *Reusable Pallet.* A pallet intended for multiple cycles of use.

2.4 Pallet Designs

2.4.1 *One-Way Pallet.* A pallet whose configuration permits retrieving or discharging from only one horizontal direction.

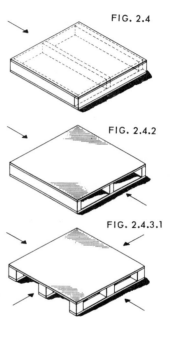

FIG. 2.4

2.4.2 *Two-Way Pallet.* A pallet whose configuration permits retrieving or discharging from opposite directions along the same horizontal axis. (Ref: ISO/TC 51 (3) ''Two-way pallet'').

FIG. 2.4.2

2.4.3 *Four-Way Pallet.* A pallet whose configuration permits retrieving or discharging from adjacent right angle directions in the same horizontal plane (Ref: ISO/TC 51 (4) ''Four way pallet'').

FIG. 2.4.3.1

2.4.3.1 Full four-way pallets (usually having block deck spacers) permit four-way entry by forks of fork lift trucks or by load wheeled forks of pallet trucks.

2.4.3.2 Partial four-way pallets (usually having notches or holes in stringers) permit four-way entry by fork lift trucks and restrict the load wheeled forks of pallet trucks to two-way entry.

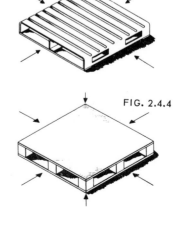

FIG. 2.4.3.2

2.4.4 *All-Way Pallet.* A pallet whose configuration permits retrieving or discharging from any direction in the same horizontal plane.

FIG. 2.4.4

2.5 Pallet Styles

2.5.1 *Single Faced Pallet.* A pallet having only one deck surface; usually a top deck. (Ref: ISO/TC 51 (5) "Single decked flat pallet").

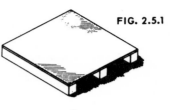

FIG. 2.5.1

2.5.2 *Double Faced Pallet.* A pallet having a top and bottom deck surface. (Ref: ISO/TC 51 (6) "Double decked flat pallet").

FIG. 2.5.2

2.5.2.1 *Reversible Pallet.* A pallet having identical top and bottom deck surfaces. (Ref: ISO/TC 51 (7) "Reversible pallet").

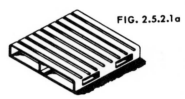

FIG. 2.5.2.1a

2.5.2.2 *Non-Reversible Pallet.* A pallet having different top and bottom deck surfaces. (See Fig. 2.2.2.3.1)

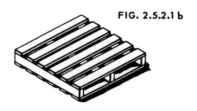

FIG. 2.5.2.1 b

2.6 Pallet Constructions

2.6.1 *Flush Pallet.* A pallet whose deck surfaces do not protrude beyond the deck spacers.

FIG. 2.6.1

2.6.2 *Wing Pallet.* A pallet whose deck surfaces protrude beyond the outer edges of the deck spacers. (Ref: ISO/TC 51 (8) "Wing pallet" and (19) "Wings").

FIG. 2.6.2.1

2.6.2.1 *Double Wing Pallet.* A pallet whose top and bottom decks protrude beyond the outer edges of the deck spacers.

2.6.2.2 *Single Wing Pallet.* A pallet whose top deck only protrudes beyond the outer edges of the deck spacers; the bottom deck is flush.

FIG. 2.6.2.2.

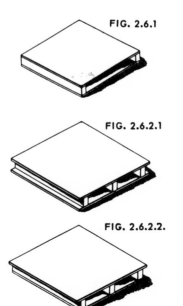

2.7 Pallet Purpose

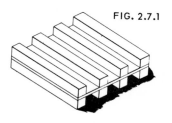

FIG. 2.7.1

2.7.1 *"Take-It-Or-Leave-It" Pallet.* A pallet whose configuration permits two choices of handling:
  a. Retrieving the pallet and unit load together, or
  b. Retrieving the unit load and leaving the pallet.

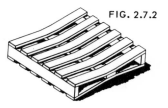

FIG. 2.7.2

2.7.2 *Special Purpose Pallet.* A pallet whose deck surfaces are planned for more expeditious handling and protection of a specific product.

FIG. 2.7.3

2.7.3 *Stacking Pallet.* A pallet having a superstructure of vertical members (fixed, removable or collapsible), to support a superimposed load. (Ref: ISO/TC 51 (10) "Post pallet").

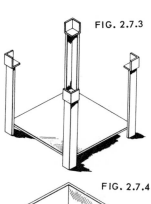

FIG. 2.7.4

2.7.4 *Container Pallet.* A pallet having a superstructure of at least three sides (fixed, removable or collapsible), with or without a lid, for use as a box. (Ref: ISO/TC 51 (9) "Box pallet").

2.8 Pallet Components

2.8.1 *Pallet Deck.* The horizontal load-carrying or load bearing surface of a pallet (Also referred to as "pallet face"). (Ref: ISO/TC 51 (13) "Deck").

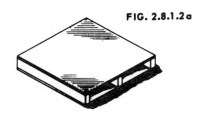

FIG. 2.8.1.2a

    2.8.1.1 TOP DECK
        Load carrying surface

    2.8.1.2 BOTTOM DECK
        Load bearing surface

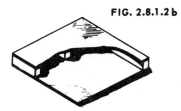

FIG. 2.8.1.2b

2.8.2 *Pallet Deck Opening.* Any void in deck faces caused by spacing of surface elements.

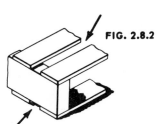

FIG. 2.8.2

2.8.3 *Deckboard.* A surface element used in the construction of a pallet deck.

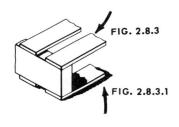

FIG. 2.8.3

FIG. 2.8.3.1

    2.8.3.1 *Edgeboard.* A deckboard assembled at right angles to and at the extreme ends of stringers or stringerboards. (Ref: ISO/TC 51 (14) "Entry members").

    2.8.3.2 *Deckboard Chamfer.* A beveled edge on the top surface of the bottom deckboards for the purpose of easing the entry or exit of pallet truck load wheels. (Ref: ISO/TC 51 (20) "Chamfer").

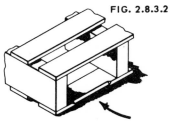

FIG. 2.8.3.2

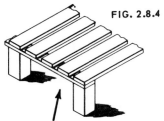

FIG. 2.8.4

    2.8.4 *Stringerboard.* A subsurface element on full four-way entry wood pallets beneath and at right angles to the deckboards but above the block, leg or post spacers. (Ref: ISO/TC 51 (18) "Stringer").

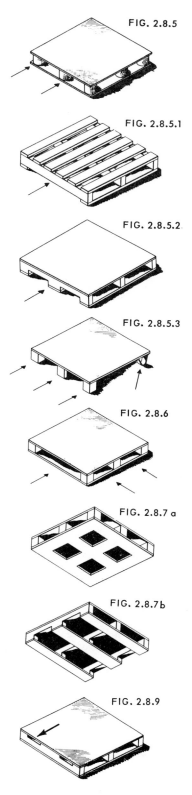

2.8.5 *Pallet Deck Spacer.* A fabricated member which supports the top deck or separates the top and bottom decks. They may be one of several types. (Ref: ISO/TC 51 (11) ''Bearers'').

FIG. 2.8.5

2.8.5.1 *Stringer.* A continuous longitudinal spacer whose length equals the pallet length in two-way and partial four-way entry pallets.

FIG. 2.8.5.1

2.8.5.2 *Notched Stringer.* A stringer which has openings cut out for insertion of pallet lifting devices.

FIG. 2.8.5.2

2.8.5.3 *Block, Leg or Column.* A spacer member used in multiples to make full four-way or all-way entry pallets. (Ref: ISO/TC 51 (15) ''Feet'').

FIG. 2.8.5.3

2.8.6 *Pallet Fork Entry.* A void in a fabricated pallet for the purpose of inserting a lifting device. (Ref: ISO/TC 51 (21) "Entry" and (22) "Free entry").

FIG. 2.8.6

2.8.7 *Pallet Truck Openings.* Voids in the bottom deck of a non-reversible double faced pallet to allow the load wheels of a pallet truck to bear on the floor. (Ref: ISO/TC 51 (23) "Openings).

FIG. 2.8.7 a

2.8.8 *Pallet Fastener.* Any device or medium used to hold together the components of an assembled pallet.

FIG. 2.8.7 b

2.8.9 *Pallet Strap Slot.* A small opening on the underside of a pallet top deck to facilitate use and prevent movement of unit load fastening material.

FIG. 2.8.9

3. PALLET SIZES

The standard sizes are given in the following Table 1.

Table 1

Rectangular

24	x	32	or	32	x	24
32	x	40	or	40	x	32
36	x	42	or	42	x	36
32	x	48	or	48	x	32
36	x	48	or	48	x	36
40	x	48	or	48	x	40
48	x	60	or	60	x	48
48	x	72	or	72	x	48
88	x	108	or	108	x	88

Square

36	x	36
42	x	42
48	x	48

(a) All dimensions are given in inches.

(b) Tolerances of plus or minus ¼ inch are permissible on all the over-all dimensions.

(c) Definition of pallet sizes shall be in accordance with Section 2.2. and Section 4.

(d) It is recommended that when pallets are to be used in International Trade, consideration should be given to specific sizes recommended by the International Organization for Standardization (ISO/R 198-1961):

ISO—International Organization for Standardization Sizes

Inches							Millimeters						
32	x	40	or	40	x	32	800	x	1000	or	1000	x	800
32	x	48	or	48	x	32	800	x	1200	or	1200	x	800
40	x	48	or	48	x	40	1000	x	1200	or	1200	x	1000

EXHIBIT B-2 Procedure for Comparative Rating for Fork Lift Trucks[2]

Some manufacturers specify a number of pounds capacity with a particular length of load while others specify a number of pounds capacity at a given number of inches from the heel of the fork. Some give an inch-pound rating based on the distance of the load center from the heel of the fork, while others base their inch-pound rating on the distance from the center of the load to the center of the front axle.

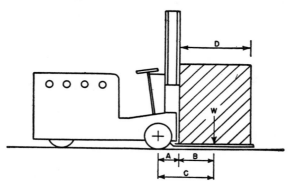

Here is one method of comparative rating:

With reference to the accompanying sketch, the symbols are interpreted as follows:

A = Distance from center of front axle to heel of fork measured in inches.

$B = \dfrac{D}{2}$ = Distance from heel of fork to center of load measured in inches.

$C = A + B$ = Distance from center of front axle to center of load measured in inches.

$D = 2 \times B$ = Length of Load.

W = Weight of load measured in pounds.

In order to calculate a load with a length other than that specified by the manufacturer, or to compare one truck with another of a different rating, it is necessary to obtain the "Inch-Pound Rating." The Inch-Pound Rating is W, the rated load; multiplied by C, the distance from the center of the front axle to the center of the load, i.e.

$$\text{Inch-Pound Rating} = W \times C$$

The inch-pound rating becomes a constant for that particular truck. Then, in order to figure (1) the maximum load length for any given load;

---

[2] Source: Material Handling Engineering, Directory & Handbook, copyright © 1960–61, Cleveland, Ohio.

or, (2) the maximum load for any given load length, the formula can be reversed to give this information, i.e.

(1) $C = \dfrac{\text{Inch-Pound Rating}}{W}$ (2) $W = \dfrac{\text{Inch-Pound Rating}}{C}$

Example: A truck has a rating of 4,000# @ 30″—which means a 4,000# load which has its center 30″ from the heel of the fork. The specifications show the distance from the center of the axle to the heel of the fork to be 15″. By applying the formulas, the inch-pound rating may be arrived at:

$C = A + B = 15 + 30 = 45″$
Inch-Pound Rating $= W \times C = 4,000 \times 45 = 180,000$ inch-pounds

The rating of 180,000 inch-pounds then becomes a constant for the truck in question. Then, to determine how long a pallet or skid which will have a gross weight of 2,500# can be made, apply the formulas:

$$C = \frac{\text{Inch-Pound Rating}}{W} = \frac{180,000}{2,500} = 72″$$

$B = C - A = 72″ - 15″ = 57″$

$D = 2 \times B = 2 \times 57″$

$= 114″$ allowable load length

Or, as another example, it is desired to know the maximum safe load for a standard 84″ rack, by applying the formulas:

$B = \dfrac{D}{2} = \dfrac{84}{2} = 42″$ $C = A + B = 15″ + 42″ = 57″$

$W = \dfrac{\text{Inch-Pound Rating}}{C} = \dfrac{180,000}{57″} = 3158\#$ gross weight allowed

EXHIBIT B-3   Typical Pallet Patterns[3]

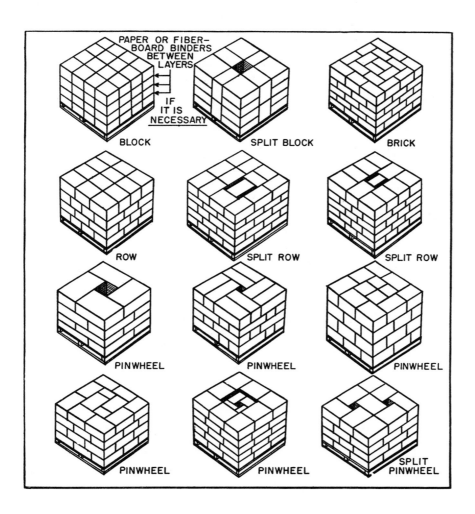

[3] Source: Material Handling Engineering, Directory & Handbook, copyright © 1960–61, Cleveland, Ohio.

Appendix C: Charts

Section 1: Planning and Control

The manager who fails to plan his work is recklessly gambling with his company's resources of manpower, equipment, facilities and money.

THE PLANNING FUNCTION

Louis A. Allen defines the planning function as "the work a manager performs to predetermine a course of action." He also defines (a) the activities of planning, (b) the characteristics of planning, and (c) the advantages of planning, as follows:

(a) The Activities of Planning[1]

[These] are forecasting, establishing objectives, programming, scheduling, budgeting, establishing procedures, developing policies.

Forecasting: the work a manager performs to estimate the future.

Establishing objectives: the work a manager performs to determine the end results to be accomplished.

Programming: the work a manager performs to establish the sequence and priority of action steps to be followed in reaching objectives.

Scheduling: the work a manager performs to establish a time sequence for program steps.

Budgeting: the work a manager performs to allocate resources necessary to accomplish objectives.

Establishing procedures: the work a manager performs to develop and apply standardized methods of performing specified work.

Developing policies: the work a manager performs to develop and interpret standing decisions that apply to repetitive questions and problems of significance to the enterprise as a whole.

---

[1] Louis A. Allen, *The Management Profession* (New York: McGraw-Hill Book Company, Copyright © 1964) by Louis A. Allen, pp. 97, 99, 100.

(b) The Characteristics of Planning

Planning is the basis for successful management action. A plan is a trap laid to capture the future. It is the process a manager follows in thinking through beforehand what he wants to accomplish and how he will do it. Neither the individual nor the enterprise need be at the beck and call of chance. We are beginning to discover that, to a surprising degree, we can master the future. We can *make* happen what we *want* to happen.

When we plan, we take time to reflect and analyze, to consider alternatives, to make sound, considered decisions about the future. We decide in advance what we are going to do, how we will do it, under what conditions we will carry it out, how we will accomplish it, and what we will require to get the results we want. Because we think ahead, we avoid the tendency to make hasty judgments and to take haphazard action.

(c) The Advantages of Planning

Proper planning greatly simplifies the task of a manager. It makes integrated and coordinated effort possible. If we know where we are going we are much more likely to get there. Once we have the whole picture in mind, it is easier to make the parts fit together and to ensure that the action that follows is timed so it will take place in the proper sequence.

Planning enables us to make most effective and economical use of manpower, equipment, facilities, and money. If we identify in advance our needs for people, we can develop individuals inside the organization so that they will be ready when the opportunity for promotion appears. Planning makes it possible to let subordinates know what is required of them and to give them an opportunity to participate in the decisions that are made. This keeps interest and enthusiasm at a high level and enables the manager to incorporate into his plans the best thinking of those closest to the point where the action will be carried out.

The importance of planning is highlighted by the emphasis well-managed companies put upon it. Industry leaders, convinced that their future success depends as much upon the thought their managers give to planning as to process and technology, expect managers at all levels, from first-line supervisors to top executives, to devote a substantial portion of their time to planning before taking action.

(d) Management's Responsibilities in Planning

Not only must every manager plan, but it is a primary responsibility of management at all levels in an organization. As the hierarchical ladder is ascended, however, the nature and scope of planning responsibilities will vary. At the highest corporate levels rests the duty of management to establish the fundamental direction of the enterprise and lay-out strategies to achieve objectives. At the lowest levels of management practically all planning efforts are very short-range and deal with matters solely under the jurisdiction of the manager. In between these extremes the nature of planning duties changes with level, scope of managerial authority, and company organization.

Exhibit C–1 shows suggested "typical" and "ideal" time allocations for executives at various levels. These are, of course, arbitrary and may not fit requirements of different companies. For the average medium-sized company they may be reasonable. For a very large company, however, it is questionable whether the emphasis is great enough for the period beyond five years. This is certainly true for companies involved in commitments beyond five years, such as aerospace companies, airlines, public utilities, and companies heavily involved in advanced technology. On the other hand, a company making women's dresses will find inappropriate the emphasis on the long range plan.

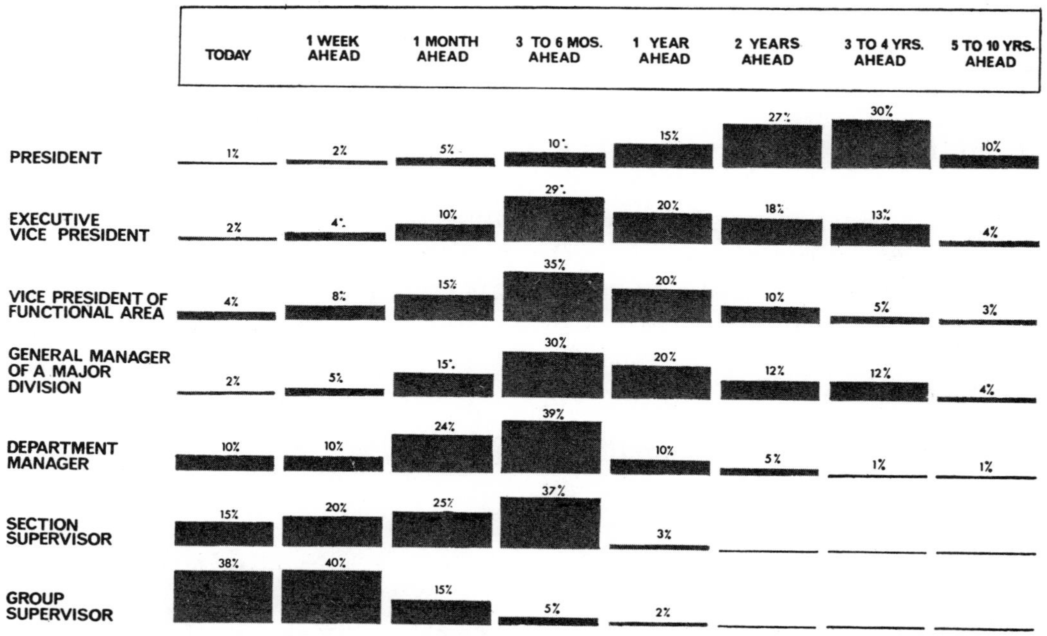

	TODAY	1 WEEK AHEAD	1 MONTH AHEAD	3 TO 6 MOS. AHEAD	1 YEAR AHEAD	2 YEARS AHEAD	3 TO 4 YRS. AHEAD	5 TO 10 YRS. AHEAD
PRESIDENT	1%	2%	5%	10%	15%	27%	30%	10%
EXECUTIVE VICE PRESIDENT	2%	4%	10%	29%	20%	18%	13%	4%
VICE PRESIDENT OF FUNCTIONAL AREA	4%	8%	15%	35%	20%	10%	5%	3%
GENERAL MANAGER OF A MAJOR DIVISION	2%	5%	15%	30%	20%	12%	12%	4%
DEPARTMENT MANAGER	10%	10%	24%	39%	10%	5%	1%	1%
SECTION SUPERVISOR	15%	20%	25%	37%	3%			
GROUP SUPERVISOR	38%	40%	15%	5%	2%			

EXHIBIT C-1 "Ideal" Allocations of Time for Planning in the "Average" Company[2]

(e) Charting for Planning and Control

In this section, only the graphical methods of planning and control will be discussed. Budgetary control and the financial approaches to managing a plant fall in the accounting orbit so will not be covered in depth here.

Production managers, whose daily operating problems consume a major share of their working day, have little time to allot to the review of long reports and involved financial statements. However, graphical reports, prepared by competent subordinates, do provide a maximum yield of intelligence in a minimum of scanning time.

---

[2] Reprinted with permission of the Macmillan Company from *Top Management Planning* by George A. Steiner. Copyright © 1969 by the Trustees of Columbia University in the City of New York, N. Y., pp. 26, 373.

Prominent among the charting systems is the Gantt chart. This technique is not new, in fact its use was initiated decades ago, but it still offers a simple, reliable and flexible method for charting operational results against established goals. Rather than devote time and space to a lengthy description of the construction of such charts, a number of practical illustrations of its utility follow. Included also are other charting methods. Most of these were used by the author both in his plant management experience and in consulting work.

PERT—PROGRAM EVALUATION AND REVIEW TECHNIQUE

PERT is basically a planning technique. It facilitates the coordination of different tasks of a project by graphic representation, with focus on satisfactory completion of a project at a predetermined date.

This provides management with another control device. By a brief review of the chart, progress can be determined, approaching bottlenecks can be avoided, and critical paths in operations, dictating the need for special attention, can be exposed.

How Does It Work?

(a) First list jobs in operational sequence. Then indicate the direction of the jobs by arrows, for example:

This represents four jobs of a project in proper operational sequence. 2 cannot start until 1 is finished, 3 can't start until 1 and 2 are finished, etc.

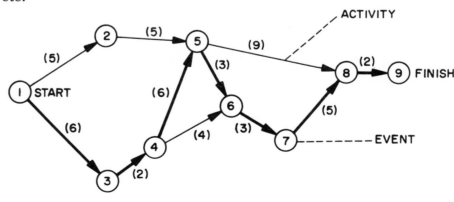

EXHIBIT C-2

(b) Exhibit C-2 illustrates a project consisting of nine jobs. The figures in parentheses represent the number of weeks needed to complete each job. Add up the estimated time values for all the jobs along each possible path. The longest one is the critical path and totals, in this instance, 27 weeks. (The heavy line on the chart indicates the longest path.) Delays in the completion of these jobs will delay completion of the project.

Determination of Critical Path:

Estimated Times			*Critical Path in Weeks*	
Activity	*Weeks*			
1–2	5		1–3 6	
1–3	6		3–4 2	
2–5	5		4–5 6	
3–4	2		5–6 3	
4–5	6		6–7 3	
4–6	4		7–8 5	
5–6	3		8–9 2	
6–7	3			
7–8	5		Total	27 weeks
5–8	9			
8–9	2			

(c) Efforts should be made to shorten the paths (time) of these jobs along the critical path. This, of course, will also result in lower costs. Close surveillance of these jobs should be exercised so that delays can be anticipated and remedial steps taken. Effective control will help to eliminate "crash" or "rush" actions which generally result in chaos, inefficiency, and added expense.

(d) In connection with the estimated time values, three estimates for each job, instead of one, can be made. This will provide more accuracy. Then apply the beta distribution principle and determine the probability that the project will be finished in 27 weeks or less. (See an explanation of this later in this section.)

PERT/Time[3]

In less certain cases it is recommended that three estimates rather than one be made for the time required to complete each activity. The three are: (a) the optimistic time, (b) the pessimistic time, and (m) the most likely time. There are several advantages to having these estimates. First, estimators may make more valid estimates if

---

[3] Reprinted with permission of the Macmillan Company from *Top Management Planning* by George A. Steiner. Copyright © 1969 by the Trustees of Columbia University in the City of New York, N. Y. pp. 464, 465.

they can express the extent of their uncertainty. Second, if a single estimate is made it is likely to be the mode, but in estimating activity time the mean is a more representative estimate than the mode.

In more complex programs the uncertainty usually involved makes one-time estimates suspect. PERT/Time solves the problem by assuming the probable duration of an activity is beta-distributed. The beta distribution has the desirable properties of being contained entirely inside a finite interval $(a-b)$ where a is the most optimistic time estimate and b is the most pessimistic. The distribution is symmetric or skewed depending on the location of the mode, m, the most likely time estimate, relative to a and b. Exhibit C–3 gives an example of one shape of the beta distribution.

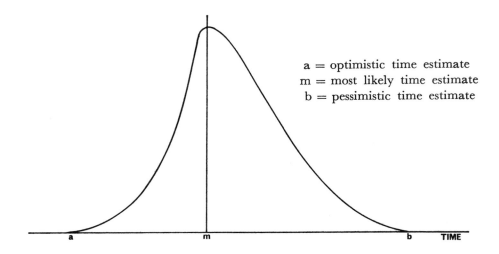

a = optimistic time estimate
m = most likely time estimate
b = pessimistic time estimate

EXHIBIT   C-3   An Example of the Beta Distribution

The use of the beta distribution serves three purposes. The first is that an "expected elapsed time" (t_e) for an activity can be determined from the three time estimates, a, b, and m. Second, probabilities of completing an activity may be computed from the three time estimates. Thus, in programs with high uncertainty, managers may be able to speak of meeting schedules in terms of probability statements. Third, the use of the beta distribution provides a statistical foundation for the PERT network. The beta distribution is, of course, dependent on subjective time estimates of people.

To make statistical inferences about the timing of future events it is necessary to typify the intervals between adjacent events in terms of their expected value and variances. The expected value is a statistical term that corresponds to the mean. Variance (the square of the standard deviation) relates to the uncertainty associated with the process. If the variance of an activity is large, there is great uncertainty connected with the activity, and vice versa. Two simple equations will produce the estimate of mean and variance for ranges of distribution encountered.

The expected elapsed time, or the mean of the distribution is given by:

$$t_e = \frac{a + 4m + b}{6}$$

where

$a = 5$ months

$m = 7$ months

$b = 15$ months

$t_e =$ elapsed time, or mean

$$t_e = \frac{5 + 4(7) + 15}{6} = 8 \text{ months}$$

The mode is, of course, 7 months, *b-m*, $15 - 7 = 8$ months. The midpoint of the range (or mid range) is $(15 + 5) \div 2 = 10$ months. Note that the mean lies one-third the distance from the mode to the midpoint of the range.

The standard deviation, which is the basis of probability statements, is given by:

$$\sigma = \frac{b - a}{6} = \frac{15 - 5}{6} = 1.67 \text{ months.}$$

In the above time estimates, the standard deviation would be 1.67 months, a fairly large deviation (relative to a mean of 8 months) indicating considerable uncertainty.

When expected elapsed times (t_e's) are calculated for each activity they are summed throughout all the network paths to determine the total expected elapsed time for every path of the network. All parallel paths are assumed to be traveled simultaneously but the completion date of the program PERT'ed is dependent on the path which takes the longest time. This path has the highest total elapsed time and is called the "critical path."

The total elapsed time of the critical path determines the expected date (T_E) for the completion of the program. Similarly, the expected date of arriving at any event in the network is calculated by summing the elapsed times for each path leading to that event and choosing the highest sum. This is because no event can occur until all activities leading into it have been completed.

PERT is an exceedingly adoptive and powerful managerial tool. When used in specific problems to which is suited, it can strengthen managerial planning and control decisions over a wide spectrum of problems concerned with scheduling activities to achieve a specified optimum of time, cost, or both.

PERT Network

Let us assume that the times given were the results of the following time estimates expressed in weeks:

Activity Code	Optimistic	Most Likely	Pessimistic	$t_e = \dfrac{a + 4m + b}{6}$
Task *a*	3	4	5	4
Task *b*	6	8	10	6
Task *c*	9	10	11	10
Task *d*	1/2	1/2	3 1/2	1

Expected delivery date of project $= 21$ weeks

Standard deviation and variances of each activity, and the project as a whole, are as follows:

Activity	$\dfrac{b-a}{6} = \sigma$	σ^2
a	1/3	1/9
b	2/3	4/9
c	1/3	1/9
d	1/2	1/4
Total		33/36

$$\sigma^2 = 33/36 = 0.917; \text{ use } 0.92, \ \sigma = \sqrt{0.92}$$
$$\sigma = 0.96 \text{ weeks}$$

What is the probability that the job will be completed in 23 weeks?

$$\frac{23 - 21}{0.96} = \frac{2}{0.96} = 2.08 = z$$

z	.00	.01	.02	.03	.04	.05	.06	.07	.08	.09
0.0	.0000	.0040	.0080	.0120	.0160	.0199	.0239	.0279	.0319	.0359
0.1	.0398	.0438	.0478	.0517	.0557	.0596	.0636	.0675	.0714	.0753
0.2	.0793	.0832	.0871	.0910	.0948	.0987	.1026	.1064	.1103	.1141
0.3	.1179	.1217	.1255	.1293	.1331	.1368	.1406	.1443	.1480	.1517
0.4	.1554	.1591	.1628	.1664	.1700	.1736	.1772	.1808	.1844	.1879
0.5	.1915	.1950	.1985	.2019	.2054	.2088	.2123	.2157	.2190	.2224
0.6	.2257	.2291	.2324	.2357	.2389	.2422	.2454	.2486	.2517	.2549
0.7	.2580	.2611	.2642	.2673	.2704	.2734	.2764	.2794	.2823	.2852
0.8	.2881	.2910	.2939	.2967	.2995	.3023	.3051	.3078	.3106	.3133
0.9	.3159	.3186	.3212	.3238	.3264	.3289	.3315	.3340	.3365	.3389
1.0	.3413	.3438	.3461	.3485	.3508	.3531	.3554	.3577	.3599	.3621
1.1	.3643	.3665	.3686	.3708	.3729	.3749	.3770	.3790	.3810	.3830
1.2	.3849	.3869	.3888	.3907	.3925	.3944	.3962	.3980	.3997	.4015
1.3	.4032	.4049	.4066	.4082	.4099	.4115	.4131	.4147	.4162	.4177
1.4	.4192	.4207	.4222	.4236	.4251	.4265	.4279	.4292	.4306	.4319
1.5	.4332	.4345	.4357	.4370	.4382	.4394	.4406	.4418	.4429	.4441
1.6	.4452	.4463	.4474	.4484	.4495	.4505	.4515	.4525	.4535	.4545
1.7	.4554	.4564	.4573	.4582	.4591	.4599	.4608	.4616	.4625	.4633
1.8	.4641	.4649	.4656	.4664	.4671	.4678	.4686	.4693	.4699	.4706
1.9	.4713	.4719	.4726	.4732	.4738	.4744	.4750	.4756	.4761	.4767
2.0	.4772	.4778	.4783	.4788	.4793	.4798	.4803	.4808	.4812	.4817
2.1	.4821	.4826	.4830	.4834	.4838	.4842	.4846	.4850	.4854	.4857
2.2	.4861	.4864	.4868	.4871	.4875	.4878	.4881	.4884	.4887	.4890
2.3	.4893	.4896	.4898	.4901	.4904	.4906	.4909	.4911	.4913	.4916
2.4	.4918	.4920	.4922	.4925	.4927	.4929	.4931	.4932	.4934	.4936
2.5	.4938	.4940	.4941	.4943	.4945	.4946	.4948	.4949	.4951	.4952
2.6	.4953	.4955	.4956	.4957	.4959	.4960	.4961	.4962	.4963	.4964
2.7	.4965	.4966	.4967	.4968	.4969	.4970	.4971	.4972	.4973	.4974
2.8	.4974	.4975	.4976	.4977	.4977	.4978	.4979	.4979	.4980	.4981
2.9	.4981	.4982	.4982	.4983	.4984	.4984	.4985	.4985	.4986	.4986
3.0	.4987	.4987	.4987	.4988	.4988	.4989	.4989	.4989	.4990	.4990

EXHIBIT C-4   Normal Curve Areas[4]

[4] Adapted from "Direct Table of the Normal Integral" from *First Course in Probability and Statistics* by J. Neyman. Copyright 1950 by Holt, Rinehart & Winston, Inc. Adapted and reproduced by permission of Holt, Rinehart & Winston, Inc., p. 345.

From the normal distribution curves it will be determined that the probability of the job being completed in 23 weeks or less is 96.2%. See Exhibit C-4, where 2.08 is .4812 × 2 = 96.2%.

PLANT SAFETY PERFORMANCE

Exhibit C-5 depicts the lost time accident record for each department in the plant compared to the total hours worked by employees. The chart should be kept current each month on an accumulated basis. Calculations are best made and the results posted on the chart within a few days after the end of the

EXHIBIT C-5   Plant Safety Performance[5]

PLANT SAFETY PERFORMANCE			
LOST TIME ACCIDENTS THIS YEAR - (9 MONTHS)			
Department	Average Number Employees	Number Of Lost Time Accidents	Total Hours Worked	Number Accidents Per 10,000 Hrs. Worked
Maintenance	17	14	25,500	5.48
Shipping & Receiving	21	5	31,500	1.59
Machine Shop	125	25	187,500	1.33
Warehouses	10	2	15,000	1.33
Tool Room	45	8	67,500	1.19
Heat Treat	7	1	10,500	0.95
Assembly	100	12	150,000	0.80
Inspection	15	1	22,500	0.44
Metal Finishing	10	0	15,000	0.00
TOTALS	350	68	525,000	1.30

[5] Lewis R. Zeyher, *Cost Reduction in the Plant* (Englewood Cliffs, N. J.: Prentice-Hall, Inc. Copyright © 1965), p. 93.

month. Delays or procrastinations of any kind should be avoided. An example of the calculations follows:

Data Required—Maintenance Department

Month	Average Number of Employees	Accumulated Hours Worked	Accumulated Lost Time Accidents	Number of Accidents Per 10,000 *Hrs. Worked*
August-8 months	14	23,300	12	5.15
September-9 months	14	25,500	14	5.48

Calculations for September:

14 employees worked an average of 157 hours each.

14 × 157 . . . . .2198 hours

Total lost time accidents 2

23,300 hours (August) + 2198 hours (September) = 25,498 hours

Number of accidents per 10,000 hours worked: 14 ÷ 2.55 = 5.48

CONFERENCE TYPE TRAINING (SCHEDULE AND ATTENDANCE)

Exhibit C-6 represents a schedule of 11 seminars as well as a record of participants' attendance. These seminars were given internally with the conference leaders drawn from company executive personnel. Two groups of supervisors attended; this chart covers Group I. A rough draft of this chart was prepared some weeks in advance of the planned starting date. Just putting the

EXHIBIT C-6   Conference Training—Schedule and Attendance[6]

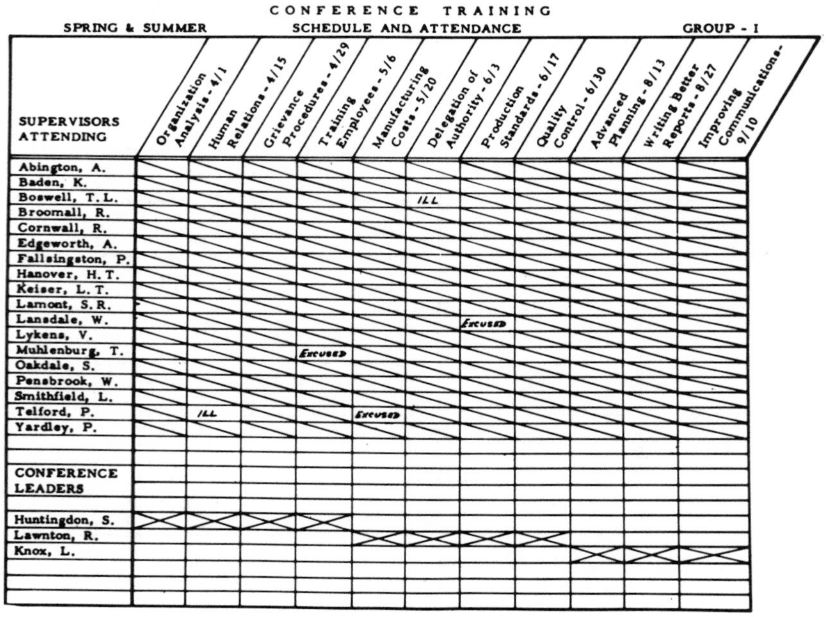

---

[6] Lewis R. Zeyher, *Cost Reduction in the Plant* (Englewood Cliffs, N. J.: Prentice-Hall, Inc. Copyright © 1965), p. 132.

facts on paper compelled some thinking, arranging and planning. The benefits of this exercise alone were well worth all the time and effort expended in the preparation of the charts. Participants, conference leaders, and their superiors received copies of this chart before the starting date. The attendance record, of course, was maintained only by the director of training. At the conclusion of the meetings copies of the chart were placed in the personnel department's file of each employee, with relevant comments concerning progress, degree of participation and related details written on the reverse side of each seminar member's individual chart.

AMERICAN MANAGEMENT ASSOCIATION SEMINARS

Exhibit C-7 is quite similar to the preceding chart (Exhibit C-6). The main difference is that this program was given externally to the plant. The planning operation, made some weeks in advance of the starting dates of seminars, was very important. No more than two supervisors were permitted to be away from the plant at the same time and were usually selected from different departments. Copies of the schedule were given in advance to each participant and his superior. At the conclusion of each participant's meeting the chart, accompanied by relevant data, was filed in the personnel department.

EXHIBIT C-7 **American Management Association Seminars**[7]

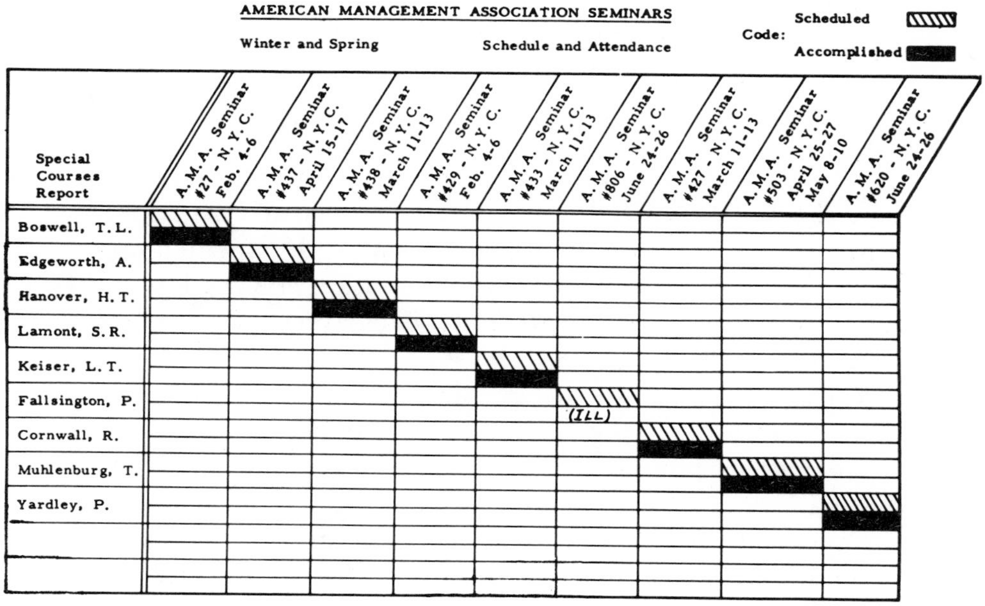

AMERICAN MANAGEMENT ASSOCIATION SEMINARS

Winter and Spring   Schedule and Attendance   Code: Scheduled / Accomplished

[7] Lewis R. Zeyher, *Cost Reduction in the Plant* (Englewood Cliffs, N. J.: Prentice-Hall, Inc. Copyright © 1965), p. 132.

PRODUCTION PERFORMANCE REPORT

Exhibit C-8 is designed for posting on department or plant bulletin boards and depicts each production employee's weekly production performance, based on measured work—time study, MTM or Work Factor systems. These reports should be posted no later than 48 hours after the end of the work week involved. In this manner the employee is kept informed of his progress, the supervisor reviews it before posting and makes notes of the poor performers, and all employees are kept informed about the records of each other.

How does plant management use this information to improve performance? The method recommended is one that had a wide acceptance and success: First, post each employee's performance every week on his individual personnel department record card, specifically designed for this purpose. Second, have this record reviewed periodically by the employee's department supervisor and also by the production manager.

In one plant within my own experience, I had my secretary review each employee's card every six weeks for performance. All those who had an average for the past six weeks of less than 100% were listed by departments, with their scores. The foremen concerned were called individually into a conference at which the offending employees were discussed. Each foreman was told to take this problem up with the employees on his particular list. These supervisors were also informed that in another six weeks this review would be repeated. If, at that time, these same employees were still performing under standard, their supervisors would be asked to provide the manager with a specific program they were going to pursue in order to improve their department's production performance.

This approach was continued until all departments showed substantial improvement. If a supervisor did not follow these instructions he was subjected to severe criticism and a possible change in status. If an offending employee could provide his supervisor with legitimate reasons for his low productivity, this information was inserted on his personnel department record card. Warnings were also indicated on this same card.

The severity and extent of disciplinary measures depends, of course, on the company's industrial relations practices and policies.

There are also occasions when a poor production performance could reflect a poor management practice or operating shortcoming, such as shortages, poor quality material, bad tooling, delays in servicing, lack of orders, poorly maintained equipment and related difficulties. The possibility of these conditions being present where poor production performance is indicated should, of course, be properly investigated before any disciplinary actions are initiated. Generally the very fact that management keeps such a record is incentive

EXHIBIT C-8   Production Performance Report[8]

PRODUCTION PERFORMANCE REPORT

Department : — WEAVING
Section : — BROADLOOM
Week Ending : — OCTOBER 19th
DEPARTMENT AVERAGE : — **98.43 %**

NAME OF EMPLOYEE *	STANDARD HOURS PRODUCED	HOURS ON STANDARDS	PERCENT PERF.
Abington, John B.	40.00	40.00	100.00
Barrington, Arthur	32.64	32.00	102.00
Brent, William R.	39.69	40.50	98.00
Charleston, Thomas	40.00	40.00	100.00
Davison, Charles T.	17.60	20.00	88.00
Edington, Peter S.	42.23	41.00	103.00
Franklin, Ernest	36.00	36.00	100.00
Hamilton, William S.	47.70	45.00	106.00
Lansdown, Lawrence, P.	22.80	38.00	60.00
Middleton, Ralph	40.20	40.00	100.50
Nettleton, John	41.58	42.00	99.00
Newton, Luther R.	39.78	39.00	102.00
Norton, William	40.00	40.00	100.00
Phillips, Howard	45.23	45.00	100.50
Parks, Arnold N.	42.84	42.00	102.00
Stanton, James	41.21	40.80	101.00
Williams, Joseph	39.90	39.90	100.00
Zimmermann, August L.	41.56	40.75	102.00
TOTALS :	690.96	701.95	1764.00
WEIGHTED AVERAGE			98.43

* *Names are fictitious*

---

[8] Lewis R. Zeyher, *Cost Reduction in the Plant* (Englewood Cliffs, N. J.: Prentice-Hall, Inc. Copyright © 1965), p. 175.

enough to encourage employees to maintain their required production performance. In order to save on clerical effort, instead of calculating and posting performances for every week you can use a work sampling process. This involves the selection of random dates throughout the year when these weekly average production performances would be calculated and reviewed, perhaps in a span of from three to six weeks or repeated about 15 times a year.

PLANT PRODUCTION PERFORMANCE

The purpose of this chart (Exhibit C-9) is to depict every department's production performance each month, complementing the preceding chart (Exhibit C–8). The weighted average of all the production employees' production performances for October 19th was 98.43. Assuming this performance was true for the entire month of October, it would appear under "Broadloom Weaving" on this chart and would also be shown in the "Under Standard" column.

These posters should be hung on plant bulletin boards around the plant. In one multi-plant company of my experience, similar posters, but depicting *plant* performances of *all* plants in the system, were compared with each other, with the best plant performances placed at the top of the list and the poorest at the bottom. This encouraged some competition among the 65 plants of the company. Both the Board Chairman and the President displayed these charts, brought up to date each month, in their offices.

MASTER SCHEDULE—WILTON CARPET

This type of chart (Exhibit C–10) provides management with an excellent control of production—from receipt of customer's order through the various production processes to warehouse or shipment. The scheduled dates, made in advance and covering every operation, establish timetables for all mill actions. The dates specify day of completion, and actual dates are written in when the task is finished. If trouble develops, the planning department or production control will immediately be made aware of it. If the trouble causes an unusual delay, rescheduling may be required and a new customer delivery date determined, or overtime work assigned. If these alternatives are not adequate, second shift work could be scheduled. By leaving some vacant spaces between the various numbered looms, as shown on this chart, flexibility is provided. A revised schedule can be inserted directly below the one abandoned.

Besides providing a "flight plan" for each order, this kind of chart offers many fringe benefits. The sales department can daily follow up progress on critical orders; a recapitulation of these schedules covering a period of weeks will provide management with data on troublesome looms, low weaver productivity, ineffective loom-fixer repair work, shoddy loom maintenance, poor

EXHIBIT C-9   Plant Production Performance[9]

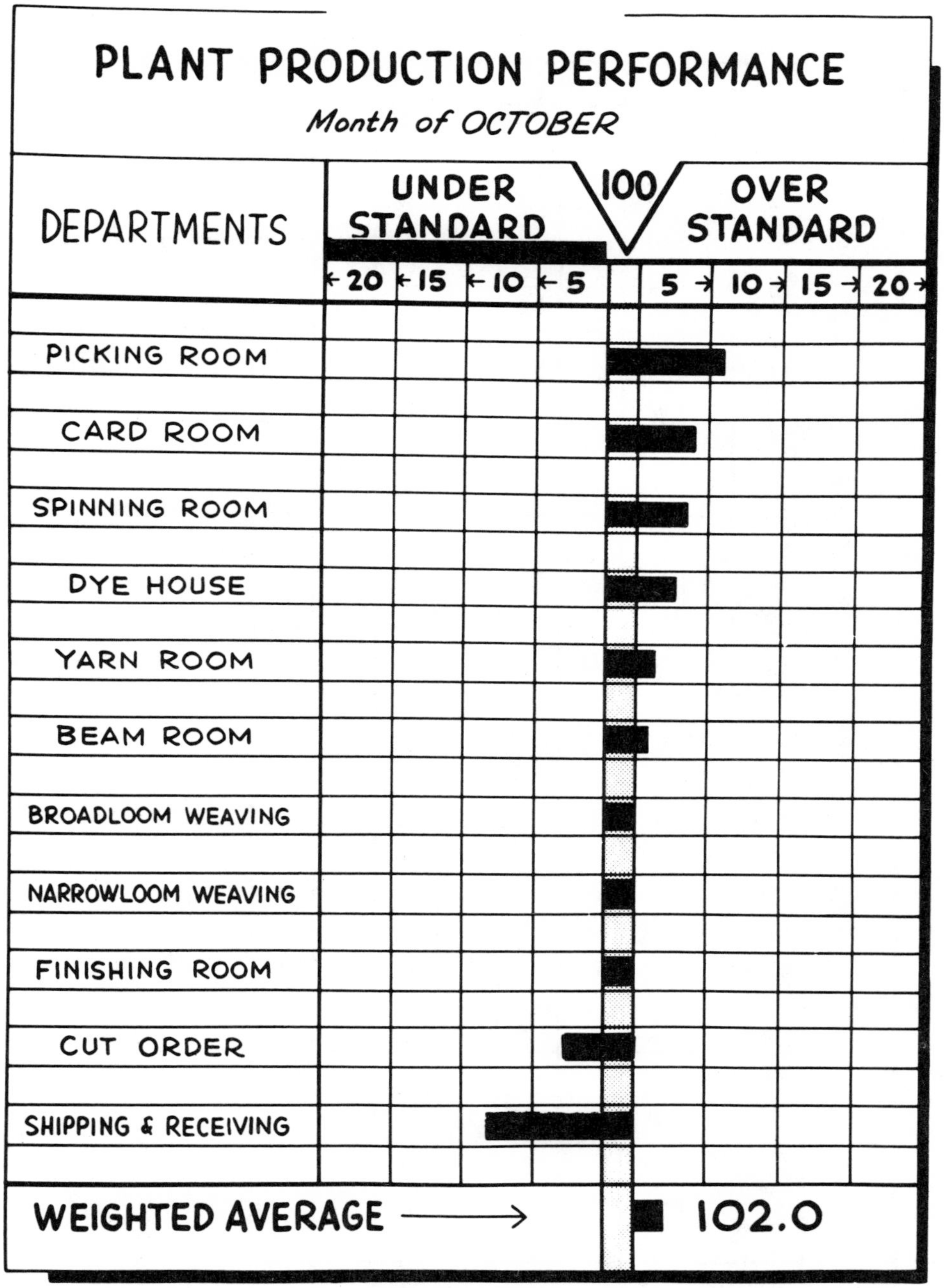

PLANT PRODUCTION PERFORMANCE
Month of OCTOBER

DEPARTMENTS	UNDER STANDARD				100	OVER STANDARD			
	←20	←15	←10	←5		5 →	10 →	15 →	20 →
PICKING ROOM									
CARD ROOM									
SPINNING ROOM									
DYE HOUSE									
YARN ROOM									
BEAM ROOM									
BROADLOOM WEAVING									
NARROWLOOM WEAVING									
FINISHING ROOM									
CUT ORDER									
SHIPPING & RECEIVING									

WEIGHTED AVERAGE ——→ ▮ 102.0

[9] Lewis R. Zeyher, *Cost Reduction in the Plant* (Englewood Cliffs, N. J.: Prentice-Hall, Inc. Copyright © 1965), p. 178.

MASTER SCHEDULE–WILTON CARPET

COMPANY ORDER NUMBER	REGISTER NUMBER	YARDS ORDERED	LOOM NUMBER	DATE ORDER RECEIVED IN PROD. CONT.	YARN REC'D FROM SPIN. MILL		SKEIN YARN WOUND IN WINDING ROOM		YARN DELIVERED TO LOOM		WEAVING STARTED		WEAVING FINISHED		CARPET FINISHED & PASSED		DELIVERED TO CUT ORDER DEPT.		CUT ORDER OPERATIONS COMPLETE		DELIVERED TO SEWING DEPT.		SEWING OPERATIONS COMPLETE		DELIVER TO WHSE. STOCK		HOLD FOR SHIPMENT		ITEM SHIPPED		COMMENTS
					SCHED.	ACT.	SCHED.	ACT.	SCHED.	ACT.	SCHED.	ACT.	SCHED.	ACT.	SCHED.	ACT.	SCHED.	ACT.	SCHED.	ACT.	SCHED.	ACT.	SCHED.	ACT.	SCHED.	ACT.	SCHED.	ACT.	SCHED.	ACT.	
2627	62100	100	22	7/8	9/3		9/10		9/7		9/8		9/12		9/13		9/15		9/17		9/19		9/22		9/22		9/23		9/23		MUST BE DELIVERED BEFORE 10/1
3824	63822	500	26	7/9	9/4		9/11		9/8		9/9		9/30		10/2		10/4		10/10		10/11		10/15		10/17		10/25		10/26		ADVISE CUSTOMER EXACT DATE OF DELIVERY
8343	63863	1,000	27	7/9	9/4		9/9		9/11		9/15		11/5		11/8		11/9		11/11		11/15		11/18		11/18		11/22		11/22		CHECK CREDIT BEFORE SHIPPING

EXHIBIT C-10 Master Schedule—Wilton Carpet[10]

[10] Lewis R. Zeyher, *Production Manager's Desk Book* (Englewood Cliffs, N. J.: Prentice-Hall, Inc. Copyright © 1969), p. 33.

quality workmanship, too frequent changes in delivery dates inspired by sales department, delays in customer orders reaching the production control department after receipt by sales department, material shortages and related operating intelligence.

LOOM LOADING CHART

This Gantt chart (Exhibit C–11) complements Exhibit C–10 and can be used for any of the mill operations. However, economically the longest operations will fit best for this approach. *Scheduled* production is usually indicated in colored pencil and *actual* production shown in black or another color. This form can easily be duplicated in the dimensions that best fit your requirements. A code is recommended to indicate various types of delays, i.e.: *A*—operator absent; *M*—machine down; *R*—repairs; *N*—no material; *S*—machine set-up; and so on. It is interesting to note that schedulers who use these charts daily design their own codes and often devise ingenious methods tailored for the type of machine and company product. Many of the fringe benefits enumerated in Exhibit C–10 are applicable here as well.

The production factor expressed in running yards per shift is usually determined by time study, MTM or Work Factor, or a similar measured work system. For example, Loom No. 136 has a production factor of 18 running yards per eight-hour shift, so 28 spaces Monday through Friday, February 1 to Tuesday, March 12th, are filled in with red pencil. This represents a customer's scheduled order for 504 yards.

MACHINE WORK LOAD CHART

The column headings make this chart (Exhibit C–12) self-explanatory. The chart is used by the scheduler for the express purpose of loading a machine to its capacity, and to avoid overloading it. This provides another excellent planning tool. For example, you have in your plant only one special purpose machine with a two 8-hour shift production capacity of 200 pieces. The sales department receives an inquiry from a customer for 4,000 pieces to be delivered in one week from the date of order and asks if you can meet this schedule. Your examination of the machine work load chart might reveal that the machine is already tied up for two weeks' work and cannot handle any more work until the third week. With this information the salesman can intelligently advise the customer. Important judgments like this can be quickly formulated only by reference to reliable production information such as supplied by this chart. However, this is a simple illustration. These charts become much more important when multi-machines are involved.

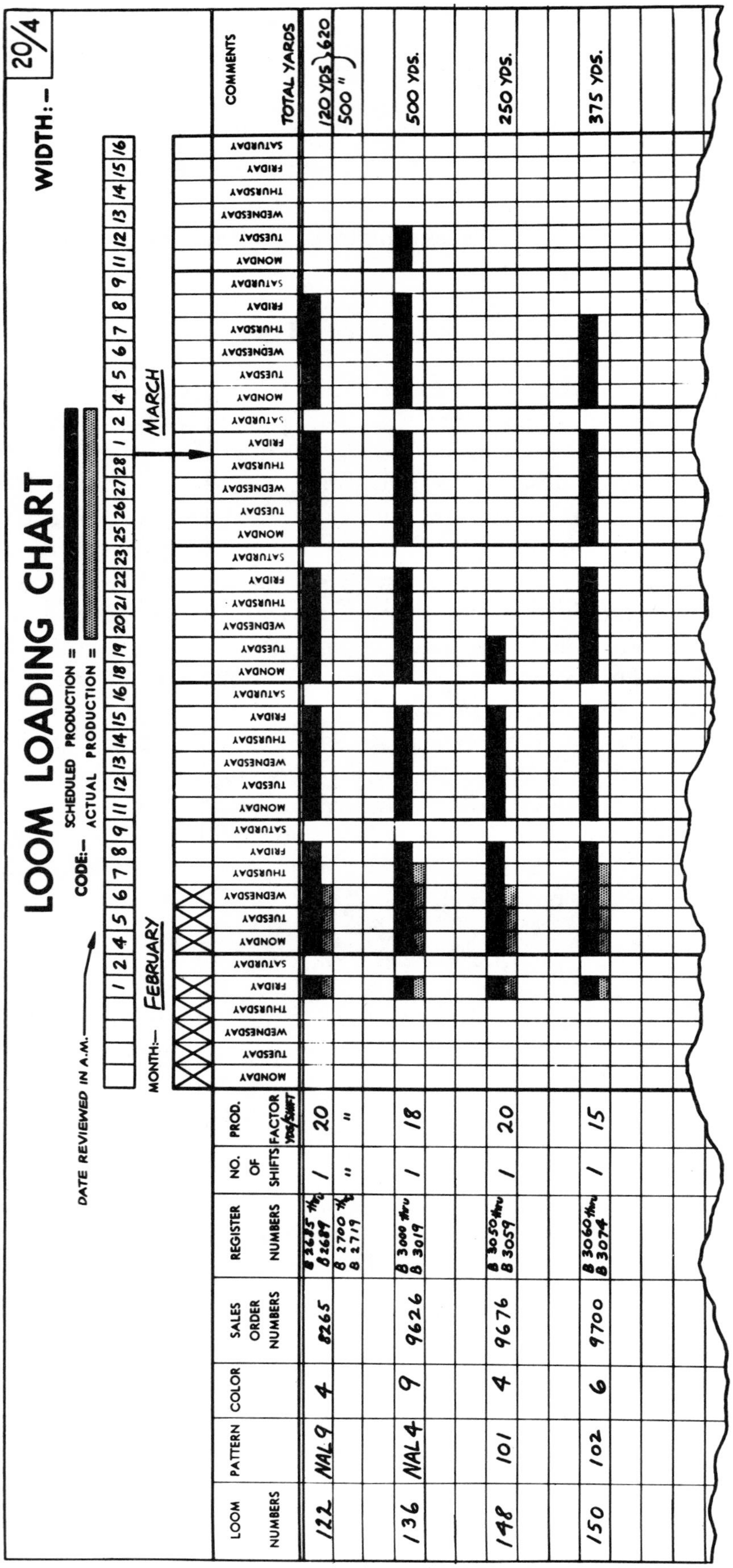

LOOM LOADING CHART

WIDTH:- 20/4

CODE:- SCHEDULED PRODUCTION = ▬▬▬   ACTUAL PRODUCTION = ▨▨▨

DATE REVIEWED IN A.M.

MONTH:- FEBRUARY — MARCH

LOOM NUMBERS	PATTERN	COLOR	SALES ORDER NUMBERS	REGISTER NUMBERS	NO. OF SHIFTS	PROD. FACTOR YDS/SHIFT	COMMENTS / TOTAL YARDS
122	NAL9	4	8265	B2635 thru B2654	1	20	120 YDS }620
				B2700 thru B2719	"	"	500 "
136	NAL4	9	9626	B3000 thru B3019	1	18	500 YDS.
148	101	4	9676	B3050 thru B3059	1	20	250 YDS.
150	102	6	9700	B3060 thru B3074	1	15	375 YDS.

EXHIBIT C-11 Loom Loading Chart[11]

[11] Lewis R. Zeyher, *Production Manager's Desk Book* (Englewood Cliffs, N. J.: Prentice-Hall, Inc. Copyright © 1969), p. 34.

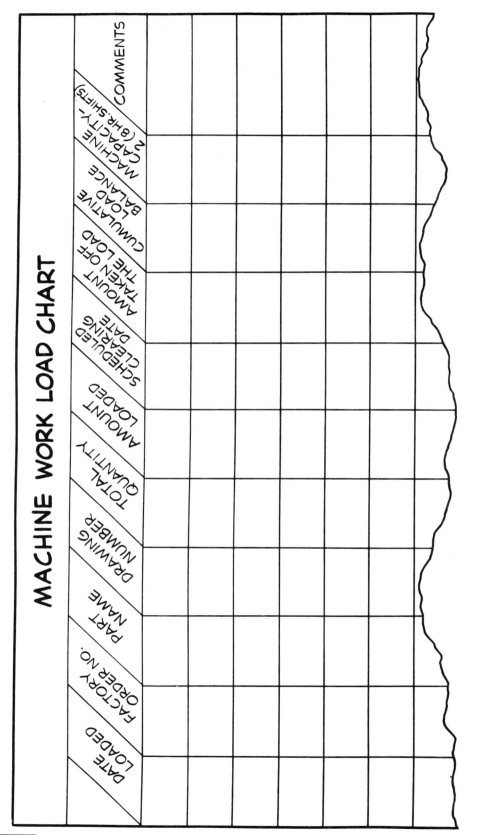

EXHIBIT C-12   Machine Work Load Chart[12]

[12] Lewis R. Zeyher, *Production Manager's Desk Book* (Englewood Cliffs, N. J.: Prentice-Hall, Inc. Copyright © 1969), p. 35.

WEEKLY STATION LOAD REPORT

This form (Exhibit C–13) is another adaptation of the Gantt chart principle. It is most often used by the scheduler in the production control department and differs from the chart in Exhibit C–12 in that it covers 12 machines rather than one. It also graphically depicts the full load, overload and underload of each machine for each week for a total of eight weeks. This is particularly useful to a jobbing type shop where specific machine time is sold to various customers. Salesmen can be advised of open machine time, suggesting the need for additional sales efforts. In addition, load and overload conditions are made known. An inspection of the chart also reveals that the borematic, hone and punch presses are currently failing to pay their way for burden expense, such as floor space, supervision, servicing, taxes and related items.

COST REDUCTION GOALS AND TIMETABLE

When initiating a plant cost reduction program, goals and timetables must be established. (See Exhibit C–14.) These should be determined by the production manager in consultation with his superior. Plant circumstances may influence these decisions but, if possible, make it cover at least a year's time and establish goals by going over the past year's operating statement with the plant controller. The goals should be realistic. Consideration also should be properly given to any reconciliation in the figures due to changes, operating modifications, design and style changes and related influences. This exhibit depicts a rough sample outline for a cost reduction program with goals and timetables. This particular program is planned on a departmental basis. Each department head is held responsible for reaching a specified goal of reduced costs as indicated in this exhibit. The chart is designed to cover a specified fiscal year and illustrates expected versus accomplished results. First, a total plant figure of cost savings should be estimated (as in the example— $170,000). Each department is then allocated a share of the total. This is followed by breaking each department's goals into quarters of the year. Actual results are inserted by the accounting department in the spaces provided. Savings in material and labor only should be considered in savings calculations. Some companies arrange to have large posters of these charts prepared and placed in conspicuous areas of the plant. This is done to solicit employee interest and cooperation in the program.

COST REDUCTION PROJECT—PROGRESS REPORT

This chart (Exhibit C–15) graphically depicts results against goals for each department head, with the program based on a *specific project* basis. Each supervisor selects a project in his department which he feels offers cost reduction

WEEKLY STATION LOAD REPORT

DATE: _____

ISSUED BY:
PRODUCTION CONTROL DEPT.

KEY: ▨ LOAD ▨ OVERLOAD ☐ UNDERLOAD

STATION WORK CENTERS	WEEKLY CAPACITY (40 HRS.)	12/20 HRS.	12/27 HRS.	1/3 HRS.	1/10 HRS.	1/17 HRS.	1/24 HRS.	1/31 HRS.	2/7 HRS.	8-WEEK RECAP TOTAL HOURS		OVER-LOAD REMARKS
SHEET METAL	480	671 / 113	176 / 93	387	369 / 111	434 / 44	472 / 8	451 / 24	480	1071	2769	
JIG BORE	40	280							40		320	
BOREMATIC	40	28 / 12	40	40	40	40	40	40	40	12	308	
BURR BENCH (2 MEN)	80			414			98	84	76	672		32
DRILL PRESS	360	731		398	200 / 111 / 58	302	246 / 114	247 / 63	360	1709	1171	
GEAR HOB	40	111		9	61	23	16 / 20	40	40	220	100	
GRINDER	40	35 / 5	37 / 3	22 / 18	37 / 3	40	40	40	40	43	227	
HONE	40	30 / 10	40	40	40	40	40	40	40	30	290	
LATHES	360	565	165 / 55 / 20	275 / 20	340 / 20	339 / 21	324 / 36	357 / 3	360	885	1995	
MILLS	320	445	82 / 63 / 50 / 23	297	281 / 39	252 / 75	281 / 53	300 / 20	320	800	1760	
PUNCH PRESS	120	61 / 59	113	118	118	117	116 / 4	120	120	79	881	
SAWS	120	113 / 7	117 / 3	120	120	120	120	120	120	10	950	

EXHIBIT C-13 Weekly Station Load Report[13]

[13] Lewis R. Zeyher, *Production Manager's Desk Book* (Englewood Cliffs, N. J.: Prentice-Hall, Inc. Copyright © 1969), p. 36.

DEPARTMENTS	Yearly Savings Expected	First Quarter		Second Quarter		Third Quarter		Fourth Quarter		Yearly Savings Accomplished
Machine Shop	$50,000	$15,000	16,000	$15,000	17,000	$10,000	10,000	$10,000	9,000	$52,000
Assembly	30,000	10,000	12,000	7,500	10,000	7,500	6,000	5,000	6,000	34,000
Toolroom	10,000	3,000	1,000	3,000	1,500	2,000	1,500	2,000	2,000	6,000
Shipping & Receiving	25,000	10,000	12,000	5,000	6,000	5,000	6,000	5,000	7,000	31,000
Plating	5,000	2,000	2,000	1,000	1,000	1,000	1,500	1,000	1,000	5,500
Maintenance	20,000	7,000	8,000	5,000	4,000	4,000	6,000	4,000	7,000	25,000
Production Control	5,000	2,000	1,000	1,000	500	1,000	500	1,000	—	2,000
Quality Control	10,000	4,000	4,000	2,000	2,200	2,000	1,800	2,000	2,000	10,000
Personnel	4,000	1,000	1,000	1,000	1,000	1,000	800	1,000	1,200	4,000
Manufacturing Engineering	11,000	4,000	5,000	3,000	4,000	2,000	1,000	2,000	2,000	12,000
TOTAL	$170,000	$58,000	$62,000	43,500	47,200	35,500	35,100	33,000	37,200	$181,500

EXHIBIT C-14   Cost Reduction Goals and Timetable[14]
Fiscal Year
(Expected vs. Accomplished Results)

[14] Lewis R. Zeyher, *Production Manager's Desk Book* (Englewood Cliffs, N. J.: Prentice-Hall, Inc. Copyright © 1969), p. 97.

COST REDUCTION PROJECT · PROGRESS REPORT

DEPT. _____

Description of project ____ — etc.

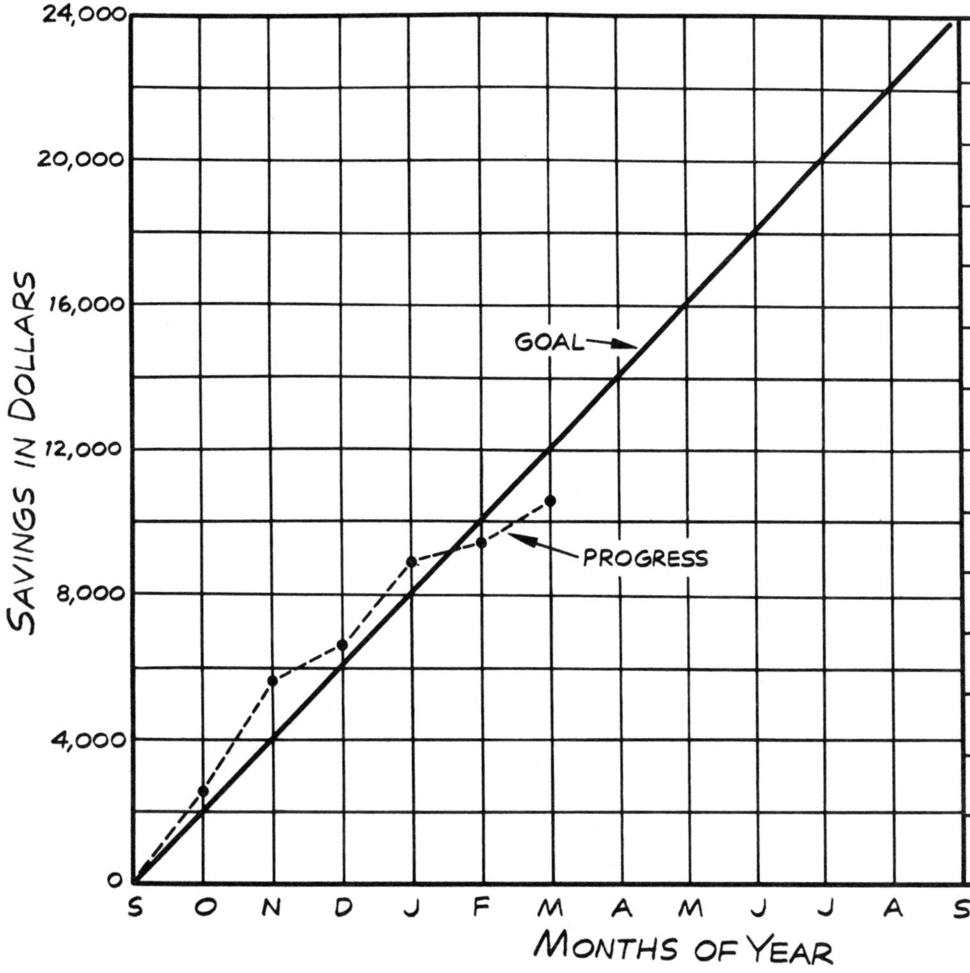

	GAIN OR LOSS FOR MONTH	GAIN/LOSS TO DATE	GOAL TO DATE
OCT.	$500	$500	$2,000
NOV.	800	1,300	4,000
DEC.	(1,000)	300	6,000
JAN.	600	900	8,000
FEB.	(1,600)	(700)	10,000
MAR.	(1,000)	(1,700)	12,000
APR.			14,000
MAY			16,000
JUNE			18,000
JULY			20,000
AUG.			22,000
SEPT.			24,000
GOAL FOR YEAR			$24,000

EXHIBIT C-15 Cost Reduction Project—Progress Report[15]

SAVINGS IN DOLLARS — MONTHS OF YEAR

[15] Lewis R. Zeyher, *Production Manager's Desk Book* (Englewood Cliffs, N. J.: Prentice-Hall, Inc. Copyright © 1969), p. 98.

possibilities. An estimate is made with staff assistance and the plant manager's approval secured. All plant departmental projects are then initiated on the same date and are expected to finish on the same date. The plant manager usually assigns one of his staff to coordinate the program and to assist the supervisors. He also has charge of developing a chart and graph for each supervisor like the subject exhibit, as well as keeping all the figures and graphs up to date. The figures in the "Gain or Loss for Month" column depict dollar savings for each month; figures in parentheses represent losses compared to monthly goals. The "Gain/Loss to Date" column is accumulative and is arrived at by algebraically adding the plus and minus figures in the first column. For example, in the first column add October, November and December figures ($500 + 800 − 1,000) and you have a gain of $300 over budget for the three months. If you examine the graph you will find a solid dot, connected to other solid dots by a broken line, with the one for December placed approximately $300 above the solid Goal line. The "Gain/Loss to Date" column and the "Goal to Date" both represent accumulative figures.

INVENTORY ANALYSIS

The construction of this chart (Exhibit C–16) to fit your plant's situation is self-explanatory. The figures usually are available in your accounting department. All figures for the different categories must be consistent—and should be based on manufactured costs.

In this instance the company was experiencing a very low yearly profit. In seeking causes for this unsatisfactory performance the chart was prepared depicting the results of the first four months of the current year. An examination of the inventory chart indicates the following:

(a) An inventory turnover ratio of 2.9 times a year is much too low. It should range between six to seven times. (No hard and fast rule can be made here unless all relevant facts are known and carefully evaluated by competent personnel.)

(b) Sales were increasing and the backlog was running approximately at the rate of four times shipments.

(c) Finished goods inventories were large, and growing. (This indicated that production was doing a poor job, which was proved true by subsequent impressive increases in shipments, brought about by a change in supervision and the installation of an improved production control plan.)

(d) Both work-in-process and material inventories were too high in relation to shipments. The inventories in both of these categories were substantially reduced by making arrangements with the electronic parts supplier to segregate and make up his shipments according to each company order covering a customer (or combination of customer orders for similar item). By this method the supplier, instead of the subject electronic company, maintained the necessary inventories of the various parts consumed. A small charge for this extra service was made by the supplier, but it eliminated a lot of costly handling time as well as lowering the costs of carrying inventories. Another feature was the practice of making shipments only when all items were available and order was complete. This eliminated part shipments.

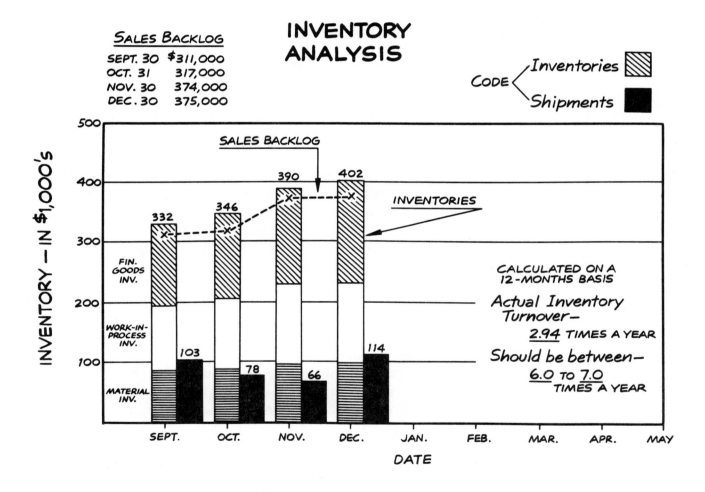

EXHIBIT C-16 Inventory Analysis[16]

[16] Lewis R. Zeyher, *Production Manager's Desk Book* (Englewood Cliffs, N. J.: Prentice-Hall, Inc. Copyright © 1969), p. 182.

(This arrangement could be used to advantage only by the small to medium sized companies.) However, I think considerable material handling and paperwork could be eliminated by a closer examination of your suppliers' methods in inventory control and shipping practices. This could result in a closer tie-in with each other's practices, resulting in mutual benefits.

(e) Material inventories included a number of obsolete items and some very slow-moving parts. Some of these were sold at reduced prices and others were scrapped. This resulted in savings in space, the cost of taking periodic physical inventories, and attendant paperwork.

SHIPPING DEPARTMENT—UNITS VS. LABOR

Labor costs began to climb in the shipping and receiving department of a company of 250 employees. The president questioned the plant manager, who had the chart shown in Exhibit C-17 prepared by an industrial engineer. Time studies had previously been made and most of the work measured. The major shipping containers were 15, 30, and 60 pound cartons with the 60 pound container (filled) counted as one unit and the others in proportion. Similar units were used for unloading trucks of supplies. All had standard hours assigned. This chart was analyzed, questions asked, and the following conclusions drawn:

(1) In this company shipments were generally heavy before most holidays. In this instance it appeared that the supervisor started to add people several weeks before Labor Day so he would not be caught short of trained help during the rush period.

(2) After Labor Day the volume of shipments decreased considerably, but *without* a corresponding reduction in weekly labor expense.

(3) The foreman had been in his present position only six months, but had ten years' experience with the company. He had not been properly trained for his new job.

(4) The supervisor apparently wanted to be popular with his subordinates—not an uncommon failing among first line supervision. The result was little, if any, control of overtime hours after the holiday rush was over.

(5) An examination of the employees' working schedules indicated that each man worked the same number of hours each day, although as depicted on the chart, Monday, Tuesday and Friday were lighter shipping days than Wednesday and Thursday. Some employees worked 12 hours a day consistently, regardless of the workload. Remedial actions taken:

(a) More training for supervisor was provided.

SHIPPING DEPT. – UNITS vs. LABOR

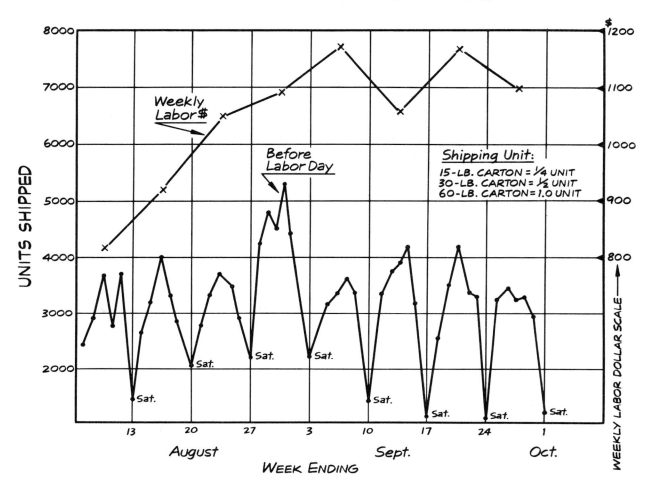

EXHIBIT C-17   Shipping Department—Units vs. Labor[17]

(b)   More surveillance over his activities was initiated.

(c)   Workers' schedules were rearranged; working hours were planned according to workload.

(d)   All 12-hour shifts were eliminated.

(e)   More assistance was given to the supervisor with his planning activities.

(f)   The supervisor was taught how to assign crews according to units (all time studied) and with knowledge of shipping patterns.

(g)   A simple paper control was prepared for use of the plant manager.

---

[17] Lewis R. Zeyher, *Production Manager's Desk Book* (Englewood Cliffs, N. J.: Prentice-Hall, Inc. Copyright © 1969), p. 183.

PURCHASING DEPARTMENT ANALYSIS

The construction of this chart (Exhibit C–18) required summarizing the dollar value of purchases and purchase commitments by months for the eight previous months, starting with July, and the dollar sales value by months of customer shipments. Several months, March and April, were estimated by using the orders then being processed in the factory that were to be delivered in this period as well as orders on file that were promised for delivery during

PURCHASING DEPT. ANALYSIS

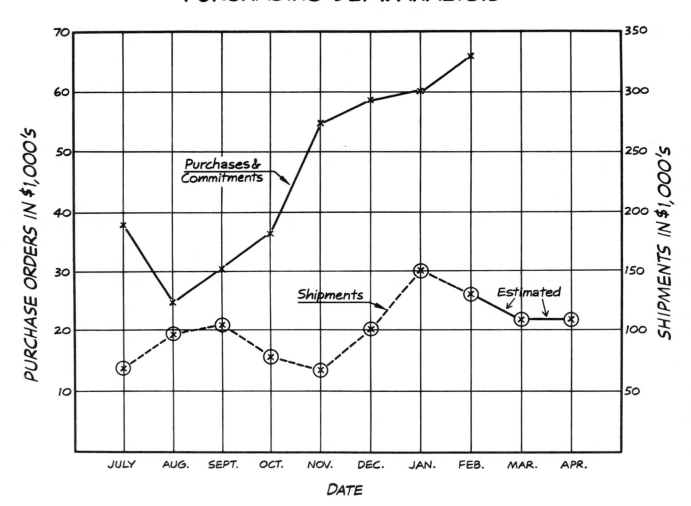

EXHIBIT C-18 Purchasing Department Analysis[18]

[18] Lewis R. Zeyher, *Production Manager's Desk Book* (Englewood Cliffs, N. J.: Prentice-Hall, Inc. Copyright © 1969), p. 185.

March and April. As depicted, two different scales were used, one for "Purchase Orders in $1,000's" and the other, "Shipments in $1,000's."

A medium size metal working plant whose monthly sales volume fluctuated experienced an uncontrolled rise in purchases and purchase order commitments. This company was in a poor cash position and needed to preserve carefully the money it had available. It also had too much money tied up in physical inventories. In addition, it was in a highly vulnerable credit position with suppliers. It was mandatory for the company to control expenditures tightly for some time. The management was at fault here. It did not maintain adequate controls. An examination of the charts reveals careless surveillance of the main function of the purchasing department. If a chart such as this one had been utilized and reviewed by plant management at the end of each month, this serious situation could have been avoided. Back in September and October a dangerous trend in both curves was apparent and dictated the need for remedial action.

If very close control is indicated, a weekly report could be submitted by the purchasing agent to his superior. This report should be prepared in detail covering such items as vendors' names, purchase order numbers, quantity ordered, units (pounds, pieces, yards, etc.), description of items, required delivery date, estimated dollars involved and, at the bottom of page, weekly totals. Depending on circumstances, a very brief weekly report depicting estimated dollars should suffice. It could also be necessary to have a weekly report depicting the type and value of items received from suppliers. When the undesirable condition improves and company finances are in good condition, the reports could be changed to monthly reporting.

AVAILABLE MACHINE HOURS

This chart (Exhibit C–19) bears a relationship to several previously discussed. It is possible that your adaptation of just one of the charts of the several described would suffice for your specific requirements. In other words, there is some duplication here, but in defense it can be stated that alternatives in choice are often desirable, and so are offered.

This particular modification of the Gantt chart should be helpful to both the production control and sales department personnel. It provides the scheduler with a facility to load his individual machines months ahead and emphasizes the soft spots. Where machine time is avi:lable, requests can be made to the sales department for more orders, which is particularly useful in a jobbing kind of business. This chart must be revised weekly and copies submitted to other interested departments. An examination of this chart also indicates a two shift operation is a distinct possibility, particularly on the No.'s 252, 201 and 256 machines, depending, of course, on backlog of orders which are not shown here.

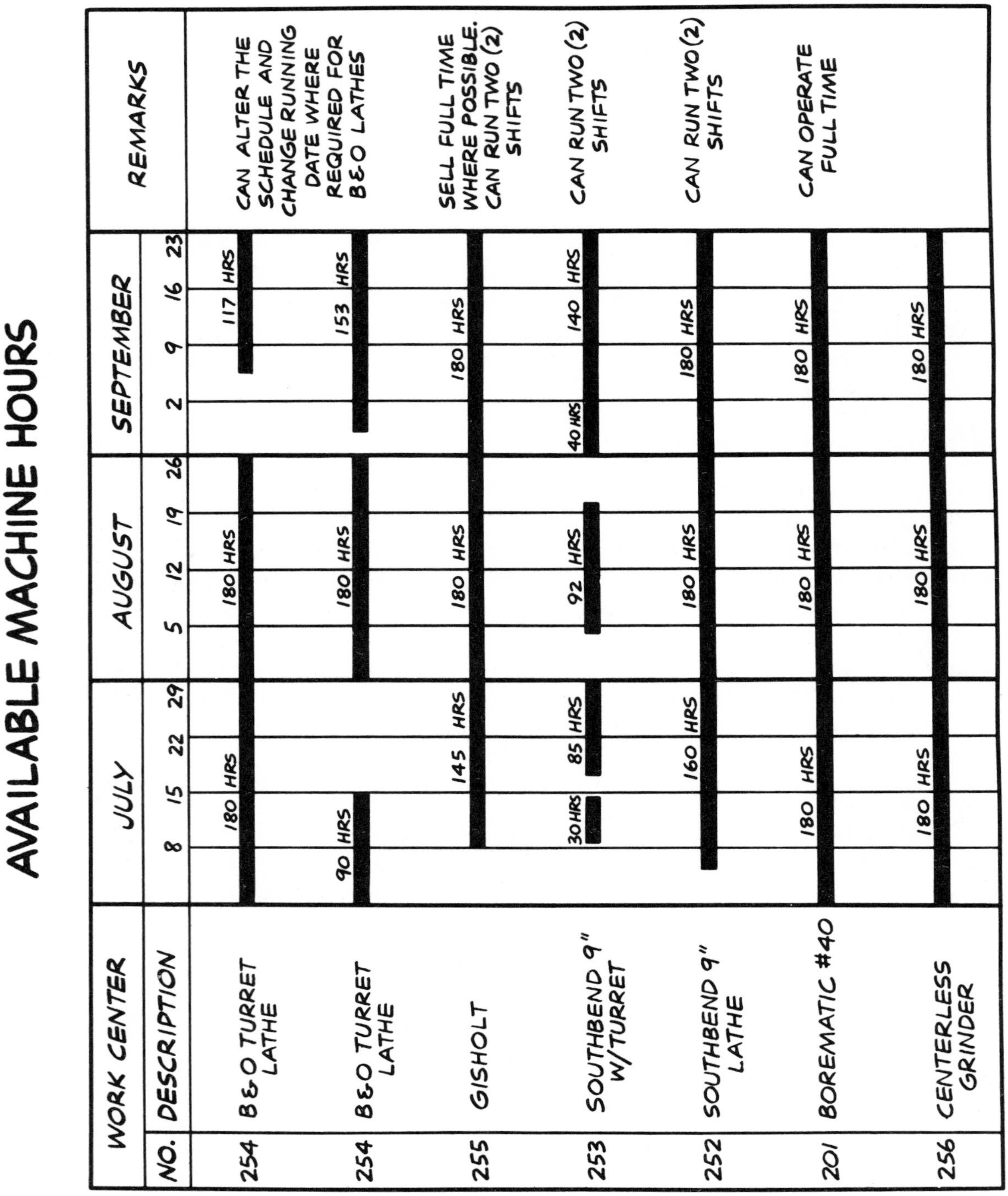

AVAILABLE MACHINE HOURS

NO.	DESCRIPTION	JULY (8 15 22 29)	AUGUST (5 12 19 26)	SEPTEMBER (2 9 16 23)	REMARKS
254	B&O TURRET LATHE	90 HRS / 180 HRS	180 HRS	117 HRS	CAN ALTER THE SCHEDULE AND CHANGE RUNNING DATE WHERE REQUIRED FOR B&O LATHES
254	B&O TURRET LATHE	145 HRS	180 HRS	153 HRS	
255	GISHOLT	180 HRS	180 HRS	180 HRS	SELL FULL TIME WHERE POSSIBLE. CAN RUN TWO (2) SHIFTS
253	SOUTHBEND 9" W/TURRET	30 HRS / 85 HRS	92 HRS	40 HRS / 140 HRS	CAN RUN TWO (2) SHIFTS
252	SOUTHBEND 9" LATHE	160 HRS / 180 HRS	180 HRS	180 HRS	CAN RUN TWO(2) SHIFTS
201	BOREMATIC #40	180 HRS	180 HRS	180 HRS	CAN OPERATE FULL TIME
256	CENTERLESS GRINDER	180 HRS	180 HRS	180 HRS	

EXHIBIT C-19   Available Machine Hours[19]

[19] Lewis R. Zeyher, *Production Manager's Desk Book* (Englewood Cliffs, N. J.: Prentice-Hall, Inc. Copyright © 1969), p. 187.

ANALYSIS OF LOOM DOWNTIME

This bar chart (Exhibit C–20) represents the production time lost by 80 looms in a textile mill. The time lost is shown by major problem areas, in percentages of total lost time. The various categories are: replace broken parts, wire motion and picking problems, smashes, bust-outs, repairs to loom while on

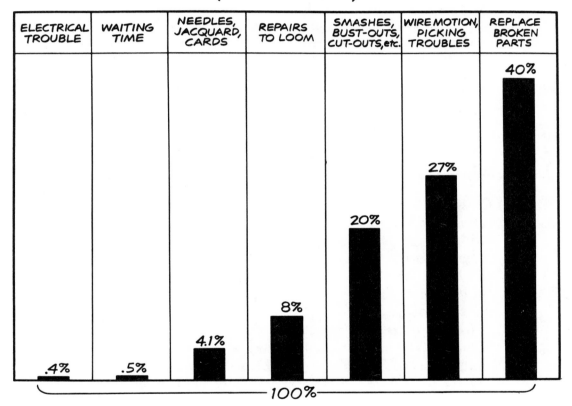

ANALYSIS OF LOOM DOWNTIME
(For 80 Looms)

EXHIBIT C-20   Analysis of Loom Downtime[20]

scheduled production, needles, Jacquard, cards, electrical trouble and waiting time. It was estimated in this actual experience that a loom was worth $20.00 per hour and the yearly loss due to these inefficiencies was $180,000. It is true that perfection in this connection is impossible, but a 50 per cent saving is a

---

[20] Lewis R. Zeyher, *Production Manager's Desk Book* (Englewood Cliffs, N. J.: Prentice-Hall, Inc. Copyright © 1969), p. 189.

realistic goal to strive for, or a saving of $90,000 which represents a substantial cost reduction in operations. Other data revealed by this analysis:

(a) Total downtime per year . . . . .9,000 hours
(b) The average loom running hours per downtime hour 16.5
(c) Poorest loom on this basis is No. 50 . . . . .2.3
(d) Best loom on this basis is No. 78 . . . . .410.0
(e) Reported total mechanical trouble as a percentage of
total loom scheduled time 6.1
(does not include unreported downtime for rest periods, coffee breaks, smoking, late starting, early quitting, etc.)

Note: The data from which this analysis was made was culled from the weavers' daily operating reports. The best approach is by the work sampling method, covered earlier in this book. It is a much more accurate method—machine operators' reports usually are very unreliable. From this it can be assumed that the losses were greater than is indicated here.

(f) Looms were down mechanically on an average of 1.0 hour every two 8–hour shifts.

Suggested Remedial Actions:

(1) Losses were excessive; definite actions are necessary.
(2) Poorly skilled weavers should be given more training (usually those with most repairs and need of maintenance).
(3) More effective supervision is indicated.
(4) A formal preventive maintenance program is needed (48 per cent of loom downtime of 9,000 hours, or 4320 hours alone, was to *replace broken parts and make repairs to looms*.
(5) More accurate operations reports are needed.
(6) Employees should be made aware of the costs of machine downtime and instructed in ways of minimizing it.
(7) The weavers with the poorest records should be given special attention by supervision, causes determined and prompt and firm actions taken to eliminate them. Then there should be consistent follow-up to check on progress. (Note: in this actual experience the analysis included each weaver's individual record based on Item 6, the average loom running hours per downtime hour. For example, the poorest weaver's figure was 2.3 and the best 410.0. In between these two extremes were numerous other weavers' poor performances, as well as good records. It is true that loom fixers contribute to these records and a study along these lines also revealed some interesting figures which will not be explored here.)

SCRAP REPORT (or WASTE)

The construction of this chart (Exhibit C–21) is self-explanatory. It is a weekly report and depicts the total dollars of scrap per 1,000 direct labor hours. If the plant is covered by measured work (time study, MTM or Work Factor systems) the direct labor hours can be changed to total dollars scrap per 1,000 *standard hours*. Where there isn't a 100 % measured work coverage, actual paid hours (time-card hours) can be used and considered the same as standard hours. This will dilute the accuracy of the results but not enough to have any appreciable affect (in relation to amount) on this kind of analysis.

In a metal working or electronics type of business, scrap reports are developed, while in the woodworking or chemical plants, waste reports are used. In woodworking the commonly employed technique is to calculate the percentage yield resulting from cutting up a known square footage of lumber. In this example, the company was a jobbing type electronics firm employing 700 people. The quality control department played a major role in cooperation with the production department supervisors in this cost reduction program. The chart was kept up to date by weekly reports depicting both losses per week (solid line) and the trend (broken line) by months. Naturally the biggest drop in costs occurred in the early months of the cost reduction program.

In order to compensate for the changing labor hours employed, the figures shown were determined through calculation by dividing the total scrap dollars experienced by the direct labor dollars.

REWORK REPORT

This report (Exhibit C–22) is very similar to the scrap report previously discussed. It is more difficult to achieve acuracy in this instance than for scrap. Poor work easily reworked could be done on unreported time and would be difficult to detect. Honesty in reporting and close supervision are important here.

NAME OF PROJECT—BUDGETARY CONTROL

The construction of this chart (Exhibit C–23) is self-explanatory. Expenditures on the project are recorded weekly. The solid black line at the top of page reflects the total dollars budgeted for this project and is inserted when chart is developed. The Average Base Expenditure (broken line) also is inserted before the start of the project. This average is calculated by dividing the total estimated cost of the project by the number of calendar weeks allowed for completing the task ($ 24,000 \div 19 = $ 1263), or rounding off figures to $1,300 per week. (An over-run allowance of about 3% is permitted.)

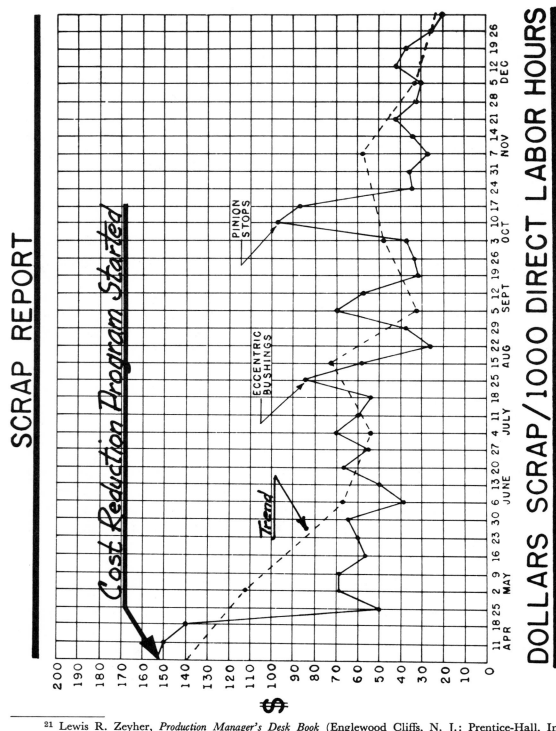

EXHIBIT C-21   Scrap Report[21]

[21] Lewis R. Zeyher, *Production Manager's Desk Book* (Englewood Cliffs, N. J.: Prentice-Hall, Inc. Copyright © 1969), p. 190.

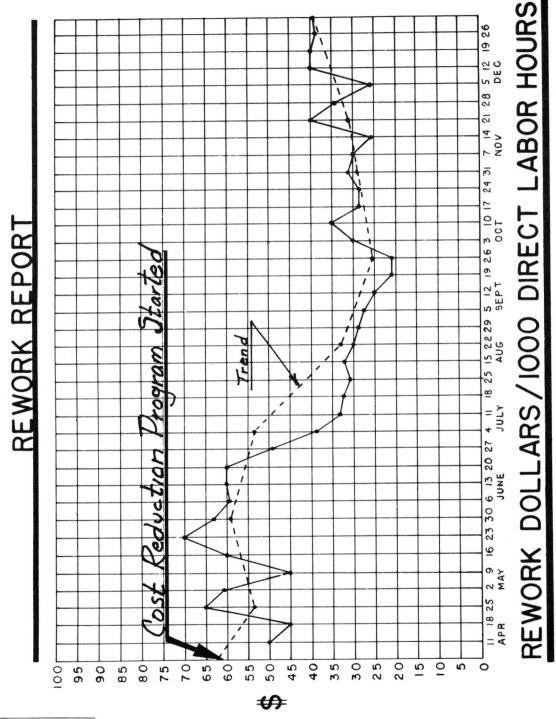

REWORK REPORT

Cost Reduction Program Started

Trend

$

100
95
90
85
80
75
70
65
60
55
50
45
40
35
30
25
20
15
10
5
0

11 18 25 | 2 9 16 23 30 | 6 13 20 27 | 4 11 18 25 | 1 8 15 22 29 | 5 12 19 26 | 3 10 17 24 31 | 7 14 21 28 | 5 12 19 26
APR | MAY | JUNE | JULY | AUG | SEPT | OCT | NOV | DEC

REWORK DOLLARS/1000 DIRECT LABOR HOURS

EXHIBIT C-22   Rework Report[22]

[22] Lewis R. Zeyher, *Production Manager's Desk Book* (Englewood Cliffs, N. J.: Prentice-Hall, Inc. Copyright © 1969), p. 191.

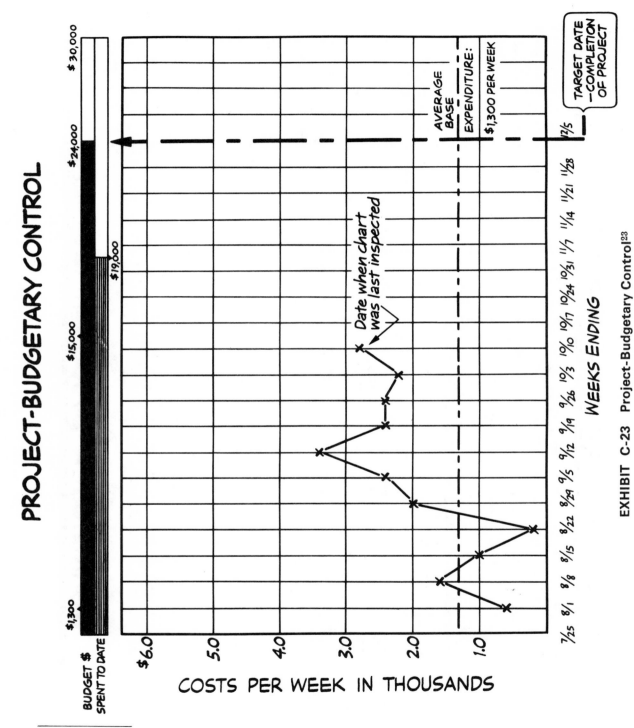

EXHIBIT C-23 Project-Budgetary Control[23]

_____

[23] Lewis R. Zeyher, *Production Manager's Desk Book* (Englewood Cliffs, N. J.: Prentice-Hall, Inc. Copyright © 1969), p. 227.

In this example the need for a special purpose piece of materials handling equipment was developed by the engineering department. It was estimated that the cost of fabricating this equipment would be $24,000 and that it would pay for itself in two years. The formal written justification for this engineering project was submitted to the production manager. He, in turn, sought approval from the vice-president of manufacturing. A number of questions were raised but eventually approval was granted. It was decided construction would be assigned to the toolroom after the design was completed in the engineering department.

The production manager knew a new project like this could become a hazardous undertaking if certain precautions were not taken. Unforeseen problems might arise and costs skyrocket unless proper control was maintained. He knew his superior would want to be periodically informed concerning progress and would insist that the allowed budget be met.

With these facts in mind this control chart was designed and responsibility designated for maintaining it. The elements in the original estimate covering design, engineering, material, labor, supplies, burden and debugging costs were presented to the department head responsible, each of whom was instructed to keep his costs within the budget figures.

Every week the accounting department submitted a *total figure* for all costs to the production department. They were entered on the chart and each week it was shown to the vice-president and progress discussed. In this manner he could follow cost results and be kept informed concerning this special activity in the plant. This type of control keeps the manager informed on where he is today and where he is headed. It also keeps him advised of progress.

GANTT CHART PRODUCTION PLANNING—10″ INDICATOR DRIVE UNIT FOR RADARSCOPE

This graphic exhibit (Exhibit C–24) represents the manufacturing department's method for the production planning of a radarscope 10″ indicator drive unit using a typical Gantt chart principle. The illustration is a copy of an actual experience drawn from the author's management background. The plant situation requiring emergency actions is described below:

The purchasing department of a very important customer had previously ordered a new assembly unit to be developed in our research and development department. The company manufactured electronic instruments and electromechanical assemblies. The development was proceeding according to plan and the delivery date for the initial ten prototypes for their engineering department had been left open. Up to this point the manufacturing department was not involved—it was still purely a research problem. Then one day the customer telephoned and asked that the current engineering design be "frozen,"

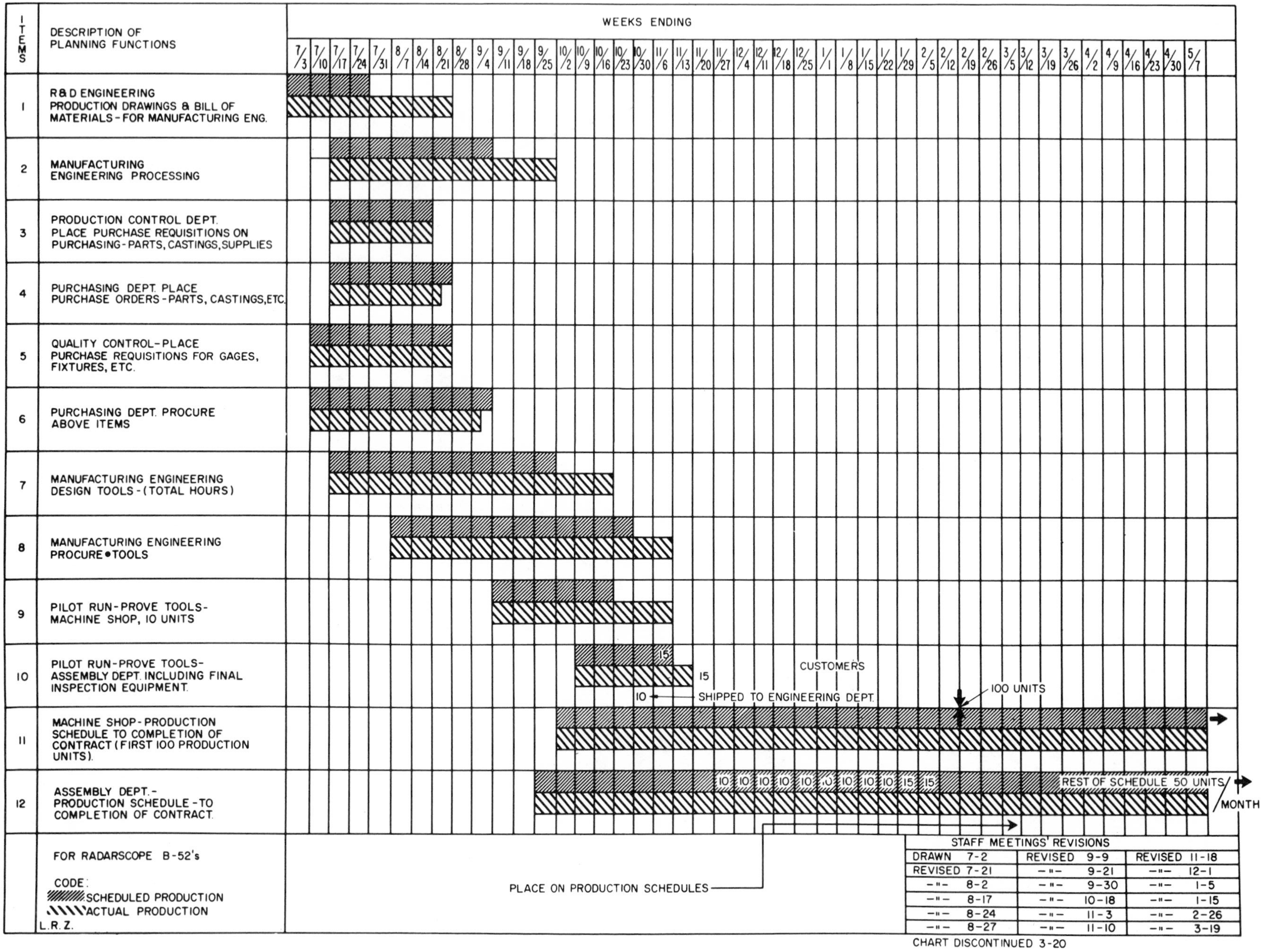
ITEMS
DESCRIPTION OF PLANNING FUNCTIONS
WEEKS ENDING
7/3 7/10 7/17 7/24 7/31 8/7 8/14 8/21 8/28 9/4 9/11 9/18 9/25 10/2 10/9 10/16 10/23 10/30 11/6 11/13 11/20 11/27 12/4 12/11 12/18 12/25 1/1 1/8 1/15 1/22 1/29 2/5 2/12 2/19 2/26 3/5 3/12 3/19 3/26 4/2 4/9 4/16 4/23 4/30 5/7
1 R&D ENGINEERING PRODUCTION DRAWINGS & BILL OF MATERIALS - FOR MANUFACTURING ENG.
2 MANUFACTURING ENGINEERING PROCESSING
3 PRODUCTION CONTROL DEPT. PLACE PURCHASE REQUISITIONS ON PURCHASING - PARTS, CASTINGS, SUPPLIES
4 PURCHASING DEPT. PLACE PURCHASE ORDERS - PARTS, CASTINGS, ETC.
5 QUALITY CONTROL - PLACE PURCHASE REQUISITIONS FOR GAGES, FIXTURES, ETC.
6 PURCHASING DEPT. PROCURE ABOVE ITEMS
7 MANUFACTURING ENGINEERING DESIGN TOOLS - (TOTAL HOURS)
8 MANUFACTURING ENGINEERING PROCURE • TOOLS
9 PILOT RUN - PROVE TOOLS - MACHINE SHOP, 10 UNITS
10 PILOT RUN - PROVE TOOLS - ASSEMBLY DEPT. INCLUDING FINAL INSPECTION EQUIPMENT.
11 MACHINE SHOP - PRODUCTION SCHEDULE TO COMPLETION OF CONTRACT (FIRST 100 PRODUCTION UNITS).
12 ASSEMBLY DEPT. - PRODUCTION SCHEDULE - TO COMPLETION OF CONTRACT.
FOR RADARSCOPE B-52's
CODE:
SCHEDULED PRODUCTION
ACTUAL PRODUCTION
L.R.Z.
15
15
10
SHIPPED TO ENGINEERING DEPT.
CUSTOMERS
100 UNITS
10 10 10 10 10 10 10 10 10 15 15
REST OF SCHEDULE 50 UNITS/MONTH
PLACE ON PRODUCTION SCHEDULES
STAFF MEETINGS' REVISIONS
DRAWN 7-2 | REVISED 9-9 | REVISED 11-18
REVISED 7-21 | —"— 9-21 | —"— 12-1
—"— 8-2 | —"— 9-30 | —"— 1-5
—"— 8-17 | —"— 10-18 | —"— 1-15
—"— 8-24 | —"— 11-3 | —"— 2-26
—"— 8-27 | —"— 11-10 | —"— 3-19
CHART DISCONTINUED 3-20

EXHIBIT C-24 Gantt Chart Production Planning—10″ Indicator-Drive Unit

and that the engineering drawings be rushed through for completion and air-mailed to them for approval. In addition, they requested that these ten pro-totypes be shipped to them air-express on a specific future date. This action was to be followed by the weekly delivery of large quantities of production units starting at a future date. They then asked for confirmation of these very tight schedules.

When this information was made known to the production control department manager he responded negatively, stating that the customer was being very unrealistic and that we shouldn't accept such impossible schedules. He was heard out, but in reply he was told that while the emergency order was a rather small one in terms of sales dollars, the customer did order millions of dollars of business from us yearly on realistic schedules. If they wanted this special order on the dates specified, we would exert every effort to satisfy them.

At this point I quickly assembled my staff, which was composed of the purchasing agent, manufacturing engineer, project engineer, general foreman of the machine shop, general foreman of the assembly department, the supervisor of the quality control department, and an engineer from the research and development department. The problem was explained, then broken up into its many facets and each element examined minutely and sparately. Individual opinions and suggestions were sought, and when these smaller elements were analyzed separately and not as a large unit, the problem began to fit into its proper perspective. Estimated dates were advanced from each supervisor for this particular part of the job. I then made concessions on overtime and labor costs and permitted other relaxations, with the department heads following by shaving their original time estimates. Many good suggestions were made regarding production shortcuts and eventually everyone accepted the unrealistic schedules as a real challenge. The final estimates were indicated on this chart, and each week a meeting was held so that progress could be discussed and the chart brought up to date. At these meetings mutual problems were discussed and obstacles eliminated or minimized. Each supervisor intitiated a follow-up program with his subordinates, having explained the problem to them previously. As a result all the tight schedules were met, and at a later date a written commendation was received from the customer.

In constructing this chart different factors were necessarily employed for keeping score. In some, such as purchasing, production control and quality control, the number of items received during each week were plotted against the number scheduled, on an accumulative basis. In manufacturing, processing—manufacturing drawings, processes and routing, hours per item were used. With Items 11 and 12, product units were used as factors. Some of the work, of course, on some items ran concurrently. As soon as one drawing and process was completed, materials, tools, gages and related items were ordered.

MANUFACTURING PLANNING CHART—BACKLOG OF SALES ORDERS

Exhibit C–25 depicts the backlog of sales orders (and eventually manufacturing orders) plotted in $1,000's against an average gross backlog for the previous year. This provides the production manager with a projected view of his work load. In this particular instance, the product consisted of different sized units of equipment costing between $100,000 to $250,000 each. This indicates that while the future schedule looked rather dim, one additional order for October and several units for November would bring the plant closer to last year's production. Manpower planning can more intelligently be implemented with accurate sales forecasts. A picture such as this should also have alerted the sales manager to firm up pending customer orders. This was an unusual situation, and rarely did this company face such a meager customer order forecast.

INDIRECT LABOR HOUR PER CENT OF DIRECT LABOR

The chart shown in Exhibit C–26, after it is initially constructed, takes little effort to maintain. It provides the production manager with a guide to his indirect cost compared to direct labor. It, of course, would not be practicable to lay off several office workers because he had a temporary dip in the direct labor load. The product manufactured here was a complex one requiring considerable engineering skills. Consistency in the shop load was very importan t. Early in the fiscal year his shop load was considerably below the normal load of 9,000 hours (the previous year's average). A pickup in sales was anticipated, which did actually occur. The production manager hesitated to arbitrarily reduce his indirect labor in view of this more encouraging forecast. This example illustrates how a plant manager's efficiency is often dependent on the sales department's effectiveness. In periods of a tight labor market, personnel cannot be indiscriminately reduced and increased at will. However, during periods of increased sales and by keeping indirect labor costs steady, certain advantages do accrue in greater absorption of burden expense.

A brief scanning of the chart reveals in retrospect that the shop labor-indirect is considerably above the standard of 30 %. This could bear further analysis and no doubt would indicate that some reduction in personnel could be made here. Even when the shop load was normal or higher than normal, such as June through October, the indirect continued over standard.

PRODUCTION PERFORMANCE CHART—MACHINE SHOP

This chart (Exhibit C–27) provides an excellent review of the entire department's expenditure for manpower in all categories. At the top of the report

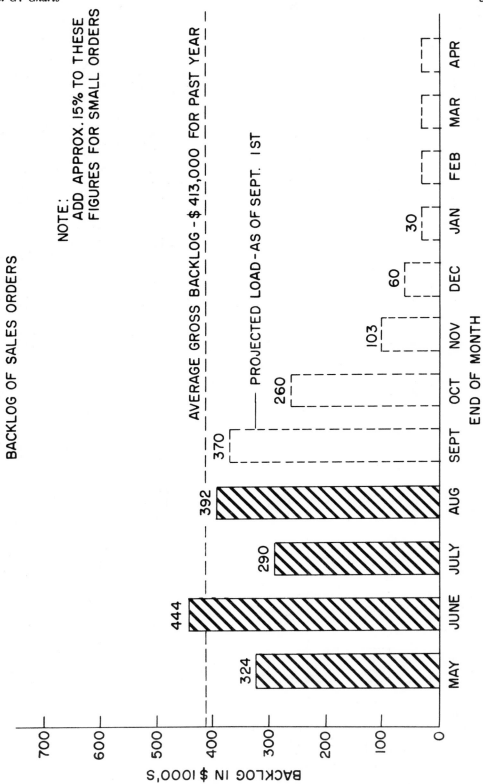

EXHIBIT C-25   Manufacturing Planning Chart—Backlog of Sales Orders

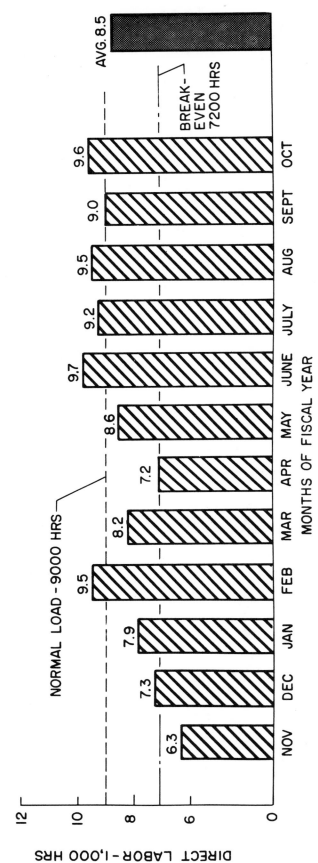

INDIRECT LABOR HOUR PERCENT OF DIRECT LABOR

												AT NORMAL LOAD – 9000 HRS	STANDARD
SHOP LABOR INDIRECT	66	58	46	34	48	57	40	42	35	30	35	30	30
OFFICE	48	40	47	30	35	49	34	31	41	40	41	42	40
SUPERVISION	29	25	30	22	23	29	21	21	28	26	21	23	20
ADMINISTRATIVE ENGINEERING	17	15	16	10	11	15	12	11	14	14	13	12	10
STAFF	11	10	11	7	7	9	6	6	9	8	7	8	7
TOTAL HOURS, INDIRECT-1,000'S	10.8	10.8	11.9	9.8	10.2	11.4	9.7	10.8	11.7	11.2	10.5	11.0	9.6

EXHIBIT C-26   Indirect Labor Hour Per Cent of Direct Labor

PRODUCTION PERFORMANCE

DISTRIBUTION:

PLANT MANAGER   DEPARTMENT: Machine Shop

DEPT. SUPT.

COST SUPT.   Week Ending: Feb. 21

	STANDARD	ACTUAL	VARIANCE	PERCENTAGE
PRODUCTION	1272.8	1315.3	(42.5)	97%
SET-UP	10.4	26.5	(16.1)	39%
SUB-TOTAL	1283.2	1341.8	(58.6)	96%
PAINTERS		225.0		
SHORT-RUN		21.9		
TOOLING		54.8		
A. TOTAL PRODUCTIVE		1643.5		
MISCELLANEOUS		26.3		
METHODS EXPERIMENTAL		4.5		
IDLE TIME				
RESET		69.3		
REWORK		9.0		
TOOL MAINTENANCE		43.5		
MACHINE MAINTENANCE		38.2		
TOTAL NON-PRODUCTIVE		190.8		
B. TOTAL PRODUCTIVE AVAILABLE		1834.3		
LEADMAN		96.8		
SET-UP		49.0		
TOOL CRIB ATTENDANT		81.0		
MISCELLANEOUS INDIRECT				
TOTAL INDIRECT		226.8		
C. ABSENCE, TARDINESS, VACATION		94.6		
D. GRAND TOTAL		2155.7		
PRODUCTIVE LABOR UTILIZED A ÷ B		90%		
ABSENCE, TARDINESS, VACATION C ÷ D		4.4%		
REMARKS:				

EXHIBIT C-27   Production Performance Chart—Machine Shop

we have actual production results (in actual time-card hours) against measured work (time study, MTM or Work Factor) standards. Then total productive work is segregated in Item A, total productive available labor in Item B, absence, tardiness and vacation hours in Item C, with a grand total figure represented in Item D. In addition, the item of productive labor utilized, which for this particular week was 90%, provides an important figure to check weekly as well as the per cent calculated for absence, tardiness and vacations. The productive labor utilized figure could be plotted on a graph each week and compared to a carefully prepared standard or budget. A periodic review of these results would reveal undesirable trends in operations dictating the need for the production manager's attention. Various modifications and changes can be made in this chart to fit your plant's specific requirements.

MACHINE TIME ANALYSIS—SUMMARY

This summary (Exhibit C–28) was compiled weekly from Exhibit C–29, which follows. It depicts hours expended for all machine operating categories. A review of these results by the production manager will reveal deficiences in need of attention. Certain items could be depicted on graphs to highlight various undesirable conditions, as shown in Exhibits C–30 and C–31. If too much manpower is required to keep such reports, a sampling could be made at random times, perhaps 15 times a year. If your company employs the use of computers it is a simple operation to program this operation. One week of operations does not present a conclusive picture; this should cover a longer period—or make use of the Work Sampling principle as was previously indicated.

XYZ INSTRUMENT CORPORATION—MACHINE TIME ANALYSIS

The figures in Exhibit C–29 were culled from time-keepers' daily operating reports and are used to make up the figures represented in Exhibit C–28. This reveals machines that are underloaded, that use excessive set-up hours, high tool and machine repair time and unexplained waiting hours. Only a sampling is illustrated here; the remainder of the machines are handled in the same manner and totals are shown in Exhibit C–28.

SET UP TIME AS A PERCENTAGE OF PRODUCTIVE TIME

The figures for this bar chart (Exhibit C–30) were obtained from Exhibit C–28, Machine Time Analysis—Summary. This reflects the set up time for each group of machines compared to the total productive time. Here again, a week's experience is not enough but is only used here to demonstrate this technique's

MACHINE TIME ANALYSIS SUMMARY	PRODUCTION HOURS	SET UP HOURS	RESET HOURS	TOOL REPAIR HOURS	MACHINE REPAIR HOURS	WAITING HOURS	TOTAL USED HOURS	AVAILABLE MACHINE HOURS	LOST MACHINE HOURS	TOTALS
AUTOMATIC SCREW MACHINES	101.8	64.6	3.6			1.4	171.4	320.0	148.6	811.4
TURRET LATHES	276.3	25.0	18.2	1.1	1.9	2.5	325.0	440.0	115.0	1205.0
PUNCH PRESSES	50.8	14.9	3.4				69.1	240.0	170.9	549.1
LATHES	207.7	6.3	1.3				215.3	440.0	224.7	1095.3
MILLING MACHINES	263.5	49.1	15.4			0.5	328.5	1200.0	871.5	2728.5
DRILL PRESSES	650.9	15.1	3.0	30.7	1.1	4.7	705.5	1400.0	694.5	3505.5
TOTALS	1551.0	175.0	44.9	31.8	3.0	9.1	1814.8	4040.0	2225.2	9894.8

EXHIBIT C-28   Machine Time Analysis—Summary

XYZ INSTRUMENT CORP.
MACHINE TIME ANALYSIS

MACH. NO.	TYPE MACHINE	PRODUCTION HOURS	SET UP HOURS	RESET HOURS	TOOL REPAIR HOURS	MACHINE REPAIR HOURS	WAITING HOURS	TOTAL USED HOURS	AVAILABLE MACHINE HOURS	LOST MACHINE HOURS
	AUTOMATIC SCREW MACHINES									
101	B & S 2–G	16.4	10.3					26.7		
102	B & S 2–G	23.0	15.0					38.0		
103	B & S 2–G	14.7						14.7		
104	B & S 0–G	21.3	1.5	3.6				26.4		
105	B & S 0–G	8.2	8.4					16.6		
106	B & S 00–G	8.8	12.8				1.4	23.0		
107	B & S 00–G	1.4	4.9					6.3		
108	B & S 00–G	8.0	11.7					19.7		
	TOTALS	101.8	64.6	3.6			1.4	171.4	320.0	148.6
	TURRET LATHES									
121	W.S. #3	37.9		2.1				40.0		
122	W.S. #3	37.8		0.9				38.7		
123	W.S. #3	37.7		0.5		0.5		38.7		
124	W.S. #2	6.7	5.3	0.6				12.6		
125	W.S. #5	31.7	3.5	5.9	0.7	1.4		43.2		
126	MIDLAND	41.7		2.2			1.1	45.0		
127	W.S. #3	25.4	2.3	5.8				33.5		
128	W.S. #3	28.6	3.4	0.2			1.4	33.6		
130	B & S #1									
131	HARDINGE									
132	HARDINGE	28.8	10.5		0.4			39.7		
	TOTALS	276.3	25.0	18.2	1.1	1.9	2.5	325.0	440.0	115.0
	PUNCH PRESSES									
301	30 TON V & O	4.6	2.8	1.1				8.5		
302	45 TON V & O	7.5	6.2	0.6				14.3		
303	20 TON BLISS	7.6	0.9					9.6		
304	30 TON BLISS	4.5	0.6					5.1		
305	100 TON MINSTER	6.4	4.4	0.6				11.4		

EXHIBIT C-29   XYZ Instrument Corporation—Machine Time Analysis

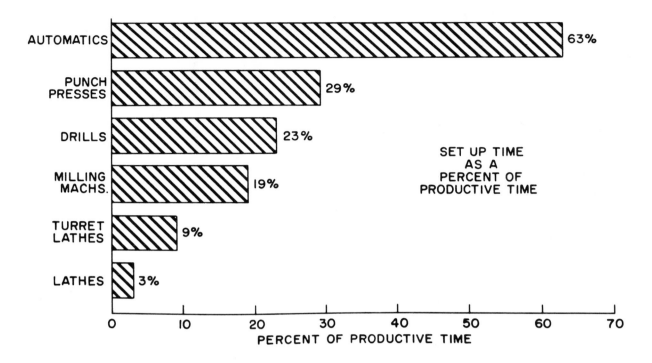

EXHIBIT C-30   Set Up Time as a Percentage of Productive Time

applicability. A review highlights the excessive time of 63% spent setting up the Automatic Screw Machines. This actual situation, upon further investigation, revealed the following: The research and development department, as well as the product development section, often visited this department and had a good-natured set up man tear down a production job to run a few hundred pieces for their use. In addition, the production runs were usually rather small quantities. This was due to the fact that the company was a jobbing type of business doing electronics and electro-mechanical assembly kinds of work with some long runs of repeat business. Many times certain screw machines were idle while the set up man was busy setting up another machine, also causing machine downtime. The two company departments that made the unprofitable demands on the set up man were instructed to give their small orders to the production control manager. He in turn would schedule these orders so they did not interfere with the efficiency of the automatics department.

USED VS. AVAILABLE MACHINE HOURS

The figures for this bar chart (Exhibit C–31) were obtained from Exhibit C–28, Machine Time Analysis—Summary, and reflect the machine time used by each group of machines compared to the available hours. This represents a week's experience, and is used here only to demonstrate this technique's utility.

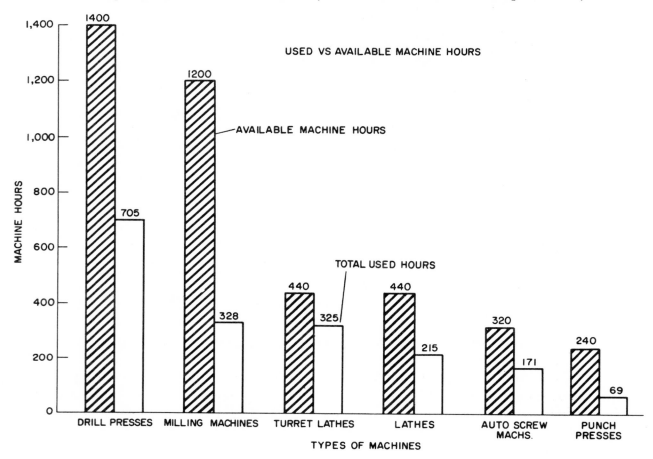

EXHIBIT C-31   Used vs. Available Machine Hours

A review of this chart discloses a very low use of machine time compared to that available, in all categories except the turret lathes. One explanation, given previously, is that these machines were employed in a jobbing type of business with many short production runs. In addition, due to the different kinds of work, a variety of types and machine capacities had to be available. Further investigations, covering a longer period of time, revealed the fact that some individual machines (on a yearly basis) were not paying their way in terms of factory floor space, capital investment, maintenance and servicing expense. A machine replacement program was initiated, some of the out-of-date machines sold, some moved to the R & D department. Others were removed from the factory floor and placed on a standby basis.

BACKLOG—MAINTENANCE WORK

One of the very costly activities of most plants is maintenance of buildings, machinery and equipment. Management control is very important—the manager must continually be made aware of the cost and inefficiencies of the maintenance department. A series of charts covering the major problem areas aids the manager in keeping informed about this department's weaknesses. Exhibits C–32, C–33, C–34, C–35 and C–36 were especially designed for this purpose. Each will be briefly discussed.

Every plant has a problem with a backlog of maintenance and repair orders. Efforts are continually expended to bring about their reduction. The chart in Exhibit C–32 depicts the number of man hours or weeks of maintenance and repair orders by months compared to the average for the previous year. It also serves a dual purpose in that it reveals the average days to start job, using a second scale on the right hand side of the chart. It also shows the past year's average figure. In this example substantial reductions are being made in both categories. The production foremen's full cooperation is vitally important. A well organized preventive maintenance program can be of great help here.

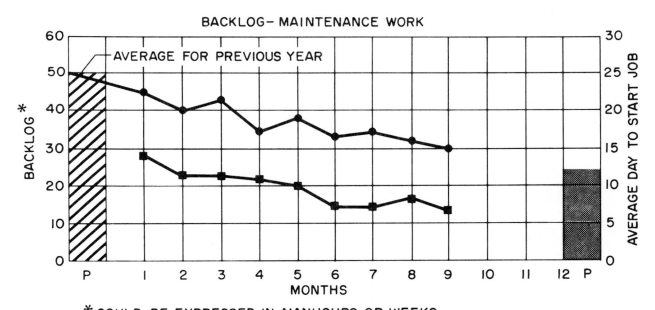

EXHIBIT C-32   Backlog—Maintenance Work[24]

---

[24] John J. Wilkinson, "How to Manage Maintenance," March-April issue, *Harvard Business Review*, Boston, Mass. Copyright © 1968, by the President and Fellows of Harvard College; all rights reserved.

MAINTENANCE—PRODUCTION PERFORMANCE OF PERSONNEL

The chart in Exhibit C–33 illustrates the production performance of maintenance personnel compared to a 100% standard. These standards should preferably be measured by time study, MTM or Work Factor systems. Both MTM and Work Factor systems provide their own basic standards based on thousands of engineered time studies. Estimated standards can be used but are not as accurate or as effective as predetermined time standards. Some companies who have had measured work installed have reported as much as a 30% improvement in personnel productivity, as well as the added advantages of providing for closer management control.

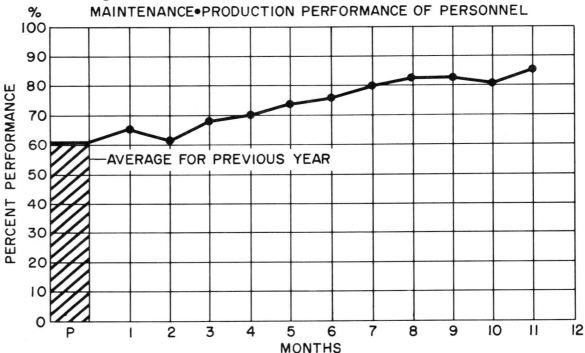

EXHIBIT C-33   Maintenance—Production Performance of Personnel[25]

DOWNTIME

The graph in Exhibit C–34 depicts the month-to-month downtime reported as a per cent of scheduled machine hours. For example, if a turret lathe is scheduled for 170 hours of work, including set up for one month, but is down for maintenance work or repairs for six hours, the per cent downtime would be

---

[25] John J. Wilkinson, "How to Manage Maintenance," March-April issue, *Harvard Business Review,* Boston, Mass. Copyright © 1968, by the President and Fellows of Harvard College; all rights reserved.

3.5% (6 ÷ 170). These percentages are compared against the past year's average. A preventive maintenance program is a great help in minimizing downtime during production runs.

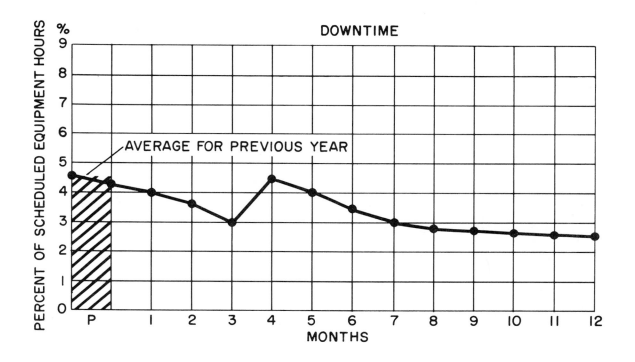

EXHIBIT C-34 Downtime[26]

MAINTENANCE—COST TRENDS

This bar chart (Exhibit C–35) is used to reflect the cost per standard hour for each month as compared to the past year's average. A standard is used where the work has been measured by one of the systems previously mentioned or by a similar method (time study, MTM or Work Factor). Generally these cost figures can be obtained from the accounting department. After the chart is drawn it takes little effort to keep it current; this responsibility can be assigned to a subordinate. The important factor is the need for the production manager

---

[26] John J. Wilkinson, "How to Manage Maintenance," March-April issue, *Harvard Business Review*, Boston, Mass. Copyright © 1968, by the President and Fellows of Harvard College; all rights reserved.

to diligently review his charts, quickly recognize the areas where his special attention is required, and *take the necessary corrective actions*. Where conditions start to deteriorate it may be necessary to run a day-to-day check. In a multi-plant company it was necessary to assign a head office staff engineer to one plant to approve or disapprove every maintenance order issued by several departments who were major offenders. After several weeks of this approach they soon fell in line and the close surveillance by the head office was discontinued.

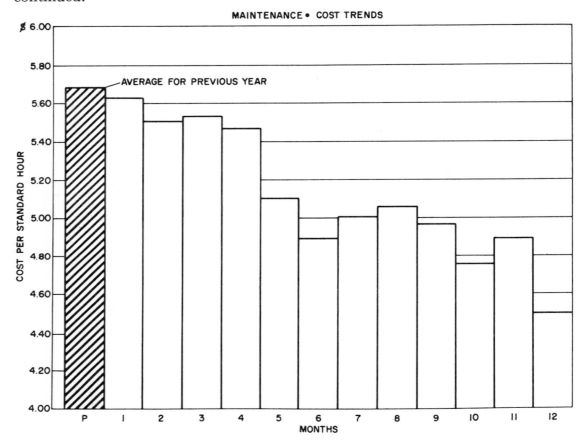

EXHIBIT C-35   Maintenance—Cost Trends[27]

MAINTENANCE—TOTAL DOLLARS EXPENDED

This bar chart in Exhibit C–36 portrays, in thousands of dollars, the total cost of maintenance by months, compared to the average for the past year.

[27] John J. Wilkinson, "How to Manage Maintenance," March-April issue, *Harvard Business Review*, Boston, Mass. Copyright © 1968, by the President and Fellows of Harvard College; all rights reserved.

The dip in expenditures for July was due to the plant closing several weeks for vacations. Charts like these can be placed on poster-size cardboard sheets and hung in conspicuous spots in the departments concerned. Most employees are interested in knowing how well they are doing, and there are few people who do not respond to goals or challenges. One interesting experience supporting this statement took place in a department where production standards were being installed. The installation was completed and the machine shop was placed on this system with each employee's daily and weekly performance being reported. The foreman of the department expected problems from a tough and troublesome lathe operator, who resented all kinds of regimentation and discipline. However, to the surprise of everyone he consistently beat every job standard he worked on and also took pride in his superior record. He was one of the informal leaders in the department and this system provided him with another opportunity to display his superiority.

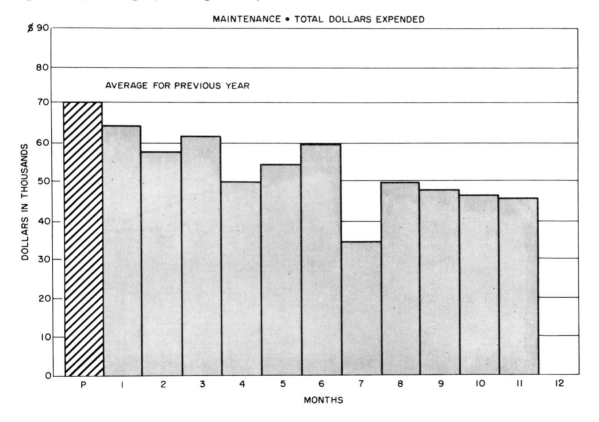

EXHIBIT C-36 Maintenance—Total Dollars Expended[28]

[28] John J. Wilkinson, "How to Manage Maintenance," March-April issue, *Harvard Business Review*, Boston, Mass. Copyright © 1968, by the President and Fellows of Harvard College; all rights reserved.

Section 2: Special Charts

(A) Units of Production—Sales Volume

See break-even chart (Exhibit C-37) with sales volume as the base.

(B) Considerations of Cost—Volume of Production and Sales

See break-even chart illustrating volume of production necessary before automated production is practical (Exhibit C–38).

(C) Surface Roughness

For surface roughness units of measurements, see Exhibit C–39.

(D) Decimal Equivalents of Common Fractions (Exhibit C–40)

(E) Industrial Engineering (Exhibits C-41, and C-42a, b, and c)

(F) Production and Inventory Control (Exhibits C-43a and b, and C-44a and b)

(G) A Tool for More Rational Planning (Exhibit C-45)

(A) Units of Production—Sales Volume

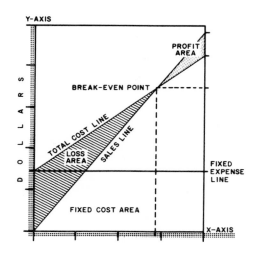

EXHIBIT C-37   Break-Even Chart with Sales Volume as the Base[1]

(B) Consideration of Cost—Volume of Production and Sales

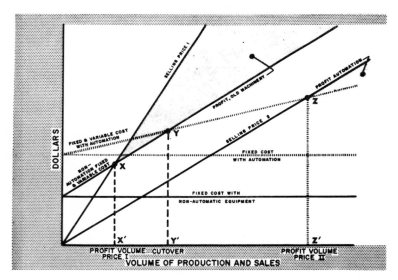

**EXHIBIT C-38   Break-Even Chart Illustrating Volume of Production Necessary
Before Automated Production Is Practical[2]**

---

[1] Used by permission of the Western Electric Company, New York, N. Y. "Planning for Manufacture,"
2nd edition. Copyright © 1965, p. 120.

[2] Used by permission of the Western Electric Company, New York, N. Y. "Planning for Manufacture,"
2nd edition. Copyright © 1965, p. 121.

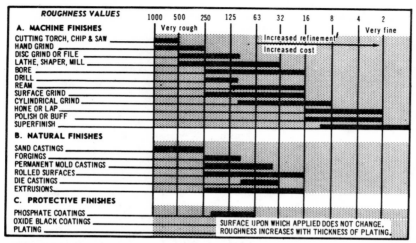

EXHIBIT C-39 Surface Roughness Units of Measurements[3]

(D) Decimal Equivalents of Common Fractions

1/64 ___.015625	25/64 ___.390625	49/64 ___.765625
1/32 _____.03125	13/32 _____.40625	25/32 _____.78125
3/64 ___.046875	27/64 ___.421875	51/64 ___.796875
1/16 _____.0625	7/16 _____.4375	13/16 _____.8125
5/64 ___.078125	29/64 ___.453125	53/64 ___.828125
3/32 _____.09375	15/32 _____.46875	27/32 _____.84375
7/64 ___.109375	31/64 ___.484375	55/64 ___.859375
1/8 _____.125	1/2 _____.500	7/8 _____.875
9/64 ___.140625	33/64 ___.515625	57/64 ___.890625
5/32 _____.15625	17/32 _____.53125	29/32 _____.90625
11/64 ___.171875	35/64 ___.546875	59/64 ___.921875
3/16 _____.1875	9/16 _____.5625	15/16 _____.9375
13/64 ___.203125	37/64 ___.578125	61/64 ___.953125
7/32 _____.21875	19/32 _____.59375	31/32 _____.96875
15/64 ___.234375	39/64 ___.609375	63/64 ___.984375
1/4 _____.250	5/8 _____.625	1 _____1.000
17/64 ___.265625	41/64 ___.640625	
9/32 _____.28125	21/32 _____.65625	
19/64 ___.296875	43/64 ___.671875	
5/16 _____.3125	11/16 _____.6875	
21/64 ___.328125	45/64 ___.703125	
11/32 _____.34375	23/32 _____.71875	
23/64 ___.359375	47/64 ___.734375	
3/8 _____.375	3/4 _____.750	

EXHIBIT C-40 Decimal Equivalents of Common Fractions

---

[3] Used by permission of the Western Electric Company, New York, N. Y. "Planning for Manufacture," 2nd edition. Copyright © 1965, p. 120.

(E) Industrial Engineering

1. What Quantity Signals Method Change [4]

You should choose other methods as quantity changes. The question is, "At what quantity?" This is answered by finding the Break-Even Point. (See Exhibit C–41.)

$$\text{Quantity} = \frac{\text{Setup } A - \text{Setup } B}{\text{Standard } B - \text{Standard } A}$$

OPERATION TIME PER PIECE (right side labeled METHOD A)

							METHOD B		
4.0	4.2	4.5	4.7	5.1	5.5	6.0			3.0
4.6	4.8	5.1	5.3	5.7	6.1	6.6			3.6
5.4	5.6	5.9	6.1	6.5	6.9	7.4	8.0	8.8	4.4
6.3	6.5	6.8	7.0	7.4	7.8	8.3	8.9	9.7	5.3
7.3	7.5	7.8	8.0	8.4	8.8	9.3	10.	10.	6.3
8.6	8.8	9.1	9.3	9.7	10.	11.	11.	12.	7.6
10.	10.	10.	11.	11.	12.	12.	13.	13.	9.1

SETUP TIME — METHOD A combined with setup values and quantity matrix

\	\	\	\	\		1.00	1.20	1.45	1.74	2.09	2.51	3.02	3.6	4.37
44	53	63	76	91										
34	43	53	66	81	10.0	10	8	7	6	5	4	3	3	2
32	41	51	64	79	12.0	12	10	8	7	6	5	4	3	3
29	38	48	61	76	14.5	15	12	10	8	7	6	5	4	3
27	36	46	59	74	17.4	17	15	12	10	8	7	6	5	4
23	32	42	55	70	20.9	21	17	15	12	10	8	7	6	5
19	28	38	51	66	25.1	25	21	17	15	12	10	8	7	6
8	23	33	46	61	30.2	30	25	21	17	15	12	10	8	7
	17	27	40	55	36.3	36	30	25	21	17	15	12	10	8
	9	19	32	47	43.7	44	36	30	25	21	17	15	12	10
METHOD B		10	23	38	52.5	53	44	36	30	25	21	17	15	12
			13	28	63.1	63	53	44	36	30	25	21	17	15

EXHIBIT C-41

[4] From *How to Chart Data* by Phil Carroll. Copyright © 1960 by Phil Carroll. Used with permission of McGraw-Hill Book Co., p. 240.

2. *Profit Changes With Sales Mix*[5]

Each of our products has a different cost—different material, different labor and, as a rule, different overhead. Probably, then, each has a different margin of profit per dollar of sales. Here are four products that have these differences (Exhibit C–42a):

SALES INCOME	VARIABLE COSTS AND MARGINS							
	SHIRTS		SHORTS		SHOES		SHEETS	
	V. COSTS	MARGIN	V. COSTS	MARGIN	V. COSTS	MARGIN	V. COSTS	MARGIN
$ 40,000	$32,000	$ 8,000	$28,000	$12,000	$ 44,000	$ –4,000	$ 36,000	$ 4,000
60,000	48,000	12,000	42,000	18,000	66,000	–6,000	54,000	6,000
80,000	64,000	16,000	65,000	24,000	88,000	–8,000	72,000	8,000
100,000	80,000	20,000	70,000	30,000	110,000	–10,000	90,000	10,000
120,000	96,000	24,000	84,000	36,000	132,000	–12,000	108,000	12,000
PLUS $30,000 CONSTANT COST								

EXHIBIT C-42a

Let's work out a chart (Exhibit C–42b) that will combine the variations in company profit or loss that are likely to result from different mixtures in customer demands of our four products.

[5] From *How to Chart Data* by Phil Carroll. Copyright © 1960 by Phil Carroll. Used with permission of McGraw-Hill Book Co., p. 188.

SHOES (right index)

	40	S
−4	40	S
−6	60	H
−8	80	O
−10	100	E
−12	120	S

SHIRTS / SHORTS (left block)

SHIRTS 40	60	80	100	120	A
40	60	80	100	120	
8	12	16	20	24	
40					
SHORTS					
	40				
60					
		40			
	60				
80		40			
	60				
		80		40	
100			60		
		80			
	100			60	
120			80		
		100			
	120			80	
			100		
		120			
				100	
			120		
22	18	−14	−10	−6	B

Shirts— $30,000

Main / SHEETS chart

A					SHEETS 40	60	80	100	120	
	40	60	80	100	120					
8	6	4	2	0	2	4	6	8		C
10	18	16	14	12	−10	−8	−6	−4	−2	
8	16	14	12	10	−8	−6	−4	−2	0	
6	14	12	10	8	−6	−4	−2	0	2	
4	12	10	8	6	−4	−2	0	2	4	
2	10	8	6	4	−2	0	2	4	6	
0	−8	−6	−4	−2	0	2	4	6	8	
2	−6	−4	−2	0	2	4	6	8	10	
4	−4	−2	0	2	4	6	8	10	12	
6	−2	0	2	4	6	8	10	12	14	
8	0	2	4	6	8	10	12	14	16	
10	2	4	6	8	10	12	14	16	18	
12	4	6	8	10	12	14	16	18	20	
14	6	8	10	12	14	16	18	20	22	
16	8	10	12	14	16	18	20	22	24	
18	10	12	14	16	18	20	22	24	26	
20	12	14	16	18	20	22	24	26	28	
22	14	16	18	20	22	24	26	28	30	
24	16	18	20	22	24	26	28	30	32	
26	18	20	22	24	26	28	30	32	34	
28	20	22	24	26	28	30	32	34	36	
30	22	24	26	28	30	32	34	36	38	

Details for building such charts may be found in Ch. 19, "How To Chart Data," McGraw-Hill Book Co.

EXHIBIT C-42b

Now we can check to see how profit or loss may change for the same total sales depending upon the several different quantities that make up our product mix (Exhibit C–42c).

Total Sales	Shirts	Shorts	Shoes	Sheets	Profit-Loss
$300,000	$ 40,000	$40,000	$120,000	$100,000	*$12,000*
300,000	40,000	60,000	100,000	100,000	
300,000	40,000	80,000	100,000	80,000	
300,000	60,000	100,000	100,000	40,000	
300,000	100,000	120,000	40,000	40,000	26,000

EXHIBIT C-42C

(F) Production and Inventory Control

1. Less Cycle Time Is Allowable[6]

To achieve a higher rate of turnover requires that you expand sales volume at a rate greater than you increase inventories, or its equivalent. You may have about the same sales with reduced inventory.

You must reduce inventories when you tie up more capital in better tools and plants or borrow or retain more dollars of working capital.

You can see how much your process cycle days must change to attain a desired turnover by completing the chart shown in Exhibit C–43a. This is based on 255 working days per year.

Per Cent Invested in SAFETY STOCK				CAPITAL %		DESIRED ANNUAL TURNOVER							
5	10	15	20	LOCK	WORK	3	4	5	6	7	8	9	10
15		Per Cent		20	80	68	51	41	34	29	25	23	20
20	15	"Fixed"		25	75	64	48	38	32	27	23	21	20
25	20	15		30	70	59	45	36	30	25	23	20	18
30	25	20	15	35	65	55	41	33	28	23	21	18	17
35	30	25	20	40	60	51	38	31	25	21	20	17	15
40	35	30	25	45	55	47	35	28	23	20	17	16	14
45	40	35	30	50	50	43	32	25	21	18	16	14	31
50	45	40	35	55	45	38	29	23	19	16	14	13	12
	50	45	40	60	40	34	25	20	17	PROCESS			10
		50	45	65	35	30	22	18	15	CYCLE DAYS			9
			50	70	30	25	19	15	13	11	9	8	

EXHIBIT C-43a

---

[6] From *Practical Production and Inventory Control* by Phil Carroll. Copyright © 1966 by Phil Carroll. Used with permission of McGraw-Hill Book Company, p. 232.

When more dollars are "locked in" to improve methods, you must reduce inventory or shrink process cycle time. Read from the chart (Exhibit C–43b) the reductions you must make.

DESIRED TURNOVER	FIXED		INVENTORY			CYCLE DAYS		
	WAS	GOES TO	WAS	MUST BE	% DROP	WERE	MUST BE	% DROP
5	20	30	20	————	————	31	————	————
6	30	40	15	————	————	23	————	————
7	40	45	10	————	————	18	————	————

EXHIBIT C-43b

2. *Work Load Changes with Product Mix*[7]

Chances are that each of our several products may take a different number of hours to produce in the shop. Hence, the work load will change with shifts in the product mix.

To "guesstimate" how much shop work there may be in a forecast, we can build a chart that combines the standard times we use in scheduling the productions of our four products (Exhibit C–44a).

From the chart, we can determine the total shop hours for any combinations of sales within the limits of the chart. Let's get answers to five product mixes (shown in Exhibit C–44b).

---

[7] From *Practical Production and Inventory Control* by Phil Carroll. Copyright © 1966 by Phil Carroll. Used with permission of McGraw-Hill Company, p. 134.

EXHIBIT C-44a

Product	SHOP HOURS $20,000 SALES
Washers	2,000
Pumps	1,000
Fans	4,000
Motors	2,000

MOTORS (staircase scale, top)

40	60	80	100	120										8	40	F
	40	60	80	100	120				MOTORS					12	60	A
		40	60	80	100	120								16	80	N
			40	60	80	100	120							20	100	S
				40	60	80	100	120						24	120	

WASHERS

40	60	80	100	120

PUMPS

Main grid:

40	60	80	100	120	A	12	14	16	18	20	22	24	26	28	30	32	34	36	C
4	6	8	10	12	A	12	14	16	18	20	22	24	26	28	30	32	34	36	C
40					6	18	20	22	24	26	28	30	32	34	36	38	40	42	
60	PUMPS				7	19	21	23	25	27	29	31	33	35	37	39	41	43	
80	40				8	20	22	24	26	28	30	32	34	36	38	40	42	44	
100	60				9	21	23	25	27	29	31	33	35	37	39	41	43	45	
120	80	40			10	22	24	26	28	30	32	34	36	38	40	42	44	46	
	100	60			11	23	25	27	29	31	33	35	37	39	41	43	45	47	
	120	80	40		12	24	26	28	30	32	34	36	38	40	42	44	46	48	
		100	60		13	25	27	29	31	33	35	37	39	41	43	45	47	49	
		120	80	40	14	26	28	30	32	34	36	38	40	42	44	46	48	50	
			100	60	15	27	29	31	33	35	37	39	41	43	45	47	49	51	
			120	80	16	28	30	32	34	36	38	40	42	SHOP			50	52	
				100	17	29	31	33	35	37	39	41	43	HOURS			51	53	
				120	18	30	32	34	36	38	40	42	44	46	48	50	52	54	

FANS scale (right):

8	40
12	60
16	80
20	100
24	120

Details for building charts may found in Ch. 19 "How to Chart Data," McGraw-Hill Book Co.

Total Sales	Washers	Pumps	Fans	Motors	Shop Hours
$300,000	$ 40,000	$ 40,000	$120,000	$100,000	40,000
300,000	40,000	60,000	100,000	100,000	_____
300,000	60,000	100,000	80,000	60,000	_____
300,000	80,000	120,000	60,000	40,000	_____
300,000	100,000	120,000	40,000	40,000	28,000

EXHIBIT C-44b

(G) A Tool for More Rational Planning

1. Return on Investment (ROI)[8]

This chart (Exhibit C–45) focuses attention on factors that ultimately determine the profitability of a venture, and displays their relationships. (In the Du Point system the complete breakdown of elements is much finer than shown in the chart.) It is clear from the chart that return on investment is highly dependent upon turnover of capital, or the amount of sales which a given dollar of investment is capable of producing. The chart also shows the significance of cost to earnings.

Given profit margin on sales, and investment turnover, the ROI can be derived. An example is worked out in Exhibit C–45.

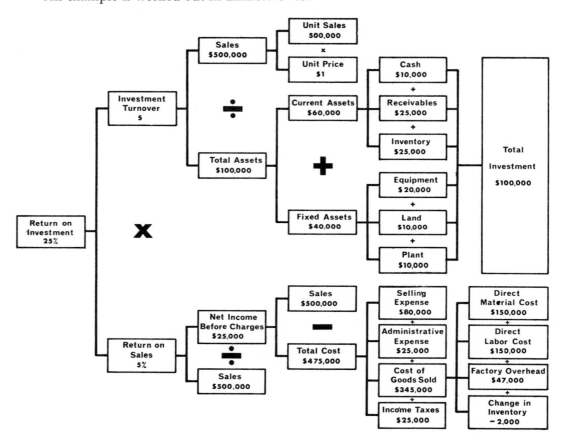

EXHIBIT C-45   Return on Investment (ROI)[8]

---

[8] Reprinted with permission of the Macmillan Company from *Top Management Planning* by George A. Steiner. Copyright © 1969 by the Trustees of Columbia University in the City of New York, N. Y., p. 373.

Appendix D: Tables

Feet per Minute

Diameter in inches	15	20	25	30	35	40	45	50	55	60	65	70	75	80	85
	Revolutions per minute														
1/16	917	1223	1528	1834	2140	2445	2751	3057	3363	3668	3974	4280	4586	4891	5197
1/8	459	611	764	917	1070	1222	1375	1528	1681	1834	1986	2139	2292	2445	2598
3/16	306	408	509	611	713	815	917	1019	1121	1222	1325	1426	1529	1630	1732
1/4	229	306	382	458	535	611	688	764	851	917	994	1070	1147	1222	1300
5/16	183	245	306	367	428	489	550	611	672	733	794	856	917	978	1039
3/8	153	204	255	306	357	408	458	509	560	611	662	713	764	815	865
7/16	131	175	218	262	306	349	393	437	481	524	568	611	656	699	743
1/2	115	153	191	229	268	306	344	382	420	459	497	535	573	611	649
9/16	102	136	170	204	238	272	306	340	373	407	441	475	509	543	577
5/8	92	123	153	184	214	245	276	306	337	367	398	428	459	489	520
11/16	83	111	138	167	194	222	249	273	300	333	360	389	416	444	472
3/4	76	102	127	153	178	203	229	254	279	306	330	357	381	408	432
13/16	71	95	119	142	166	190	213	237	261	284	308	332	356	379	403
7/8	66	87	109	131	153	175	196	219	241	262	285	306	329	349	372
15/16	61	81	101	122	142	163	183	204	224	244	265	285	305	326	346
1	57	76	96	115	134	153	172	191	210	229	258	267	287	306	325
1 1/16	54	72	90	108	126	144	162	180	197	215	233	251	269	287	305
1 1/8	51	68	85	102	119	136	153	170	187	204	221	238	255	272	289
1 3/16	48	64	81	97	113	129	145	161	177	193	209	225	242	258	274
1 1/4	46	61	76	92	107	123	137	153	168	183	199	214	230	245	260
1 5/16	44	58	73	87	102	116	131	146	160	175	189	204	218	233	247
1 3/8	42	56	70	83	97	111	125	139	153	167	180	195	208	222	236
1 7/16	40	53	66	80	93	106	119	133	146	159	172	186	199	212	225
1 1/2	38	51	64	76	89	102	115	127	140	153	165	178	191	204	216
1 9/16	37	49	61	73	85	98	110	122	134	146	159	171	183	195	207
1 5/8	35	47	59	71	82	94	106	117	129	141	152	165	176	188	199
1 11/16	34	45	57	68	79	90	102	113	124	136	147	158	170	181	192
1 3/4	33	44	55	66	76	87	98	109	120	131	142	153	164	175	185
1 13/16	32	42	53	63	74	84	95	106	116	127	137	148	158	169	179
1 7/8	31	41	51	61	71	82	92	102	112	122	133	143	153	163	173
1 15/16	30	39	49	59	69	79	89	99	108	118	128	138	148	158	167
2	29	38	48	57	67	76	86	96	105	115	124	134	143	153	162
2 1/8	27	36	45	54	63	72	81	90	99	108	117	126	135	144	153
2 1/4	25	34	42	51	59	68	76	86	94	102	111	119	128	136	145
2 3/8	24	32	40	48	56	64	73	81	89	97	105	113	121	129	137
2 1/2	23	31	38	46	54	61	69	76	84	92	100	107	114	122	130
2 5/8	22	29	36	44	51	58	65	73	80	87	94	102	109	116	123
2 3/4	21	28	35	42	49	56	63	70	77	83	90	97	104	111	118
2 7/8	20	26	33	40	46	53	59	66	73	79	86	92	99	106	112
3	19	26	32	38	45	51	57	64	70	76	83	89	95	102	108
3 1/8	18	24	31	37	43	49	55	61	67	73	79	85	92	98	104
3 1/4	18	23	29	35	41	47	53	59	64	70	76	82	88	94	100
3 3/8	17	23	28	34	40	45	51	57	62	68	74	79	85	90	96
3 1/2	16	22	27	33	38	44	49	55	60	66	71	76	82	87	93
3 5/8	16	21	26	32	37	42	47	53	58	63	68	74	79	84	89
3 3/4	15	20	26	31	36	41	46	51	56	61	66	71	77	82	87
3 7/8	15	20	25	30	35	39	44	49	54	59	64	69	74	79	84
4	14	19	24	29	33	38	43	48	53	57	62	67	72	76	81
4 1/4	14	18	23	27	31	36	40	45	49	54	58	63	67	72	76
4 1/2	13	17	21	25	30	34	38	42	47	51	55	59	64	68	72
4 3/4	12	16	20	24	28	32	36	40	44	48	52	56	60	64	68
5	12	15	19	23	27	31	34	38	42	46	50	54	57	61	65
5 1/4	11	15	18	22	25	29	33	36	40	44	47	51	55	58	62
5 1/2	10	14	17	21	24	28	31	35	38	42	45	49	52	56	59
5 3/4	10	13	17	20	23	27	30	33	37	40	43	47	50	53	56
6	10	13	16	19	22	26	29	32	35	38	41	45	48	51	54
6 1/4	9	12	15	18	21	24	28	31	34	37	40	43	46	49	52
6 1/2	9	12	15	18	21	24	26	29	32	35	38	41	44	47	50
6 3/4	9	11	14	17	20	23	26	28	31	34	37	40	43	45	48
7	8	11	14	16	19	22	25	27	30	33	36	38	41	44	46
7 1/4	8	11	13	16	18	21	24	26	29	32	34	37	40	42	45
7 1/2	8	10	13	15	18	20	23	25	28	31	33	36	38	41	43
7 3/4	7	10	12	15	17	20	22	25	27	30	32	34	37	39	43
8	7	10	12	14	17	19	21	24	26	29	31	33	36	38	41

Feet per Minute

90	95	100	110	120	125	130	140	150	160	170	175	180	190	200	Diameter in inches
Revolutions per minute															
5502	5808	6114	6725	7337	7643	7948	8560	9171	9782						1/16
2750	2903	3056	3362	3667	3820	3973	4278	4584	4890	5195	5348	5501	5806	6112	1/8
1834	1936	2038	2242	2446	2548	2649	2853	3057	3261	3465	3567	3668	3872	4076	3/16
1376	1453	1528	1681	1734	1910	1986	2139	2292	2445	2598	2674	2750	2903	3056	1/4
1100	1161	1222	1344	1466	1527	1589	1711	1833	1955	2077	2139	2200	2322	2444	5/16
916	967	1018	1121	1222	1273	1323	1425	1527	1629	1731	1782	1832	1934	2036	3/8
786	830	874	961	1049	1093	1136	1224	1311	1398	1486	1530	1573	1661	1748	7/16
688	726	764	840	917	955	993	1070	1146	1222	1229	1337	1375	1452	1528	1/2
611	645	679	747	813	869	883	951	1019	1086	1154	1188	1222	1290	1358	9/16
552	581	612	673	736	765	796	857	918	979	1040	1071	1102	1163	1224	5/8
500	527	555	611	666	692	722	770	833	888	944	971	999	1054	1110	11/16
458	483	508	559	610	635	661	711	762	813	864	889	914	965	1016	3/4
427	450	474	521	569	593	616	664	711	758	806	830	853	901	948	13/16
392	416	438	482	526	548	569	613	657	701	745	767	788	832	876	7/8
366	387	407	448	488	509	529	570	611	651	692	712	733	773	814	15/16
344	363	382	420	458	478	497	535	573	611	649	669	688	726	764	1
323	341	359	395	431	449	467	503	539	579	610	628	646	682	718	1 1/16
306	324	340	374	408	425	442	476	510	544	578	595	612	646	680	1 1/8
290	306	322	354	386	403	419	451	483	515	547	564	580	612	644	1 3/16
274	291	306	337	367	383	398	428	459	490	520	536	551	581	612	1 1/4
262	276	291	320	349	351	378	407	437	466	495	509	524	553	582	1 5/16
250	264	278	306	334	348	361	389	417	445	472	487	500	528	556	1 3/8
239	251	265	292	318	331	345	371	398	424	451	464	477	504	530	1 7/16
230	241	254	279	305	318	330	356	381	406	432	445	457	483	508	1 1/2
220	232	244	268	293	305	317	342	366	390	415	427	439	464	488	1 9/16
212	222	234	257	281	293	304	328	351	374	398	410	421	445	468	1 5/8
203	215	223	249	271	283	294	316	339	362	384	396	407	429	452	1 11/16
196	207	218	240	262	273	283	305	327	349	371	382	392	414	436	1 3/4
190	200	211	232	253	264	274	295	317	338	359	369	380	401	422	1 13/16
184	194	204	224	244	255	265	286	306	326	347	357	367	388	408	1 7/8
177	187	197	217	236	246	256	276	296	315	335	345	355	374	394	1 15/16
172	181	191	210	229	239	248	267	287	306	325	334	344	363	382	2
162	171	180	198	216	225	234	252	270	288	306	315	324	342	360	2 1/8
153	162	170	187	204	213	221	238	255	272	289	298	306	313	340	2 1/4
145	153	161	177	193	201	209	225	242	258	274	282	290	306	322	2 3/8
138	145	153	168	184	191	199	213	230	245	260	268	275	291	306	2 1/2
131	138	145	160	174	181	189	203	218	232	247	254	261	276	290	2 5/8
125	132	139	153	167	174	181	195	209	222	236	242	250	264	278	2 3/4
119	125	132	145	158	165	172	185	198	211	224	231	238	251	264	2 7/8
114	121	127	140	152	159	165	178	191	203	216	222	228	241	254	3
110	116	122	134	146	153	159	171	183	195	207	214	219	232	244	3 1/8
105	111	117	129	140	146	152	164	176	188	199	205	211	222	234	3 1/4
102	108	113	124	136	141	147	158	170	181	192	198	203	215	226	3 3/8
98	104	109	120	131	136	142	153	164	174	186	191	196	207	218	3 1/2
95	100	105	116	126	131	137	147	158	168	179	184	189	200	210	3 5/8
92	97	102	112	122	128	133	143	154	163	175	179	184	194	205	3 3/4
87	94	99	108	118	123	128	138	148	158	167	172	177	186	197	3 7/8
86	91	96	105	115	120	124	134	143	153	163	167	172	182	191	4
81	85	90	99	108	112	117	126	135	144	153	157	162	171	180	4 1/4
76	81	85	93	102	106	110	119	127	136	144	148	153	161	170	4 1/2
72	76	80	88	97	101	105	113	121	129	137	141	145	153	161	4 3/4
69	73	76	84	92	96	99	107	115	122	130	134	138	145	153	5
65	69	73	80	87	91	95	102	109	116	124	127	131	138	145	5 1/4
63	66	69	76	83	87	90	97	104	111	118	121	125	132	139	5 1/2
60	63	66	73	80	83	86	93	100	106	113	116	120	126	133	5 3/4
57	61	64	70	76	80	83	89	95	102	108	111	114	121	127	6
55	58	61	67	73	76	79	86	92	98	104	107	110	116	122	6 1/4
53	56	59	65	70	73	76	82	88	94	100	103	106	112	117	6 1/2
51	54	57	62	68	71	74	79	85	91	96	99	102	108	113	6 3/4
49	52	55	60	66	68	71	76	82	87	93	96	98	104	109	7
47	50	53	58	63	66	69	74	79	84	90	92	95	100	105	7 1/4
46	48	51	56	61	64	66	71	76	81	87	89	92	97	102	7 1/2
44	47	49	54	59	62	64	69	74	79	84	86	89	93	98	7 3/4
43	45	48	53	57	60	62	67	72	77	81	84	86	91	96	8

EXHIBIT D-1 Cutting Speed Conversion Table: Surface Feet to R.P.M. (Feet per Minute)[1]

[1] From *Timestudy for Cost Control* by Phil Carroll, 3rd edition. Copyright © 1954 by McGraw-Hill, Inc. Used with permission of McGraw-Hill Book Company, New York, N. Y., pp. 280–281.

Substance	Specific gravity	Avg. weight, pounds per cu. ft.
Metals, Alloys, Ores		
Aluminum, cast-hammered	2.55-2.80	165
Brass, cast-rolled	7.7	481
Bronze, aluminum	8.4-8.7	534
Bronze, 7.9 to 14% Sn	7.4-8.9	509
Bronze, phosphor	8.88	554
Copper, cast-rolled	8.8-8.95	556
Copper ore, pyrites	4.1-4.3	262
German silver	8.58	536
Gold, cast-hammered	19.25-19.35	1205
Gold coin (U. S.)	17.18-17.2	1073
Iridium	21.78-22.42	1383
Iron, gray cast	7.03-7.13	442
Iron, cast, pig	7.2	450
Iron, wrought	7.6-7.9	485
Iron, spiegel-eisen	7.5	468
Iron, ferro-silicon	6.7-7.3	437
Iron ore, hematite	5.2	325
Iron ore, limonite	3.6-4.0	237
Iron ore, magnetite	4.9-5.2	315
Iron slag	2.5-3.0	172
Lead	11.34	710
Lead ore, galena	7.3-7.6	465
Manganese	7.42	475
Manganese ore, pyro-lusite	3.7-4.6	259
Mercury	13.6	849
Monel metal, rolled	8.97	555
Nickel	8.9	537
Platinum, cast-hammered	21.5	1330
Silver, cast-hammered	10.4-10.6	656
Steel, cold-drawn	7.83	489
Steel, machine	7.80	487
Steel, tool	7.70-7.73	481
Tin, cast-hammered	7.2-7.5	459
Tin ore, cassiterite	6.4-7.0	418
Tungsten	19.22	1200
Zinc, cast-rolled	6.9-7.2	440
Zinc ore, blende	3.9-4.2	253
Various Solids		
Cereals, oats, bulk	0.51	26
Cereals, barley, bulk	0.62	39
Cereals, corn, rye, bulk	0.73	45
Cereals, wheat, bulk	0.77	48
Cork	0.22-0.26	15
Cotton, flax, hemp	1.47-1.50	93
Fats	0.90-0.97	58
Flour, loose	0.40-0.50	28
Flour, pressed	0.70-0.80	47
Glass, common	2.40-2.80	162
Glass, plate or crown	2.45-2.72	161
Glass, crystal	2.90-3.00	184
Glass, flint	3.2-4.7	247
Hay and straw, bales	0.32	20
Leather	0.86-1.02	59
Paper	0.70-1.15	58
Potatoes, piled	0.67	44
Rubber, caoutchouc	0.92-0.96	59
Rubber goods	1.0-2.0	94
Salt, granulated, piled	0.77	48

Substance	Specific gravity	Avg. weight, pounds per cu. ft.
Saltpeter	1.07	67
Starch	1.53	96
Sulphur	1.93-2.07	125
Wool	1.32	82
Timber, air-dry		
Apple	0.66-0.74	44
Ash, black	0.55	34
Ash, white	0.64-0.71	42
Birch, sweet yellow	0.71-0.72	44
Cedar, white, red	0.35	22
Cherry, wild red	0.43	27
Chestnut	0.48	30
Cypress	0.45-0.48	29
Fir, Douglas	0.48-0.55	32
Fir, balsam	0.40	25
Elm, white	0.56	35
Hemlock	0.45-0.50	29
Hickory	0.74-0.80	48
Locust	0.67-0.77	45
Mahogany	0.56-0.85	44
Maple, sugar	0.68	43
Maple, white	0.53	33
Oak, chestnut	0.74	46
Oak, live	0.87	54
Oak, red, black	0.64-0.71	45
Oak, white	0.77	48
Pine, Oregon	0.51	32
Pine, red	0.48	30
Pine, white	0.43	27
Pine, Southern	0.61-0.67	38-42
Pine, Norway	0.55	34
Poplar	0.43	27
Redwood, California	0.42	26
Spruce, white, red	0.45	28
Teak, African	0.99	62
Teak, Indian	0.66-0.88	48
Walnut, black	0.59	37
Willow	0.42-0.50	28
Various Liquids		
Alcohol, ethyl (100%)	0.789	49
Alcohol, methyl(100%)	0.796	50
Acid, muriatic, 40%	1.20	75
Acid, nitric, 91%	1.50	94
Acid, sulphuric, 87%	1.80	112
Chloroform	1.500	95
Ether	0.736	46
Lye, soda, 66%	1.70	106
Oils, vegetable	0.91-0.94	58
Oils, mineral, lubricants	0.88-0.94	57
Turpentine	0.861-0.867	54
Water, 4°C., max. density	1.0	62.428
Water, 100°C.	0.9584	59.830
Water, ice	0.88-0.92	56
Water, snow, fresh fallen	0.125	8
Water, sea water	1.02-1.03	64
Gases, see pp. 319 and 383.		
Ashlar Masonry		
Granite, syenite, gneiss	2.4-2.7	159
Limestone	2.1-2.8	153
Marble	2.4-2.8	162

Substance	Specific gravity	Avg. weight, pounds per cu. ft.
Sandstone	2.0-2.6	143
Bluestone	2.3-2.6	153
Rubble Masonry		
Granite, syenite, gneiss	2.3-2.6	153
Limestone	2.0-2.7	147
Sandstone	1.9-2.5	137
Bluestone	2.3-2.5	147
Marble	2.3-2.7	156
Dry Rubble Masonry		
Granite, syenite, gneiss	1.9-2.3	130
Limestone, marble	1.9-2.1	125
Sandstone, bluestone	1.8-1.9	110
Brick Masonry		
Hard brick	1.8-2.3	128
Medium brick	1.6-2.0	112
Soft brick	1.4-1.9	103
Sand-lime brick	1.4-2.2	112
Concrete Masonry		
Cement, stone, sand	2.2-2.4	144
Cement, slag, etc.	1.9-2.3	130
Cement, cinder, etc.	1.5-1.7	100
Building Materials		
Ashes, cinders	0.64-0.72	40-45
Cement, Portland, loose	1.5	94
Portland cement	3.1-3.2	196
Lime, gypsum, loose	0.85-1.00	53-64
Mortar, lime, set	1.4-1.9	103
Mortar, Portland cement	2.08-2.25	135
Slags, bank slag	1.1-1.2	67-72
Slags, bank screenings	1.5-1.9	98-117
Slags, machine slag	1.5	96
Slags, slag sand	0.8-0.9	49-55
Earth, etc., Excavated		
Clay, dry	1.0	63
Clay, damp plastic	1.76	110
Clay and gravel, dry	1.6	100
Earth, dry, loose	1.2	76
Earth, dry, packed	1.5	95
Earth, moist, loose	1.3	78
Earth, moist, packed	1.6	96
Earth, mud, flowing	1.7	108
Earth, mud, packed	1.8	115
Riprap, limestone	1.3-1.4	80-85
Riprap, sandstone	1.4	90
Riprap, shale	1.7	105
Sand, gravel, dry, loose	1.4-1.7	90-105
Sand, gravel, dry, packed	1.6-1.9	100-120
Sand, gravel, wet	1.89-2.16	126
Excavations in Water		
Sand or gravel	0.96	60
Sand or gravel and clay	1.00	65
Clay	1.28	80
River mud	1.44	90
Soil	1.12	70
Stone riprap	1.00	65
Minerals		
Asbestos	2.1-2.8	153
Barytes	4.50	281

Substance	Specific gravity	Avg. weight, pounds per cu. ft.
Basalt	2.7-3.2	184
Bauxite	2.55	159
Bluestone	2.5-2.6	159
Borax	1.7-1.8	109
Chalk	1.8-2.8	143
Clay, marl	1.8-2.6	137
Dolomite	2.9	181
Feldspar, orthoclase	2.5-2.7	162
Gneiss	2.6-2.9	175
Granite	2.6-2.7	165
Greenstone, trap	2.8-3.2	187
Gypsum, alabaster	2.3-2.8	159
Hornblende	3.0	187
Limestone	2.1-2.86	155
Marble	2.6-2.86	170
Magnesite	3.0	187
Phosphate rock, apatite	3.2	200
Porphyry	2.6-2.9	172
Pumice, natural	0.37-0.90	40
Quartz, flint	2.5-2.8	165
Sandstone	2.0-2.6	143
Serpentine	2.7-2.8	171
Shale, slate	2.6-2.9	172
Soapstone, talc	2.6-2.8	169
Syenite	2.6-2.7	165
Stone, Quarried, Piled		
Basalt, granite, gneiss	1.5	96
Limestone, marble, quartz	1.5	95
Sandstone	1.3	82
Shale	1.5	92
Greenstone, hornblende	1.7	107
Bituminous Substances		
Asphaltum	1.1-1.5	81
Coal, anthracite	1.4-1.8	97
Coal, bituminous	1.2-1.5	84
Coal, lignite	1.1-1.4	78
Coal, peat, turf, dry	0.65-0.85	47
Coal, charcoal, pine	0.28-0.44	23
Coal, charcoal, oak	0.47-0.57	33
Coal, coke	1.0-1.4	75
Graphite	1.64-2.7	135
Paraffin	0.87-0.91	56
Petroleum	0.87	54
Petroleum, refined (kerosene)	0.78-0.82	50
Petroleum, benzine	0.73-0.75	46
Petroleum, gasoline	0.70-0.75	45
Pitch	1.07-1.15	69
Tar, bituminous	1.20	75
Coal and Coke, Piled		
Coal, anthracite	0.75-0.93	47-58
Coal, bituminous, lignite	0.64-0.87	40-54
Coal, peat, turf	0.32-0.42	20-26
Coal, charcoal	0.16-0.23	10-14
Coal, coke	0.37-0.51	23-32

(Water at 4 deg. cent. and normal atmospheric pressure taken as unity.)

EXHIBIT D-2   Approximate Specific Gravities and Densities[2]

[2] From *Timestudy for Cost Control* by Phil Carroll, 3rd edition. Copyright © 1954 by McGraw-Hill, Inc. Used with permission of McGraw-Hill Book Company, New York, N. Y., pp. 284–285.

1–50

Number	Area circle	Circumference circle	Reciprocal	Logarithm	Square	Square root
1	.79	3.14	1.00000	.000	1	1.00
2	3.14	6.28	.50000	.301	4	1.41
3	7.07	9.43	.33333	.477	9	1.73
4	12.57	12.57	.25000	.602	16	2.00
5	19.64	15.71	.20000	.699	25	2.24
6	28.27	18.85	.16667	.778	36	2.45
7	38.48	21.99	.14286	.845	49	2.65
8	50.27	25.13	.12500	.903	64	2.83
9	63.62	28.27	.11111	.954	81	3.00
10	78.54	31.42	.10000	1.000	100	3.16
11	95.03	34.56	.09091	1.041	121	3.32
12	113.10	37.70	.08333	1.079	144	3.46
13	132.73	40.84	.07692	1.114	169	3.61
14	153.94	43.98	.07143	1.146	196	3.74
15	176.72	47.12	.06667	1.176	225	3.87
16	201.06	50.27	.06250	1.204	256	4.00
17	226.98	53.41	.05882	1.230	289	4.12
18	254.47	56.55	.05556	1.255	324	4.24
19	283.53	59.69	.05263	1.279	361	4.36
20	314.16	62.83	.05000	1.301	400	4.47
21	346.36	65.97	.04762	1.322	441	4.58
22	380.13	69.12	.04545	1.342	484	4.69
23	415.48	72.26	.04348	1.362	529	4.80
24	452.39	75.40	.04167	1.380	576	4.90
25	490.87	78.54	.04000	1.398	625	5.00
26	530.93	81.68	.03846	1.415	676	5.10
27	572.56	84.82	.03704	1.431	729	5.20
28	615.75	87.97	.03571	1.447	784	5.29
29	660.52	91.11	.03448	1.462	841	5.39
30	706.86	94.25	.03333	1.477	900	5.48
31	754.77	97.39	.03226	1.491	961	5.57
32	804.25	100.53	.03125	1.505	1024	5.66
33	855.30	103.67	.03030	1.519	1089	5.74
34	907.92	106.81	.02941	1.531	1156	5.83
35	962.11	109.96	.02857	1.544	1225	5.92
36	1017.9	113.10	.02778	1.556	1296	6.00
37	1075.2	116.24	.02703	1.568	1369	6.08
38	1134.1	119.38	.02632	1.580	1444	6.16
39	1194.6	122.52	.02564	1.591	1521	6.25
40	1256.6	125.66	.02500	1.602	1600	6.32
41	1320.3	128.81	.02439	1.613	1681	6.40
42	1385.4	131.95	.02381	1.623	1764	6.48
43	1452.2	135.09	.02326	1.633	1849	6.56
44	1520.5	138.23	.02273	1.643	1936	6.63
45	1590.4	141.37	.02222	1.653	2025	6.71
46	1661.9	144.51	.02174	1.663	2116	6.78
47	1734.9	147.65	.02128	1.672	2209	6.86
48	1809.6	150.80	.02083	1.681	2304	6.93
49	1885.7	153.94	.02041	1.690	2401	7.00
50	1963.5	157.08	.02000	1.699	2500	7.07

50–100

Square root	Square	Logarithm	Reciprocal	Circumference circle	Area circle	Number
7.07	2500	1.699	.02000	157.1	1963.5	50
7.14	2601	1.708	.01961	160.2	2042.8	51
7.21	2704	1.716	.01923	163.4	2123.7	52
7.28	2809	1.724	.01887	166.5	2206.2	53
7.35	2916	1.732	.01852	169.7	2290.2	54
7.42	3025	1.740	.01818	172.8	2375.8	55
7.48	3136	1.748	.01786	175.9	2463.0	56
7.55	3249	1.756	.01754	179.1	2551.8	57
7.62	3364	1.763	.01724	182.2	2642.1	58
7.68	3481	1.771	.01695	185.4	2734.0	59
7.75	3600	1.778	.01667	188.5	2827.4	60
7.81	3721	1.785	.01639	191.6	2922.5	61
7.87	3844	1.792	.01613	194.8	3019.1	62
7.94	3969	1.799	.01587	197.9	3117.3	63
8.00	4096	1.806	.01563	201.1	3217.0	64
8.06	4225	1.813	.01538	204.2	3318.3	65
8.12	4356	1.820	.01515	207.4	3421.2	66
8.19	4489	1.826	.01493	210.5	3525.7	67
8.25	4624	1.833	.01471	213.6	3631.7	68
8.31	4761	1.839	.01449	216.8	3739.3	69
8.37	4900	1.845	.01429	219.9	3848.5	70
8.43	5041	1.851	.01408	223.1	3959.2	71
8.49	5184	1.857	.01389	226.2	4071.5	72
8.54	5329	1.863	.01370	229.3	4185.4	73
8.60	5476	1.869	.01351	232.5	4300.8	74
8.66	5625	1.875	.01333	235.6	4417.9	75
8.72	5776	1.881	.01316	238.8	4536.5	76
8.78	5929	1.886	.01299	241.9	4656.6	77
8.83	6084	1.892	.01282	245.0	4778.4	78
8.89	6241	1.898	.01266	248.2	4901.7	79
8.94	6400	1.903	.01250	251.3	5026.6	80
9.00	6561	1.908	.01235	254.5	5153.0	81
9.06	6724	1.914	.01220	257.6	5281.0	82
9.11	6889	1.919	.01205	260.8	5410.6	83
9.17	7056	1.924	.01190	263.9	5541.8	84
9.22	7225	1.929	.01176	267.0	5674.5	85
9.27	7396	1.935	.01163	270.2	5808.8	86
9.33	7569	1.940	.01149	273.3	5944.7	87
9.38	7744	1.944	.01136	276.5	6082.1	88
9.43	7921	1.949	.01124	279.6	6221.1	89
9.49	8100	1.954	.01111	282.7	6361.7	90
9.54	8281	1.959	.01099	285.9	6503.9	91
9.59	8464	1.964	.01087	289.0	6647.6	92
9.64	8649	1.968	.01075	292.2	6792.9	93
9.70	8836	1.973	.01064	295.3	6939.8	94
9.75	9025	1.978	.01053	298.5	7088.2	95
9.80	9216	1.982	.01042	301.6	7238.2	96
9.85	9409	1.987	.01031	304.7	7389.8	97
9.90	9604	1.991	.01020	307.9	7543.0	98
9.95	9801	1.996	.01010	311.0	7697.7	99
10.00	10000	2.000	.01000	314.2	7854.0	100

EXHIBIT D-3   Functions of Numbers 1–999[3]

[3] From *Timestudy for Cost Control* by Phil Carroll, 3rd edition. Copyright © 1954 by McGraw-Hill, Inc. Used with permission of McGraw-Hill Book Company, New York, N. Y., pp. 288–293.

Number	Area circle	Circumference circle	10 × Reciprocal	Logarithm	Square	Square root
200	31416	628	.05000	2.301	40000	14.14
202	32047	635	.04951	2.305	40804	14.21
204	32685	641	.04902	2.310	41616	14.28
206	33329	647	.04854	2.314	42436	14.35
208	33980	653	.04808	2.318	43264	14.42
210	34636	660	.04762	2.322	44100	14.49
212	35299	666	.04717	2.326	44944	14.56
214	35968	672	.04673	2.330	45796	14.63
216	36644	679	.04630	2.334	46656	14.70
218	37325	685	.04587	2.338	47524	14.76
220	38013	691	.04545	2.342	48400	14.83
222	38708	697	.04505	2.346	49284	14.90
224	39408	704	.04464	2.350	50176	14.97
226	40115	710	.04425	2.354	51076	15.03
228	40828	716	.04386	2.358	51984	15.10
230	41548	723	.04348	2.362	52900	15.17
232	42273	729	.04310	2.365	53824	15.23
234	43005	735	.04274	2.369	54756	15.30
236	43744	741	.04237	2.373	55696	15.36
238	44488	748	.04202	2.377	56644	15.43
240	45239	754	.04167	2.380	57600	15.49
242	45996	760	.04132	2.384	58564	15.56
244	46760	767	.04098	2.387	59536	15.62
246	47529	773	.04065	2.391	60516	15.68
248	48305	779	.04032	2.394	61504	15.75
250	49087	785	.04000	2.398	62500	15.81
252	49876	792	.03968	2.401	63504	15.87
254	50671	798	.03937	2.405	64516	15.94
256	51472	804	.03906	2.408	65536	16.00
258	52279	811	.03876	2.412	66564	16.06
260	53093	817	.03846	2.415	67600	16.12
262	53913	823	.03817	2.418	68644	16.19
264	54739	829	.03788	2.422	69696	16.25
266	55572	836	.03759	2.425	70756	16.31
268	56410	842	.03731	2.428	71824	16.37
270	57256	848	.03704	2.431	72900	16.43
272	58107	855	.03676	2.435	73984	16.49
274	58965	861	.03650	2.438	75076	16.55
276	59829	867	.03623	2.441	76176	16.61
278	60699	873	.03597	2.444	77284	16.67
280	61575	880	.03571	2.447	78400	16.73
282	62458	886	.03546	2.450	79524	16.79
284	63347	892	.03521	2.453	80656	16.85
286	64242	899	.03497	2.456	81796	16.91
288	65144	905	.03472	2.459	82944	16.97
290	66052	911	.03448	2.462	84100	17.03
292	66966	917	.03425	2.465	85264	17.09
294	67887	924	.03401	2.468	86436	17.15
296	68813	930	.03378	2.471	87616	17.20
298	69747	936	.03356	2.474	88804	17.26
300	70686	942	.03333	2.477	90000	17.32
Number	Area circle	Circumference circle	10 × Reciprocal	Logarithm	Square	Square root

Number	Area circle	Circumference circle	10 × Reciprocal	Logarithm	Square	Square root
100	7854	314.2	.10000	2.000	10000	10.00
102	8171	320.4	.09804	2.009	10404	10.10
104	8495	326.7	.09615	2.017	10816	10.20
106	8825	333.0	.09434	2.025	11236	10.30
108	9161	339.3	.09259	2.033	11664	10.39
110	9503	345.6	.09091	2.041	12100	10.49
112	9852	351.9	.08929	2.049	12544	10.58
114	10207	358.1	.08772	2.057	12996	10.68
116	10568	364.4	.08621	2.064	13456	10.77
118	10936	370.7	.08475	2.072	13924	10.86
120	11310	377.0	.08333	2.079	14400	10.95
122	11690	383.3	.08197	2.086	14884	11.05
124	12076	389.6	.08065	2.093	15376	11.14
126	12469	395.8	.07937	2.100	15876	11.23
128	12868	402.1	.07813	2.107	16384	11.31
130	13273	408.4	.07692	2.114	16900	11.40
132	13685	414.7	.07576	2.121	17424	11.49
134	14103	421.0	.07463	2.127	17956	11.58
136	14527	427.3	.07353	2.134	18496	11.66
138	14957	433.5	.07246	2.140	19044	11.75
140	15394	439.8	.07143	2.146	19600	11.83
142	15837	446.1	.07042	2.152	20164	11.92
144	16286	452.4	.06944	2.158	20736	12.00
146	16742	458.7	.06849	2.164	21316	12.08
148	17203	465.0	.06757	2.170	21904	12.17
150	17672	471.2	.06667	2.176	22500	12.25
152	18146	477.5	.06579	2.182	23104	12.33
154	18627	483.8	.06494	2.188	23716	12.41
156	19113	490.1	.06410	2.193	24336	12.49
158	19607	496.4	.06329	2.199	24964	12.57
160	20106	502.7	.06250	2.204	25600	12.65
162	20612	508.9	.06173	2.210	26244	12.73
164	21124	515.2	.06098	2.215	26896	12.81
166	21642	521.5	.06024	2.220	27556	12.88
168	22167	527.8	.05952	2.225	28224	12.96
170	22698	534.1	.05882	2.230	28900	13.04
172	23235	540.4	.05814	2.236	29584	13.11
174	23779	546.6	.05747	2.241	30276	13.19
176	24328	552.9	.05682	2.246	30976	13.27
178	24885	559.2	.05618	2.250	31684	13.34
180	25447	565.5	.05556	2.255	32400	13.42
182	26016	571.8	.05495	2.260	33124	13.49
184	26590	578.1	.05435	2.265	33856	13.56
186	27172	584.3	.05376	2.270	34596	13.64
188	27759	590.6	.05319	2.274	35344	13.71
190	28353	596.9	.05263	2.279	36100	13.78
192	28953	603.2	.05208	2.283	36864	13.86
194	29559	609.5	.05155	2.288	37636	13.93
196	30172	615.8	.05102	2.292	38416	14.00
198	30791	622.0	.05051	2.297	39204	14.07
200	31416	628.3	.05000	2.301	40000	14.14
Number	Area circle	Circumference circle	10 × Reciprocal	Logarithm	Square	Square root

EXHIBIT D-3 (Continued)

Number	Area circle	Circumference circle	10 × Reciprocal	Logarithm	Square	Square root
550	237583	1728	.01818	2.740	302500	23.45
555	241922	1744	.01802	2.744	308025	23.56
560	246301	1759	.01786	2.748	313600	23.66
565	250719	1775	.01770	2.752	319225	23.77
570	255176	1791	.01754	2.756	324900	23.87
575	259672	1806	.01739	2.760	330625	23.98
580	264208	1822	.01724	2.763	336400	24.08
585	268783	1838	.01709	2.767	342225	24.19
590	273397	1854	.01695	2.771	348100	24.29
595	278051	1869	.01681	2.775	354025	24.39
600	282743	1885	.01667	2.778	360000	24.49
610	292247	1916	.01639	2.785	372100	24.70
620	301907	1948	.01613	2.792	384400	24.90
630	311725	1979	.01587	2.799	396900	25.10
640	321699	2011	.01563	2.806	409600	25.30
650	331831	2042	.01538	2.813	422500	25.50
660	342119	2074	.01515	2.820	435600	25.69
670	352565	2105	.01493	2.826	448900	25.88
680	363168	2136	.01471	2.833	462400	26.08
690	373928	2168	.01449	2.839	476100	26.27
700	384845	2199	.01429	2.845	490000	26.46
710	395919	2231	.01408	2.851	504100	26.65
720	407150	2262	.01389	2.857	518400	26.83
730	418539	2293	.01370	2.863	532900	27.02
740	430084	2325	.01351	2.869	547600	27.20
750	441786	2356	.01333	2.875	562500	27.39
760	453646	2388	.01316	2.881	577600	27.57
770	465663	2419	.01299	2.886	592900	27.75
780	477836	2450	.01282	2.892	608400	27.93
790	490167	2482	.01266	2.898	624100	28.11
800	502655	2513	.01250	2.903	640000	28.28
810	515300	2545	.01235	2.908	656100	28.46
820	528102	2576	.01220	2.914	672400	28.64
830	541061	2608	.01205	2.919	688900	28.81
840	554177	2639	.01190	2.924	705600	28.98
850	567450	2670	.01176	2.929	722500	29.15
860	580880	2702	.01163	2.935	739600	29.33
870	594468	2733	.01149	2.940	756900	29.50
880	608212	2765	.01136	2.944	774400	29.66
890	622114	2796	.01124	2.949	792100	29.83
900	636173	2827	.01111	2.954	810000	30.00
910	650388	2859	.01099	2.959	828100	30.17
920	664761	2890	.01087	2.964	846400	30.33
930	679291	2922	.01075	2.968	864900	30.50
940	693978	2953	.01064	2.973	883600	30.66
950	708822	2985	.01053	2.978	902500	30.82
960	723823	3016	.01042	2.982	921600	30.98
970	738981	3047	.01031	2.987	940900	31.14
980	754296	3079	.01020	2.991	960400	31.31
990	769769	3110	.01010	2.996	980100	31.46
999	783828	3139	.01001	2.999	998001	31.61

Number	Area circle	Circumference circle	10 × Reciprocal	Logarithm	Square	Square root
300	70686	942	.03333	2.477	90000	17.32
305	73062	958	.03279	2.484	93025	17.46
310	75477	974	.03226	2.491	96100	17.61
315	77931	990	.03175	2.498	99225	17.75
320	80425	1005	.03125	2.505	102400	17.89
325	82958	1021	.03077	2.512	105625	18.03
330	85530	1037	.03030	2.519	108900	18.17
335	88141	1052	.02985	2.525	112225	18.30
340	90792	1068	.02941	2.531	115600	18.44
345	93482	1084	.02899	2.538	119025	18.57
350	96211	1100	.02857	2.544	122500	18.71
355	98980	1115	.02817	2.550	126025	18.84
360	101788	1131	.02778	2.556	129600	18.97
365	104635	1147	.02740	2.562	133225	19.11
370	107521	1162	.02703	2.568	136900	19.24
375	110447	1178	.02667	2.574	140625	19.36
380	113411	1194	.02632	2.580	144400	19.49
385	116416	1210	.02597	2.585	148225	19.62
390	119459	1225	.02564	2.591	152100	19.75
395	122542	1241	.02532	2.597	156025	19.87
400	125664	1257	.02500	2.602	160000	20.00
405	128825	1272	.02469	2.607	164025	20.12
410	132025	1288	.02439	2.613	168100	20.25
415	135265	1304	.02410	2.618	172225	20.37
420	138544	1320	.02381	2.623	176400	20.49
425	141863	1335	.02353	2.628	180625	20.62
430	145220	1351	.02326	2.633	184900	20.74
435	148617	1367	.02299	2.638	189225	20.86
440	152053	1382	.02273	2.643	193600	20.98
445	155528	1398	.02247	2.648	198025	21.10
450	159043	1414	.02222	2.653	202500	21.21
455	162597	1429	.02198	2.658	207025	21.33
460	166190	1445	.02174	2.663	211600	21.45
465	169823	1461	.02151	2.667	216225	21.56
470	173494	1477	.02128	2.672	220900	21.68
475	177205	1492	.02105	2.677	225625	21.79
480	180956	1508	.02083	2.681	230400	21.91
485	184745	1524	.02062	2.686	235225	22.02
490	188574	1539	.02041	2.690	240100	22.14
495	192442	1555	.02020	2.695	245025	22.25
500	196350	1571	.02000	2.699	250000	22.36
505	200296	1587	.01980	2.703	255025	22.47
510	204282	1602	.01961	2.708	260100	22.58
515	208307	1618	.01942	2.712	265225	22.69
520	212372	1634	.01923	2.716	270400	22.80
525	216475	1649	.01905	2.720	275625	22.91
530	220618	1665	.01887	2.724	280900	23.02
535	224801	1681	.01869	2.728	286225	23.13
540	229022	1697	.01852	2.732	291600	23.24
545	233283	1712	.01835	2.736	297025	23.35
550	237583	1728	.01818	2.740	302500	23.45

EXHIBIT D-3 (Continued)

EXHIBIT D-4   Random Sampling Numbers[4]

90	78	82	54	47	20	83	80	10	41	35	22	23	03	98	79	74	41	35	05	78	73	95	47	83
78	58	68	87	41	11	08	81	29	89	71	23	10	01	79	25	06	00	45	80	64	70	95	34	29
51	42	21	03	88	20	05	35	93	00	68	12	09	55	09	36	54	95	22	82	48	30	09	56	87
93	15	07	60	86	67	37	94	24	35	82	44	19	92	96	21	84	29	04	29	83	32	05	10	48
27	12	31	66	62	09	54	17	31	23	27	30	37	36	79	75	50	39	57	12	67	23	22	09	33
79	44	83	55	47	96	50	93	56	82	58	16	35	18	87	64	08	22	47	93	86	43	43	30	17
89	73	43	91	03	57	91	35	40	64	13	61	94	37	16	09	93	96	25	87	30	23	42	54	31
29	30	90	00	58	15	99	93	33	67	80	08	59	21	66	13	54	56	85	25	05	32	03	52	52
97	33	17	26	25	04	73	18	10	05	34	40	32	65	07	28	68	29	31	97	89	57	95	55	16
07	15	44	92	47	28	50	93	03	53	37	70	19	68	59	95	39	87	90	46	98	64	46	24	71
82	50	35	50	80	23	67	81	25	02	83	08	12	70	00	25	31	33	80	06	19	86	14	59	27
59	21	86	16	30	27	85	16	26	34	50	15	87	22	69	71	36	95	90	76	90	99	79	63	21
04	19	60	33	05	29	02	33	74	56	38	84	21	07	35	93	54	70	18	47	14	62	75	45	02
96	91	44	09	94	06	89	50	88	83	82	50	11	82	51	30	68	91	06	28	86	65	17	45	20
31	71	03	53	38	94	02	52	72	15	44	49	53	42	43	00	36	97	67	64	12	27	46	00	18
03	70	22	67	59	98	10	64	68	08	79	06	89	48	41	85	72	10	87	24	96	04	20	68	00
08	45	79	46	89	74	73	67	60	15	70	37	61	44	07	67	89	81	54	26	57	17	63	27	74
37	80	05	75	64	48	51	68	68	27	71	75	45	32	27	76	35	26	58	88	67	74	48	90	94
90	63	56	69	37	19	74	48	63	31	52	36	84	40	66	72	66	03	41	87	65	29	12	36	64
22	69	38	02	88	89	71	43	01	87	41	79	42	99	29	41	08	47	32	19	45	29	59	69	90
05	79	69	67	64	36	14	82	65	26	40	51	63	42	48	85	48	34	12	04	33	26	52	26	52
48	91	53	03	82	64	24	06	31	03	97	44	82	24	89	88	48	66	54	10	41	27	09	11	61
94	64	97	27	25	62	23	94	40	54	56	32	97	78	90	58	86	41	75	19	42	90	85	36	68
15	85	82	52	08	52	96	26	92	88	93	11	03	23	52	78	23	57	85	43	53	90	42	22	22
09	81	37	66	56	99	08	59	19	48	29	69	21	64	95	12	08	15	24	45	59	25	22	76	96
43	83	99	02	76	12	16	45	52	66	35	70	93	09	52	75	40	34	35	62	65	42	27	20	59
31	98	09	80	62	75	26	64	57	26	46	41	47	90	97	99	46	10	51	42	73	28	98	89	91
81	35	42	62	84	37	02	59	78	16	17	96	05	71	39	88	05	34	05	92	22	43	89	66	89
97	95	56	39	75	65	47	61	86	33	14	88	55	33	69	70	87	79	94	46	17	61	72	27	01
37	63	35	93	23	17	30	14	51	51	17	28	21	74	67	12	11	57	19	27	38	70	73	82	92
39	22	96	00	48	52	49	62	09	40	08	30	27	54	70	96	06	52	12	80	36	12	38	68	05
61	29	84	34	51	60	19	77	82	16	64	45	02	27	04	65	55	90	95	04	20	39	29	96	28
38	84	18	10	29	19	09	66	06	78	37	09	60	50	21	52	72	01	52	70	29	65	05	37	16
64	29	48	04	08	55	72	25	25	77	54	26	27	24	39	66	67	06	40	00	99	35	70	69	58
64	02	32	99	63	62	42	89	32	20	81	14	08	40	45	22	15	37	49	38	96	51	19	08	27
13	83	39	51	30	31	49	94	83	66	02	50	95	18	98	58	84	90	58	81	00	40	91	12	46
83	30	90	09	35	41	12	87	93	66	85	96	20	65	34	23	13	05	41	01	91	48	95	59	45
46	63	53	97	63	18	86	37	56	20	35	62	66	11	37	30	91	89	97	51	64	78	06	95	65
54	43	40	02	41	55	70	52	96	87	02	82	61	21	88	60	65	98	42	09	03	61	20	83	01
27	18	65	62	01	97	45	79	51	37	74	47	20	11	48	97	93	73	86	50	46	61	95	01	24
45	42	16	13	20	34	51	08	71	52	39	17	71	39	84	97	27	72	49	42	81	62	32	87	22
35	92	97	02	34	93	32	95	81	13	92	05	40	70	95	71	66	61	24	08	77	32	73	66	79
60	55	35	57	24	52	95	84	90	64	38	39	72	70	17	98	42	85	96	67	41	11	83	17	78
43	17	21	09	60	58	86	12	31	11	66	61	43	96	00	93	97	00	15	20	37	96	73	56	63
07	85	74	58	28	38	74	68	32	61	87	14	71	83	47	90	11	96	70	08	67	04	34	46	08
33	00	29	08	87	42	59	40	24	97	44	99	13	56	87	95	02	47	97	89	23	51	45	37	83
97	14	00	42	23	72	03	19	02	41	11	23	36	98	32	19	91	42	03	58	62	23	74	45	06
68	58	32	80	82	40	49	71	83	37	93	49	99	60	72	88	14	26	88	95	48	69	35	40	63
39	87	38	16	06	82	92	62	32	75	67	64	50	49	39	29	55	53	92	97	04	48	60	53	90
37	73	01	84	87	42	88	30	93	75	01	18	34	73	30	28	44	28	18	01	00	38	26	38	57

[4] Reprinted from A. Hald, *Statistical Tables,* John Wiley & Sons, Inc., New York, N. Y. Copyright © 1952.

EXHIBIT D-5   Table for Finding Rate of Return on Machine Tool Investments[5]

Gross 1st Year % Return	True Rate of Return after Taxes*	Equivalent Rate of Return before Taxes#
50%	30%	60%
45	27	54
40	24	48
35	21	42
30	18	36
25	15	30
22	13	26
20	11.5	23
18	10	20
16	8.5	17
14	7	14
12	5.5	11
10	4	8
9	3	6
8	2.25	4.5
7	1.5	3
6	0	0
5	0	0

HOW TO USE THIS TABLE

1. Divide first year earnings by amount of investment to find Gross First Year % Return of proposed new machine(s).

2. Read across to find *True Rate of Return after Taxes* and *Equivalent Rate of Return before Taxes*.

*True Rate of Return after Taxes is based on:
12-17-yr life of new machine
Accelerated depreciation for tax purposes
New machine to be worth 15% of original cost
  when it's eventually replaced
Deterioration to 75% efficiency in 15 years
Nominal present market value for machines being
  replaced

EXAMPLE:

$20,000 machine with $4,000 earnings
$ 4,000 ÷ 20,000 = 20%
True rate of return
   after taxes ....= 11.5%
Equivalent rate of return
   before taxes ....= 23.0%

#Investment return is usually considered before taxes.
This is the return from the machine tool investment to
compare with the before-tax return of other investments

---

[5] Baxter T. Fullerton, *Machine Replacement for the Shop Manager,* Huebner Publications, Inc., copyright © 1961, Cleveland, Ohio, p. 51.

EXHIBIT D-6   Adjustments for Off-Standard Cases[6]

ADJUSTMENTS FOR OFF-STANDARD CASES[6]

Add to or subtract from True Rate of Return after Taxes as indicated

True Rate of Return After Taxes	Different length of life of new machine		Different depreciation	Different "end of life"
	Under $10,000 *machines with* 8-11 *yr life*	*Over* $50,000 *machines with more than* 17 *yr life*	*Straight line depreciation*	*Final value* 5% *of original cost*
30%	−0.5%	0%	−2%	+0.5%
27	−0.5	0	−2	+0.5
24	−0.5	0	−2	+0.5
21	−1	0	−2	+0.5
18	−1	0	−1.5	+0.5
15	−1.5	0	−1.5	+0.25
13	−2	0	−1	+0.25
11.5	−2	+0.5	−1	+0.25
10	−2	+0.5	−1	+0.25
8.5	−2	+1	−0.5	0
7	−2	+1	−0.5	0
5.5	−2.5	+1	−0.5	0
4	−4	+1	−0.5	0
3	−3	+1	−0.5	0
2.25	−2.25	+0.75	−0.5	0
1.5	−1.5	+0.75	−0.25	0
0	0	+1.5	0	0
0	0	+0.5	0	0

Machine Tools Are Good Investments

On this basis, machine tool investments are earning more money than many of us have realized! We don't normally think of a "10–yr. payoff" as being much of a machine investment, but this may now be viewed in a different light. A 10% first-year gross return (10–yr. payoff) is actually a 4% after-tax and an 8% equivalent before-tax return. There are times when a machine tool with an 8% return will be a much safer and better investment than some of the 8% return investments available elsewhere.

A Standard Pattern of Results

Where do we get this simple table?

It is not the result of a formula. It's the actual recorded pattern of the answers

---

[6] Baxter T. Fullerton, *Machine Replacement for the Shop Manager*, Huebner Publications, Inc. Copyright © 1961, Cleveland, Ohio, pp. 52–53.

value of new machine	Different rate of declining earnings because of new machine deterioration			*No Present market value for machines being replaced*
Final value 25% of original cost	*50% of "normal" deterioration*	*150% of "normal" deterioration*	*Straight line deterioration*	
−0.5%	+0.5%	−0.5%	−0.5%	−1%
−0.5	+0.5	−0.5	−0.5	−1
−0.5	+0.5	−0.5	−0.5	−1
−0.5	+0.5	−0.5	−0.5	−1
−0.5	+0.5	−0.5	−0.5	−1
−0.5	+0.5	−0.5	−0.5	−1
−0.25	+0.5	−0.5	−0.5	−1
−0.25	+0.5	−0.5	−0.5	−1
0	+1	−0.5	−0.5	−1
0	+1	−0.5	−0.5	−1
0	+1	−0.5	−0.5	−1
0	+1	−0.5	−0.5	−1
0	+0.5	−0.5	−0.5	−1
+0.25	+0.5	−0.5	−0.25	−1
+0.25	+0.75	−0.75	−0.25	−1
+0.5	+0.75	−1.5	−0.5	−1.5
+1	+1.5	0	0	0
0	+0.75	0	0	0

from many dozens of trial replacement calculations similar to the longhand calculation from the last chapter. These calculations have been solved for different kinds and prices of machine tools, different rates of first-year earnings, different lengths of life for the new machine, different depreciation methods, different end of life values for the new machine, different rates of deterioration of the new machine, differences in the loss of market value of the old equipment if it is not replaced, and for combinations of these differences.

Make all of these calculations, and you'll find that the answers fall into a definite and simple pattern such as shown in the table, and that the rate of first-year earnings is the only thing that changes the answer to a great extent.

To determine this pattern, we can establish a set of standard conditions of machine tool life, and work out a set of answers based on these conditions. We can also show how much difference (or how little difference) a deviation will make in any given situation.

EXHIBIT D-7   Interest and the Time Value of Money[7]

Why must interest be considered in many economy studies? Interest, from the point of view of the business enterprise or the individual, can be considered as a rental for the use of money. If you have saved some money and I would like to borrow it from you to use in my business, you will agree to lend or rent it to me only if I agree to pay you interest for the use of the money. If I were not willing to pay you a rental (interest) for the use of your money, there would be no incentive for you to lend me the money and your incentive to save money would be reduced.

If you were to place $1,000 in an insured savings bank account paying 4 per cent interest compounded annually, at the end of 10 years you would have $1,480 in the account, assuming you made no further deposits or withdrawals. You would thus not be willing to give me $1,000 in return for a promise to give you $1,000 ten years from now. You would consider that the equivalent value of a present sum of $1,000 ten years from now is $1,480 at 4 per cent interest. You would also consider that you should possibly receive more than a 4 per cent return if you lend the money to me rather than deposit it in an insured savings bank account because of the possibly greater risks involved in an uninsured loan. You may therefore consider that you should receive a 5 per cent return on the $1,000 you lend me. At 5 per cent interest, the equivalent value of a present sum of $1,000 ten years from now is $1,629.

The values of sums of money at different points in time depend upon the interest rate and the time spans. A sum of money today is not equivalent to the same sum ten years later unless interest were assumed at the unrealistic rate of zero. We shall consider in this chapter the methods of calculating these equivalent values of money at different points in time.

Simple Interest. If I lend you $100 for three years at 5 per cent interest payable annually, you will pay me $5 at the end of each of three years, or a total of $15, in addition to returning the principal sum of $100 at the end of the three years.

Simple interest on any investment is computed by multiplying the present investment by the interest rate per period (in this case per year) by the number of periods (in this case years). If we let

[7] From *Economic Analysis for Engineering and Managerial Decision-Making* by Barish, copyright © 1962 by McGraw-Hill, Inc. Used with permission of McGraw-Hill Book Company, pp. 49–55.

P = sum of money at a time designated as the present

i = interest rate per interest period

n = number of interest periods

Then, the simple-interest charge equals Pin.

In the example previously cited, simple interest for the three-year period equals ($100)(0.05)(3), or $15.

Compound Interest. If, in the previous example, I lent you the $100 with the understanding that you would credit me with interest at 5 per cent each year, but would retain the interest until the termination of the loan at the end of three years, paying me interest on the interest you retained, the process is called compounding interest. This is similar to depositing money in a savings bank account and letting the interest accumulate in the account so that interest is earned on the interest.

An arithmetic calculation of the amount that I should receive at the end of three years is shown below.

Year	Principal amount at beginning of year (A)	Interest earned during year at 5% interest (B)	Total amount at end of year (A) + (B)
1	$100.00	$100.00(0.05) = $5.00	$105.00
2	105.00	105.00(0.05) = 5.25	110.25
3	110.25	110.25(0.05) = 5.51	115.76

When you pay me $115.76 at the end of three years, $0.76 of the sum is interest on the interest which you retained until the end of the loan period. Another way of looking at these calculations is that $100.00 today is equivalent to $115.76 three years later, calculating interest at 5 per cent.

Let us designate S as the future sum which is equivalent to present sum P. Thus, S is a sum of money n interest periods from the present which is equivalent to P with interest rate i. We can develop a compound-interest formula to compute S.

Year	Principal amount at beginning of year (A)	Interest earned during year (B)	Total amount at end of year (A) + (B)
1	P	Pi	$P + Pi = P(1 + i)$
2	$P(1 + i)$	$P(1 + i)i$	$P(1 + i) + P(1 + i)i = P(1 + i)(1 + i) = P(1 + i)^2$
3	$P(1 + i)^2$	$P(1 + i)^2 i$	$P(1 + i)^2 + P(1 + i)^2 i = P(1 + i)^2(1 + i) = P(1 + i)^3$
.	.	.	.
.	.	.	.
.	.	.	.
n	$P(1 + i)^{n-1}$	$P(1 + i)^{n-1}i$	$P(1 + i)^{n-1} + P(1 + i)^{n-1}i = P(1 + i)^{n-1}(1 + i) = P(1 + i)^n$

Thus, the total amount S, at the end of year n, equals $P(1 + i)^n$.

The formula $S = P(1 + i)^n$ was developed using interest compounded annually, so that n equaled the number of years and i the annual interest rate. If interest is compounded more frequently, then n will be equal to the number of periods of compounding and i will be equal to the rate per interest period, not the nominal rate per year.

Let us consider how much \$100 would be worth in five years at 4 per cent interest compounded semiannually. In this case, $n = 10$ half-year periods and $i = 0.02$, or 2 per cent per half-year period. Using our formula, the solution would be

$$S = P(1 + i)^n$$
$$S = \$100(1 + 0.02)^{10}$$
$$S = \$121.90$$

Present Worth. If I desire to receive \$100 three years from now, how much money must I invest now if the money will earn 5 per cent interest? Another way of asking this same question is: What is the present worth of \$100 three years from now, with interest at 5 per cent?

This question can be answered using the compound-interest formula, $S = P(1 + i)^n$. In this case, P is the unknown quantity and S is the \$100 which I desire to receive three years hence. Solving, $S = P(1 + i)^n$ for P,

$$P = S\left(\frac{1}{1 + i}\right)^n$$

To answer the original question,

$$P = S\left(\frac{1}{1 + i}\right)^n$$
$$P = \$100\left(\frac{1}{1 + 0.05}\right)^3$$
$$P = \$86.38$$

Uniform-series Payments. If I deposit \$100 at the end of each year for four years in a savings bank account which earns interest at the rate of 4 per cent annually and make no withdrawals during the period, how much money will I have in the account at the end of the four years?

The \$100 deposited at the end of the fourth year will earn no interest and will, therefore, contribute only \$100 to the total in the account. The \$100 deposited at the end of the third year will earn interest for one year and will, therefore, contribute $\$100(1 + 0.04)$. The \$100 deposited at the end of the second year will earn interest for two years and will, therefore, contribute $\$100(1 + 0.04)^2$. The \$100 deposited at the end of the first year will earn interest for three years and will, therefore, contribute $\$100(1 + 0.04)^3$. The total in the account at the end of the four years will be $100 + 100(1 + 0.04) + 100(1 + 0.04)^2 + 100(1 + 0.04)^3$.

Solving, we find

$$S = \$100[1 + (1 + 0.04) + (1 + 0.04)^2 + (1 + 0.04)^3]$$
$$= \$424.60$$

These calculations can be graphically represented by a simple line diagram to show the time relationships, as follows:

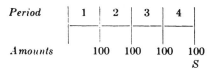

In this problem, we desire to determine the value of S which will be equal to the compound amounts of the four \$100 payments. Using our compound-interest formula for the value of each \$100 payment at the end of period 4 gives us the previously stated formulation.

Formula for Compound Amount in Fund. Let us say that $R =$ a single end-of-period payment and let us expand our definition of S so that it is a sum of money n interest periods from the present which is equivalent either to P with interest rate i or to a uniform series of end-of-period payments, R, with interest rate i.

We want to derive a formula for the value of a fund in which R dollars are deposited at the end of each period for n periods and in which all monies in the fund earn interest at a rate of i. Graphically, we can represent the model as follows:

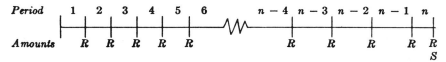

We want to find that value of S which will be equivalent to all of the n end-of-period R's at interest rate i.

The R dollars deposited at the end of the nth period earn no interest and, therefore, contribute R dollars to the fund. The R dollars deposited at the end of the $(n - 1)$ period earn interest for 1 year and will, therefore, contribute $R(1 + i)$ dollars to the fund. The R dollars deposited at the end of the $(n - 2)$ period earn interest for 2 years and will, therefore contribute $R(1 + i)^2$. These years of earned interest in the contributions will continue to increase in this manner, and the R deposited at the end of the first period will have earned interest for $(n - 1)$ periods. The total in the fund S is, thus, equal to $R + R(1 + i) + R(1 + i)^2 + R(1 + i)^3 + R(1 + i)^4 + \cdots + R(1 + i)^{n-2} + R(1 + i)^{n-1}$. Factoring out R, we obtain

$$S = R[1 + (1 + i) + (1 + i)^2 + \cdots + (1 + i)^{n-2} + (1 + i)^{n-1}] \quad (1)$$

If we multiply both sides of this equation by $(1 + i)$, we obtain

$$(1 + i)S = R[(1 + i) + (1 + i)^2 + (1 + i)^3 + \cdots + (1 + i)^{n-1}$$
$$+ (1 + i)^n] \quad (2)$$

If we now subtract equation (1) from (2), we obtain

$$(1 + i)S - S = R[(1 + i) + (1 + i)^2 + (1 + i)^3 + \cdots$$
$$+ (1 + i)^{n-1} + (1 + i)^n] - R[1 + (1 + i)$$
$$+ (1 + i)^2 + \cdots + (1 + i)^{n-2} + (1 + i)^{n-1}]$$
$$iS = R[(1 + i)^n - 1]$$
$$S = R\left[\frac{(1 + i)^n - 1}{i}\right]$$

This formula enables us to compute the compound amount of a uniform periodic series. Using it to verify the value of S in the immediately preceding illustration, we obtain

$$S = \$100\left[\frac{(1 + 0.04)^4 - 1}{0.04}\right] = \$424.60$$

Sinking-fund Deposit. I desire to deposit a uniform sum at the end of each year for the next 10 years to provide a sum of $5,000 at the end of the 10-year period for my son's education. If the fund will earn a 4 per cent return on its monies, how much should I deposit at the end of each year? (This type of fund is frequently called a sinking fund.)

To answer this question, let us take the previously developed formula for the compound amount of a uniform series and solve for R, the end-of-period deposit.

$$S = R\left[\frac{(1 + i)^n - 1}{i}\right]$$
$$R = S\left[\frac{i}{(1 + i)^n - 1}\right]$$

Then, to answer my question

$$R = \$5,000\left[\frac{0.04}{(1 + 0.04)^{10} - 1}\right]$$
$$R = \$416.45$$

Capital Recovery. If I lend you $10,000 with the understanding that you will repay this principal sum, including interest at 4 per cent on all unpaid balances, in uniform annual payments, how much should you pay me at the end of each year for five years?

Let us refer to our previous formula for end-of-period payments, R

$$R = S\left[\frac{i}{(1+i)^n - 1}\right]$$

and since $\qquad S = P(1+i)^n$

$$R = P(1+i)^n\left[\frac{i}{(1+i)^n - 1}\right]$$

$$R = P\left[\frac{i(1+i)^n}{(1+i)^n - 1}\right]$$

To answer our original question,

$$R = \$10,000\left[\frac{0.04(1+0.04)^5}{(1+0.04)^5 - 1}\right]$$

$$R = \$2,246.30$$

Present Value of a Uniform Series. How much should you be willing to pay me now for my agreement to pay you \$100 at the end of each year for six years if you desire 6 per cent return on the payment you make? (In other words, you would like to receive your principal plus 6 per cent interest on unpaid balances.)

Referring to our previous formula for capital recovery,

$$R = P\left[\frac{i(1+i)^n}{(1+i)^n - 1}\right]$$

Let us solve for P:

$$P = R\left[\frac{(1+i)^n - 1}{i(1+i)^n}\right]$$

To answer our question,

$$P = \$100\left[\frac{(1+0.06)^6 - 1}{0.06(1+0.06)^6}\right]$$

$$P = \$491.70$$

Interest Tables. To evaluate numerically the terms with exponents in the answers to our previous questions required the use of logarithms or a calculator or slide rule and is somewhat tedious. To facilitate the solution of interest calculation problems, tables such as those in the Appendix of this book are convenient. These give values for each of the following six interest factors.

Factor name	*Formula*	*Abbreviation*
Single-payment compound-amount factor	$(1+i)^n$	SPCA $i =$ $n =$
Single-payment present-worth factor	$\dfrac{1}{(1+i)^n}$	SPPW $i =$ $n =$
Uniform-series compound-amount factor	$\dfrac{(1+i)^n - 1}{i}$	USCA $i =$ $n =$

Factor name	*Formula*	*Abbreviation*
Sinking-fund-payment factor	$\dfrac{i}{(1 + i)^n - 1}$	SFP $i =$ $n =$
Capital-recovery factor	$\dfrac{i(1 + i)^n}{(1 + i)^n - 1}$	CR $i =$ $n =$
Uniform-series present-worth factor	$\dfrac{(1 + i)^n - 1}{i(1 + i)^n}$	USPW $i =$ $n =$

The values of these factors are computed for the following interest rates: $\frac{1}{2}\%$, 1%, 2%, 3%, 4%, 5%, 6%, 8%, 10%, 12%, 15%, 20%, 25%, 30%, 40%, and 50%. We shall use these interest tables in the solution of most problems in this book requiring the use of the interest formula.

When the values of interest factors are desired for interest rates for which tables are not available, approximations of the desired values can be obtained by linear interpolation between the values for available interest rates on each side of the desired rate. For most of the types of economy calculations with which we are concerned, such approximate solutions are adequate and can obviate the necessity for the more arduous calculations using the formula.

EXHIBIT D-8   Table of Interest Factors[8]

Period n	Single-payment compound-amount (SPCA) — Future value of $1 $(1+i)^n$	Single-payment present-worth (SPPW) — Present value of $1 $\dfrac{1}{(1+i)^n}$	Uniform-series compound-amount (USCA) — Future value of uniform series of $1 $\dfrac{(1+i)^n-1}{i}$	Sinking-fund payment (SFP) — Uniform series whose future value is $1 $\dfrac{i}{(1+i)^n-1}$	Capital recovery (CR) — Uniform series with present value of $1 $\dfrac{i(1+i)^n}{(1+i)^n-1}$	Uniform-series present-worth (USPW) — Present value of uniform series of $1 $\dfrac{(1+i)^n-1}{i(1+i)^n}$
1	1.005	0.9950	1.000	1.00000	1.00500	0.995
2	1.010	0.9901	2.005	0.49875	0.50375	1.985
3	1.015	0.9851	3.015	0.33167	0.33667	2.970
4	1.020	0.9802	4.030	0.24813	0.25313	3.950
5	1.025	0.9754	5.050	0.19801	0.20301	4.926
6	1.030	0.9705	6.076	0.16460	0.16960	5.896
7	1.036	0.9657	7.106	0.14073	0.14573	6.862
8	1.041	0.9609	8.141	0.12283	0.12783	7.823
9	1.046	0.9561	9.182	0.10891	0.11391	8.779
10	1.051	0.9513	10.228	0.09777	0.10277	9.730
11	1.056	0.9466	11.279	0.08866	0.09366	10.677
12	1.062	0.9419	12.336	0.08107	0.08607	11.619
13	1.067	0.9372	13.397	0.07464	0.07964	12.556
14	1.072	0.9326	14.464	0.06914	0.07414	13.489
15	1.078	0.9279	15.537	0.06436	0.06936	14.417
16	1.083	0.9233	16.614	0.06019	0.06519	15.340
17	1.088	0.9187	17.697	0.05651	0.06151	16.259
18	1.094	0.9141	18.786	0.05323	0.05823	17.173
19	1.099	0.9096	19.880	0.05030	0.05530	18.082
20	1.105	0.9051	20.979	0.04767	0.05267	18.987
21	1.110	0.9006	22.084	0.04528	0.05028	19.888
22	1.116	0.8961	23.194	0.04311	0.04811	20.784
23	1.122	0.8916	24.310	0.04113	0.04613	21.676
24	1.127	0.8872	25.432	0.03932	0.04432	22.563
25	1.133	0.8828	26.559	0.03765	0.04265	23.446
26	1.138	0.8784	27.692	0.03611	0.04111	24.324
27	1.144	0.8740	28.830	0.03469	0.03969	25.198
28	1.150	0.8697	29.975	0.03336	0.03836	26.068
29	1.156	0.8653	31.124	0.03213	0.03713	26.933
30	1.161	0.8610	32.280	0.03098	0.03598	27.794
35	1.191	0.8398	38.145	0.02622	0.03122	32.035
40	1.221	0.8191	44.159	0.02265	0.02765	36.172
45	1.252	0.7990	50.324	0.01987	0.02487	40.207
50	1.283	0.7793	56.645	0.01765	0.02265	44.143
55	1.316	0.7601	63.126	0.01584	0.02084	47.981
60	1.349	0.7414	69.770	0.01433	0.01933	51.726
65	1.383	0.7231	76.582	0.01306	0.01806	55.377
70	1.418	0.7053	83.566	0.01197	0.01697	58.939
75	1.454	0.6879	90.727	0.01102	0.01602	62.414
80	1.490	0.6710	98.068	0.01020	0.01520	65.802
85	1.528	0.6545	105.594	0.00947	0.01447	69.108
90	1.567	0.6383	113.311	0.00883	0.01383	72.331
95	1.606	0.6226	121.222	0.00825	0.01325	75.476
100	1.647	0.6073	129.334	0.00773	0.01273	78.543

1% Interest Factors

Period n	Single-payment compound-amount (SPCA) Future value of \$1 $(1+i)^n$	Single-payment present-worth (SPPW) Present value of \$1 $\dfrac{1}{(1+i)^n}$	Uniform-series compound-amount (USCA) Future value of uniform series of \$1 $\dfrac{(1+i)^n-1}{i}$	Sinking-fund payment (SFP) Uniform series whose future value is \$1 $\dfrac{i}{(1+i)^n-1}$	Capital recovery (CR) Uniform series with present value of \$1 $\dfrac{i(1+i)^n}{(1+i)^n-1}$	Uniform-series present-worth (USPW) Present value of uniform series of \$1 $\dfrac{(1+i)^n-1}{i(1+i)^n}$
1	1.010	0.9901	1.000	1.00000	1.01000	0.990
2	1.020	0.9803	2.010	0.49751	0.50751	1.970
3	1.030	0.9706	3.030	0.33002	0.34002	2.941
4	1.041	0.9610	4.060	0.24628	0.25628	3.902
5	1.051	0.9515	5.101	0.19604	0.20604	4.853
6	1.062	0.9420	6.152	0.16255	0.17255	5.795
7	1.072	0.9327	7.214	0.13863	0.14863	6.728
8	1.088	0.9235	8.286	0.12069	0.13069	7.652
9	1.094	0.9143	9.369	0.10674	0.11674	8.566
10	1.105	0.9053	10.462	0.09558	0.10558	9.471
11	1.116	0.8963	11.567	0.08645	0.09645	10.368
12	1.127	0.8874	12.683	0.07885	0.08885	11.255
13	1.138	0.8787	13.809	0.07241	0.08241	12.134
14	1.149	0.8700	14.947	0.06690	0.07690	13.004
15	1.161	0.8613	16.097	0.06212	0.07212	13.865
16	1.173	0.8528	17.258	0.05794	0.06794	14.718
17	1.184	0.8444	18.430	0.05426	0.06426	15.562
18	1.196	0.8360	19.615	0.05098	0.06098	16.398
19	1.208	0.8277	20.811	0.04805	0.05805	17.226
20	1.220	0.8195	22.019	0.04542	0.05542	18.046
21	1.232	0.8114	23.239	0.04303	0.05303	18.857
22	1.245	0.8034	24.472	0.04086	0.05086	19.660
23	1.257	0.7954	25.716	0.03889	0.04889	20.456
24	1.270	0.7876	26.973	0.03707	0.04707	21.243
25	1.282	0.7798	28.243	0.03541	0.04541	22.023
26	1.295	0.7720	29.526	0.03387	0.04387	22.795
27	1.308	0.7644	30.821	0.03245	0.04245	23.560
28	1.321	0.7568	32.129	0.03112	0.04112	24.316
29	1.335	0.7493	33.450	0.02990	0.03990	25.066
30	1.348	0.7419	34.785	0.02875	0.03875	25.808
35	1.417	0.7059	41.660	0.02400	0.03400	29.409
40	1.489	0.6717	48.886	0.02046	0.03046	32.835
45	1.565	0.6391	56.481	0.01771	0.02771	36.095
50	1.645	0.6080	64.463	0.01551	0.02551	39.196
55	1.729	0.5785	72.852	0.01373	0.02373	42.147
60	1.817	0.5504	81.670	0.01224	0.02224	44.955
65	1.909	0.5237	90.937	0.01100	0.02100	47.627
70	2.007	0.4983	100.676	0.00993	0.01993	50.169
75	2.109	0.4741	110.913	0.00902	0.01902	52.587
80	2.217	0.4511	121.672	0.00822	0.01822	54.888
85	2.330	0.4292	132.979	0.00752	0.01752	57.078
90	2.449	0.4084	144.863	0.00690	0.01690	59.161
95	2.574	0.3886	157.354	0.00636	0.01636	61.143
100	2.705	0.3697	170.481	0.00587	0.01587	63.029

2% Interest Factors

Period *n*	Single-payment compound-amount (SPCA)	Single-payment present-worth (SPPW)	Uniform-series compound-amount (USCA)	Sinking-fund payment (SFP)	Capital recovery (CR)	Uniform-series present-worth (USPW)
	Future value of $1 $(1+i)^n$	Present value of $1 $\dfrac{1}{(1+i)^n}$	Future value of uniform series of $1 $\dfrac{(1+i)^n-1}{i}$	Uniform series whose future value is $1 $\dfrac{i}{(1+i)^n-1}$	Uniform series with present value of $1 $\dfrac{i(1+i)^n}{(1+i)^n-1}$	Present value of uniform series of $1 $\dfrac{(1+i)^n-1}{i(1+i)^n}$
1	1.020	0.9804	1.000	1.00000	1.02000	0.980
2	1.040	0.9612	2.020	0.49505	0.51505	1.942
3	1.061	0.9423	3.060	0.32675	0.34675	2.884
4	1.082	0.9238	4.122	0.24262	0.26262	3.808
5	1.104	0.9057	5.204	0.19216	0.21216	4.713
6	1.126	0.8880	6.308	0.15853	0.17853	5.601
7	1.149	0.8706	7.434	0.13451	0.15451	6.472
8	1.172	0.8535	8.583	0.11651	0.13651	7.325
9	1.195	0.8368	9.755	0.10252	0.12252	8.162
10	1.219	0.8203	10.950	0.09133	0.11133	8.983
11	1.243	0.8043	12.169	0.08218	0.10218	9.787
12	1.268	0.7885	13.412	0.07456	0.09456	10.575
13	1.294	0.7730	14.680	0.06812	0.08812	11.348
14	1.319	0.7579	15.974	0.06260	0.08260	12.106
15	1.346	0.7430	17.293	0.05783	0.07783	12.849
16	1.373	0.7284	18.639	0.05365	0.07365	13.578
17	1.400	0.7142	20.012	0.04997	0.06997	14.292
18	1.428	0.7002	21.412	0.04670	0.06670	14.992
19	1.457	0.6864	22.841	0.04378	0.06378	15.678
20	1.486	0.6730	24.297	0.04116	0.06116	16.351
21	1.516	0.6598	25.783	0.03878	0.05878	17.011
22	1.546	0.6468	27.299	0.03663	0.05663	17.658
23	1.577	0.6342	28.845	0.03467	0.05467	18.292
24	1.608	0.6217	30.422	0.03287	0.05287	18.914
25	1.641	0.6095	32.030	0.03122	0.05122	19.523
26	1.673	0.5976	33.671	0.02970	0.04970	20.121
27	1.707	0.5859	35.344	0.02829	0.04829	20.707
28	1.741	0.5744	37.051	0.02699	0.04699	21.281
29	1.776	0.5631	38.792	0.02578	0.04578	21.844
30	1.811	0.5521	40.568	0.02465	0.04465	22.396
35	2.000	0.5000	49.994	0.02000	0.04000	24.999
40	2.208	0.4529	60.402	0.01656	0.03656	27.355
45	2.438	0.4102	71.893	0.01391	0.03391	29.490
50	2.692	0.3715	84.579	0.01182	0.03182	31.424
55	2.972	0.3365	98.587	0.01014	0.03014	33.175
60	3.281	0.3048	114.052	0.00877	0.02877	34.761
65	3.623	0.2761	131.126	0.00763	0.02763	36.197
70	4.000	0.2500	149.978	0.00667	0.02667	37.499
75	4.416	0.2265	170.792	0.00586	0.02586	38.677
80	4.875	0.2051	193.772	0.00516	0.02516	39.745
85	5.383	0.1858	219.144	0.00456	0.02456	40.711
90	5.943	0.1683	247.157	0.00405	0.02405	41.587
95	6.562	0.1524	278.085	0.00360	0.02360	42.380
100	7.245	0.1380	312.232	0.00320	0.02320	43.098

3% Interest Factors

Period n	Single-payment compound-amount (SPCA) Future value of $1 $(1 + i)^n$	Single-payment present-worth (SPPW) Present value of $1 $\dfrac{1}{(1 + i)^n}$	Uniform-series compound-amount (USCA) Future value of uniform series of $1 $\dfrac{(1 + i)^n - 1}{i}$	Sinking-fund payment (SFP) Uniform series whose future value is $1 $\dfrac{i}{(1 + i)^n - 1}$	Capital recovery (CR) Uniform series with present value of $1 $\dfrac{i(1 + i)^n}{(1 + i)^n - 1}$	Uniform-series present-worth (USPW) Present value of uniform series of $1 $\dfrac{(1 + i)^n - 1}{i(1 + i)^n}$
1	1.030	0.9709	1.000	1.00000	1.03000	0.971
2	1.061	0.9426	2.030	0.49261	0.52261	1.913
3	1.093	0.9151	3.091	0.32353	0.35353	2.829
4	1.126	0.8885	4.184	0.23903	0.26903	3.717
5	1.159	0.8626	5.309	0.18835	0.21835	4.580
6	1.194	0.8375	6.468	0.15460	0.18460	5.417
7	1.230	0.8131	7.662	0.13051	0.16051	6.230
8	1.267	0.7894	8.892	0.11246	0.14246	7.020
9	1.305	0.7664	10.159	0.09843	0.12843	7.786
10	1.344	0.7441	11.464	0.08723	0.11723	8.530
11	1.384	0.7224	12.808	0.07808	0.10808	9.253
12	1.426	0.7014	14.192	0.07046	0.10046	9.954
13	1.469	0.6810	15.618	0.06403	0.09403	10.635
14	1.513	0.6611	17.086	0.05853	0.08853	11.296
15	1.558	0.6419	18.599	0.05377	0.08377	11.938
16	1.605	0.6232	20.157	0.04961	0.07961	12.561
17	1.653	0.6050	21.762	0.04595	0.07595	13.166
18	1.702	0.5874	23.414	0.04271	0.07271	13.754
19	1.754	0.5703	25.117	0.03981	0.06981	14.324
20	1.806	0.5537	26.870	0.03722	0.06722	14.877
21	1.860	0.5375	28.676	0.03487	0.06487	15.415
22	1.916	0.5219	30.537	0.03275	0.06275	15.937
23	1.974	0.5067	32.453	0.03081	0.06081	16.444
24	2.033	0.4919	34.426	0.02905	0.05905	16.936
25	2.094	0.4776	36.459	0.02743	0.05743	17.413
26	2.157	0.4637	38.553	0.02594	0.05594	17.877
27	2.221	0.4502	40.710	0.02456	0.05456	18.327
28	2.288	0.4371	42.931	0.02329	0.05329	18.764
29	2.357	0.4243	45.219	0.02211	0.05211	19.188
30	2.427	0.4120	47.575	0.02102	0.05102	19.600
35	2.814	0.3554	60.462	0.01654	0.04654	21.487
40	3.262	0.3066	75.401	0.01326	0.04326	23.115
45	3.782	0.2644	92.720	0.01079	0.04079	24.519
50	4.384	0.2281	112.797	0.00887	0.03887	25.730
55	5.082	0.1968	136.072	0.00735	0.03735	26.774
60	5.892	0.1697	163.053	0.00613	0.03613	27.676
65	6.830	0.1464	194.333	0.00515	0.03515	28.453
70	7.918	0.1263	230.594	0.00434	0.03434	29.123
75	9.179	0.1089	272.631	0.00367	0.03367	29.702
80	10.641	0.0940	321.363	0.00311	0.03311	30.201
85	12.336	0.0811	377.857	0.00265	0.03265	30.631
90	14.300	0.0699	443.349	0.00226	0.03226	31.002
95	16.578	0.0603	519.272	0.00193	0.03193	31.323
100	19.219	0.0520	607.288	0.00165	0.03165	31.599

4% **Interest Factors**

Period n	Single-payment compound-amount (SPCA) — Future value of $1 $(1 + i)^n$	Single-payment present-worth (SPPW) — Present value of $1 $\dfrac{1}{(1 + i)^n}$	Uniform-series compound-amount (USCA) — Future value of uniform series of $1 $\dfrac{(1 + i)^n - 1}{i}$	Sinking-fund payment (SFP) — Uniform series whose future value is $1 $\dfrac{i}{(1 + i)^n - 1}$	Capital recovery (CR) — Uniform series with present value of $1 $\dfrac{i(1 + i)^n}{(1 + i)^n - 1}$	Uniform-series present-worth (USPW) — Present value of uniform series of $1 $\dfrac{(1 + i)^n - 1}{i(1 + i)^n}$
1	1.040	0.9615	1.000	1.00000	1.04000	0.962
2	1.082	0.9246	2.040	0.49020	0.53020	1.886
3	1.125	0.8890	3.122	0.32035	0.36035	2.775
4	1.170	0.8548	4.246	0.23549	0.27549	3.630
5	1.217	0.8219	5.416	0.18463	0.22463	4.452
6	1.265	0.7903	6.633	0.15076	0.19076	5.242
7	1.316	0.7599	7.898	0.12661	0.16661	6.002
8	1.369	0.7307	9.214	0.10853	0.14853	6.733
9	1.423	0.7026	10.583	0.09449	0.13449	7.435
10	1.480	0.6756	12.006	0.08329	0.12329	8.111
11	1.539	0.6496	13.486	0.07415	0.11415	8.760
12	1.601	0.6246	15.026	0.06655	0.10655	9.385
13	1.665	0.6006	16.627	0.06014	0.10014	9.986
14	1.732	0.5775	18.292	0.05467	0.09467	10.563
15	1.801	0.5553	20.024	0.04994	0.08994	11.118
16	1.873	0.5339	21.825	0.04582	0.08582	11.652
17	1.948	0.5134	23.698	0.04220	0.08220	12.166
18	2.026	0.4936	25.645	0.03899	0.07899	12.659
19	2.107	0.4746	27.671	0.03614	0.07614	13.134
20	2.191	0.4564	29.778	0.03358	0.07358	13.590
21	2.279	0.4388	31.969	0.03128	0.07128	14.029
22	2.370	0.4220	34.248	0.02920	0.06920	14.451
23	2.465	0.4057	36.618	0.02731	0.06731	14.857
24	2.563	0.3901	39.083	0.02559	0.06559	15.247
25	2.666	0.3751	41.646	0.02401	0.06401	15.622
26	2.772	0.3607	44.312	0.02257	0.06257	15.983
27	2.883	0.3468	47.084	0.02124	0.06124	16.330
28	2.999	0.3335	49.968	0.02001	0.06001	16.663
29	3.119	0.3207	52.966	0.01888	0.05888	16.984
30	3.243	0.3083	56.085	0.01783	0.05783	17.292
35	3.946	0.2534	73.652	0.01358	0.05358	18.665
40	4.801	0.2083	95.026	0.01052	0.05052	19.793
45	5.841	0.1712	121.029	0.00826	0.04826	20.720
50	7.107	0.1407	152.667	0.00655	0.04655	21.482
55	8.646	0.1157	191.159	0.00523	0.04523	22.109
60	10.520	0.0951	237.991	0.00420	0.04420	22.623
65	12.799	0.0781	294.968	0.00339	0.04339	23.047
70	15.572	0.0642	364.290	0.00275	0.04275	23.395
75	18.945	0.0528	448.631	0.00223	0.04223	23.680
80	23.050	0.0434	551.245	0.00181	0.04181	23.915
85	28.044	0.0357	676.090	0.00148	0.04148	24.109
90	34.119	0.0293	827.983	0.00121	0.04121	24.267
95	41.511	0.0241	1012.785	0.00099	0.04099	24.398
100	50.505	0.0198	1237.624	0.00081	0.04081	24.505

5% Interest Factors

Period n	Single-payment compound-amount (SPCA)	Single-payment present-worth (SPPW)	Uniform-series compound-amount (USCA)	Sinking-fund payment (SFP)	Capital recovery (CR)	Uniform-series present-worth (USPW)
	Future value of $1 $(1 + i)^n$	Present value of $1 $\dfrac{1}{(1 + i)^n}$	Future value of uniform series of $1 $\dfrac{(1 + i)^n - 1}{i}$	Uniform series whose future value is $1 $\dfrac{i}{(1 + i)^n - 1}$	Uniform series with present value of $1 $\dfrac{i(1 + i)^n}{(1 + i)^n - 1}$	Present value of uniform series of $1 $\dfrac{(1 + i)^n - 1}{i(1 + i)^n}$
1	1.050	0.9524	1.000	1.00000	1.05000	0.952
2	1.103	0.9070	2.050	0.48780	0.53780	1.859
3	1.158	0.8638	3.153	0.31721	0.36721	2.723
4	1.216	0.8227	4.310	0.23201	0.28201	3.546
5	1.276	0.7835	5.526	0.18097	0.23097	4.329
6	1.340	0.7462	6.802	0.14702	0.19702	5.076
7	1.407	0.7107	8.142	0.12282	0.17282	5.786
8	1.477	0.6768	9.549	0.10472	0.15472	6.463
9	1.551	0.6446	11.027	0.09069	0.14069	7.108
10	1.629	0.6139	12.578	0.07950	0.12950	7.722
11	1.710	0.5847	14.207	0.07039	0.12039	8.306
12	1.796	0.5568	15.917	0.06283	0.11283	8.863
13	1.886	0.5303	17.713	0.05646	0.10646	9.394
14	1.980	0.5051	19.599	0.05102	0.10102	9.899
15	2.079	0.4810	21.579	0.04634	0.09634	10.380
16	2.183	0.4581	23.657	0.04227	0.09227	10.838
17	2.292	0.4363	25.840	0.03870	0.08870	11.274
18	2.407	0.4155	28.132	0.03555	0.08555	11.690
19	2.527	0.3957	30.539	0.03275	0.08275	12.085
20	2.653	0.3769	33.066	0.03024	0.08024	12.462
21	2.786	0.3589	35.719	0.02800	0.07800	12.821
22	2.925	0.3418	38.505	0.02597	0.07597	13.163
23	3.072	0.3256	41.430	0.02414	0.07414	13.489
24	3.225	0.3101	44.502	0.02247	0.07247	13.799
25	3.386	0.2953	47.727	0.02095	0.07095	14.094
26	3.556	0.2812	51.113	0.01956	0.06956	14.375
27	3.733	0.2678	54.669	0.01829	0.06829	14.643
28	3.920	0.2551	58.403	0.01712	0.06712	14.898
29	4.116	0.2429	62.323	0.01605	0.06605	15.141
30	4.322	0.2314	66.439	0.01505	0.06505	15.372
35	5.516	0.1813	90.320	0.01107	0.06107	16.374
40	7.040	0.1420	120.800	0.00828	0.05828	17.159
45	8.985	0.1113	159.700	0.00626	0.05626	17.774
50	11.467	0.0872	209.348	0.00478	0.05478	18.256
55	14.636	0.0683	272.713	0.00367	0.05367	18.633
60	18.679	0.0535	353.584	0.00283	0.05283	18.929
65	23.840	0.0419	456.798	0.00219	0.05219	19.161
70	30.426	0.0329	588.529	0.00170	0.05170	19.343
75	38.833	0.0258	756.654	0.00132	0.05132	19.485
80	49.561	0.0202	971.229	0.00103	0.05103	19.596
85	63.254	0.0158	1245.087	0.00080	0.05080	19.684
90	80.730	0.0124	1594.607	0.00063	0.05063	19.752
95	103.035	0.0097	2040.694	0.00049	0.05049	19.806
100	131.501	0.0076	2610.025	0.00038	0.05038	19.848

6% Interest Factors

Period n	Single-payment compound-amount (SPCA)	Single-payment present-worth (SPPW)	Uniform-series compound-amount (USCA)	Sinking-fund payment (SFP)	Capital recovery (CR)	Uniform-series present-worth (USPW)
	Future value of \$1 $(1 + i)^n$	Present value of \$1 $\dfrac{1}{(1 + i)^n}$	Future value of uniform series of \$1 $\dfrac{(1 + i)^n - 1}{i}$	Uniform series whose future value is \$1 $\dfrac{i}{(1 + i)^n - 1}$	Uniform series with present value of \$1 $\dfrac{i(1 + i)^n}{(1 + i)^n - 1}$	Present value of uniform series of \$1 $\dfrac{(1 + i)^n - 1}{i(1 + i)^n}$
1	1.060	0.9434	1.000	1.00000	1.06000	0.943
2	1.124	0.8900	2.060	0.48544	0.54544	1.833
3	1.191	0.8396	3.184	0.31411	0.37411	2.673
4	1.262	0.7921	4.375	0.22859	0.28859	3.465
5	1.338	0.7473	5.637	0.17740	0.23740	4.212
6	1.419	0.7050	6.975	0.14336	0.20336	4.917
7	1.504	0.6651	8.394	0.11914	0.17914	5.582
8	1.594	0.6274	9.897	0.10104	0.16104	6.210
9	1.689	0.5919	11.491	0.08702	0.14702	6.802
10	1.791	0.5584	13.181	0.07587	0.13587	7.360
11	1.898	0.5268	14.972	0.06679	0.12679	7.887
12	2.012	0.4970	16.870	0.05928	0.11928	8.384
13	2.133	0.4688	18.882	0.05296	0.11296	8.853
14	2.261	0.4423	21.015	0.04758	0.10758	9.295
15	2.397	0.4173	23.276	0.04296	0.10296	9.712
16	2.540	0.3936	25.673	0.03895	0.09895	10.106
17	2.693	0.3714	28.213	0.03544	0.09544	10.477
18	2.854	0.3503	30.906	0.03236	0.09236	10.828
19	3.026	0.3305	33.760	0.02962	0.08962	11.158
20	3.207	0.3118	36.786	0.02718	0.08718	11.470
21	3.400	0.2942	39.993	0.02500	0.08500	11.764
22	3.604	0.2775	43.392	0.02305	0.08305	12.042
23	3.820	0.2618	46.996	0.02128	0.08128	12.303
24	4.049	0.2470	50.816	0.01968	0.07968	12.550
25	4.292	0.2330	54.865	0.01823	0.07823	12.783
26	4.549	0.2198	59.156	0.01690	0.07690	13.003
27	4.822	0.2074	63.706	0.01570	0.07570	13.211
28	5.112	0.1956	68.528	0.01459	0.07459	13.406
29	5.418	0.1846	73.640	0.01358	0.07358	13.591
30	5.743	0.1741	79.058	0.01265	0.07265	13.765
35	7.686	0.1301	111.435	0.00897	0.06897	14.498
40	10.286	0.0972	154.762	0.00646	0.06646	15.046
45	13.765	0.0727	212.744	0.00470	0.06470	15.456
50	18.420	0.0543	290.336	0.00344	0.06344	15.762
55	24.650	0.0406	394.172	0.00254	0.06254	15.991
60	32.988	0.0303	533.128	0.00188	0.06188	16.161
65	44.145	0.0227	719.083	0.00139	0.06139	16.289
70	59.076	0.0169	967.932	0.00103	0.06103	16.385
75	79.057	0.0126	1300.949	0.00077	0.06077	16.456
80	105.796	0.0095	1746.600	0.00057	0.06057	16.509
85	141.579	0.0071	2342.982	0.00043	0.06043	16.549
90	189.465	0.0053	3141.075	0.00032	0.06032	16.579
95	253.546	0.0039	4209.104	0.00024	0.06024	16.601
100	339.302	0.0029	5638.368	0.00018	0.06018	16.618

8% Interest Factors

Period n	Single-payment compound-amount (SPCA) Future value of \$1 $(1 + i)^n$	Single-payment present-worth (SPPW) Present value of \$1 $\dfrac{1}{(1 + i)^n}$	Uniform-series compound-amount (USCA) Future value of uniform series of \$1 $\dfrac{(1 + i)^n - 1}{i}$	Sinking-fund payment (SFP) Uniform series whose future value is \$1 $\dfrac{i}{(1 + i)^n - 1}$	Capital recovery (CR) Uniform series with present value of \$1 $\dfrac{i(1 + i)^n}{(1 + i)^n - 1}$	Uniform-series present-worth (USPW) Present value of uniform series of \$1 $\dfrac{(1 + i)^n - 1}{i(1 + i)^n}$
1	1.080	0.9259	1.000	1.00000	1.08000	0.926
2	1.166	0.8573	2.080	0.48077	0.56077	1.783
3	1.260	0.7938	3.246	0.30803	0.38803	2.577
4	1.360	0.7350	4.506	0.22192	0.30192	3.312
5	1.469	0.6806	5.867	0.17046	0.25046	3.993
6	1.587	0.6302	7.336	0.13632	0.21632	4.623
7	1.714	0.5835	8.923	0.11207	0.19207	5.206
8	1.851	0.5403	10.637	0.09401	0.17401	5.747
9	1.999	0.5002	12.488	0.08008	0.16008	6.247
10	2.159	0.4632	14.487	0.06903	0.14903	6.710
11	2.332	0.4289	16.645	0.06008	0.14008	7.139
12	2.518	0.3971	18.977	0.05270	0.13270	7.536
13	2.720	0.3677	21.495	0.04652	0.12652	7.904
14	2.937	0.3405	24.215	0.04130	0.12130	8.244
15	3.172	0.3152	27.152	0.03683	0.11683	8.559
16	3.426	0.2919	30.324	0.03298	0.11298	8.851
17	3.700	0.2703	33.750	0.02963	0.10963	9.122
18	3.996	0.2502	37.450	0.02670	0.10670	9.372
19	4.316	0.2317	41.446	0.02413	0.10413	9.604
20	4.661	0.2145	45.762	0.02185	0 10185	9.818
21	5.034	0.1987	50.423	0.01983	0.09983	10.017
22	5.437	0.1839	55.457	0.01803	0.09803	10.201
23	5.871	0.1703	60.893	0.01642	0.09642	10.371
24	6.341	0.1577	66.765	0.01498	0.09498	10.529
25	6.848	0.1460	73.106	0.01368	0.09368	10.675
26	7.396	0.1352	79.954	0.01251	0.09251	10.810
27	7.988	0.1252	87.351	0.01145	0.09145	10.935
28	8.627	0.1159	95.339	0.01049	0.09049	11.051
29	9.317	0.1073	103.966	0.00962	0.08962	11.158
30	10.063	0.0994	113.283	0.00883	0.08883	11.258
35	14.785	0.0676	172.317	0.00580	0.08580	11.655
40	21.725	0.0460	259.057	0.00386	0.08386	11.925
45	31.920	0.0313	386.506	0.00259	0.08259	12 108
50	46.902	0.0213	573.770	0.00174	0.08174	12.233
55	68.914	0.0145	848.923	0.00118	0.08118	12 319
60	101.257	0.0099	1253.213	0.00080	0.08080	12.377
65	148.780	0.0067	1847.248	0.00054	0.08054	12.416
70	218.606	0.0046	2720.080	0.00037	0.08037	12.443
75	321.205	0.0031	4002.557	0.00025	0.08025	12.461
80	471.955	0.0021	5886.935	0.00017	0.08017	12.474
85	693.456	0.0014	8655.706	0.00012	0.08012	12.482
90	1018.915	0.0010	12723.939	0.00008	0.08008	12.488
95	1497.121	0.0007	18701.507	0.00005	0.08005	12.492
100	2199.761	0.0005	27484.516	0.00004	0.08004	12.494

10% Interest Factors

Period n	Single-payment compound-amount (SPCA)	Single-payment present-worth (SPPW)	Uniform-series compound-amount (USCA)	Sinking-fund payment (SFP)	Capital recovery (CR)	Uniform-series present-worth (USPW)
	Future value of $1 $(1 + i)^n$	Present value of $1 $\dfrac{1}{(1 + i)^n}$	Future value of uniform series of $1 $\dfrac{(1 + i)^n - 1}{i}$	Uniform series whose future value is $1 $\dfrac{i}{(1 + i)^n - 1}$	Uniform series with present value of $1 $\dfrac{i(1 + i)^n}{(1 + i)^n - 1}$	Present value of uniform series of $1 $\dfrac{(1 + i)^n - 1}{i(1 + i)^n}$
1	1.100	0.9091	1.000	1.00000	1.10000	0.909
2	1.210	0.8264	2.100	0.47619	0.57619	1.736
3	1.331	0.7513	3.310	0.30211	0.40211	2.487
4	1.464	0.6830	4.641	0.21547	0.31547	3.170
5	1.611	0.6209	6.105	0.16380	0.26380	3.791
6	1.772	0.5645	7.716	0.12961	0.22961	4.355
7	1.949	0.5132	9.487	0.10541	0.20541	4.868
8	2.144	0.4665	11.436	0.08744	0.18744	5.335
9	2.358	0.4241	13.579	0.07364	0.17364	5.759
10	2.594	0.3855	15.937	0.06275	0.16275	6.144
11	2.853	0.3505	18.531	0.05396	0.15396	6.495
12	3.138	0.3186	21.384	0.04676	0.14676	6.814
13	3.452	0.2897	24.523	0.04078	0.14078	7.103
14	3.797	0.2633	27.975	0.03575	0.13575	7.367
15	4.177	0.2394	31.772	0.03147	0.13147	7.606
16	4.595	0.2176	35.950	0.02782	0.12782	7.824
17	5.054	0.1978	40.545	0.02466	0.12466	8.022
18	5.560	0.1799	45.599	0.02193	0.12193	8.201
19	6.116	0.1635	51.159	0.01955	0.11955	8.365
20	6.727	0.1486	57.275	0.01746	0.11746	8.514
21	7.400	0.1351	64.002	0.01562	0.11562	8.649
22	8.140	0.1228	71.403	0.01401	0.11401	8.772
23	8.954	0.1117	79.543	0.01257	0.11257	8.883
24	9.850	0.1015	88.497	0.01130	0.11130	8.985
25	10.835	0.0923	98.347	0.01017	0.11017	9.077
26	11.918	0.0839	109.182	0.00916	0.10916	9.161
27	13.110	0.0763	121.100	0.00826	0.10826	9.237
28	14.421	0.0693	134.210	0.00745	0.10745	9.307
29	15.863	0.0630	148.631	0.00673	0.10673	9.370
30	17.449	0.0573	164.494	0.00608	0.10608	9.427
35	28.102	0.0356	271.024	0.00369	0.10369	9.644
40	45.259	0.0221	442.593	0.00226	0.10226	9.779
45	72.890	0.0137	718.905	0.00139	0.10139	9.863
50	117.391	0.0085	1163.909	0.00086	0.10086	9.915
55	189.059	0.0053	1880.591	0.00053	0.10053	9.947
60	304.482	0.0033	3034.816	0.00033	0.10033	9.967
65	490.371	0.0020	4893.707	0.00020	0.10020	9.980
70	789.747	0.0013	7887.470	0.00013	0.10013	9.987
75	1271.895	0.0008	12708.954	0.00008	0.10008	9.992
80	2048.400	0.0005	20474.002	0.00005	0.10005	9.995
85	3298.969	0.0003	32979.690	0.00003	0.10003	9.997
90	5313.023	0.0002	53120.226	0.00002	0.10002	9.998
95	8556.676	0.0001	85556.760	0.00001	0.10001	9.999

12% Interest Factors

Period n	Single-payment compound-amount (SPCA) Future value of $1 $(1 + i)^n$	Single-payment present-worth (SPPW) Present value of $1 $\dfrac{1}{(1 + i)^n}$	Uniform-series compound-amount (USCA) Future value of uniform series of $1 $\dfrac{(1 + i)^n - 1}{i}$	Sinking-fund payment (SFP) Uniform series whose future value is $1 $\dfrac{i}{(1 + i)^n - 1}$	Capital recovery (CR) Uniform series with present value of $1 $\dfrac{i(1 + i)^n}{(1 + i)^n - 1}$	Uniform-series present-worth (USPW) Present value of uniform series of $1 $\dfrac{(1 + i)^n - 1}{i(1 + i)^n}$
1	1.120	0.8929	1.000	1.00000	1.12000	0.893
2	1.254	0.7972	2.120	0.47170	0.59170	1.690
3	1.405	0.7118	3.374	0.29635	0.41635	2.402
4	1.574	0.6355	4.779	0.20923	0.32923	3.037
5	1.762	0.5674	6.353	0.15741	0.27741	3.605
6	1.974	0.5066	8.115	0.12323	0.24323	4.111
7	2.211	0.4523	10.089	0.09912	0.21912	4.564
8	2.476	0.4039	12.300	0.08130	0.20130	4.968
9	2.773	0.3606	14.776	0.06768	0.18768	5.328
10	3.106	0.3220	17.549	0.05698	0.17698	5.650
11	3.479	0.2875	20.655	0.04842	0.16842	5.938
12	3.896	0.2567	24.133	0.04144	0.16144	6.194
13	4.363	0.2292	28.029	0.03568	0.15568	6.424
14	4.887	0.2046	32.393	0.03087	0.15087	6.628
15	5.474	0.1827	37.280	0.02682	0.14682	6.811
16	6.130	0.1631	42.753	0.02339	0.14339	6.974
17	6.866	0.1456	48.884	0.02046	0.14046	7.120
18	7.690	0.1300	55.750	0.01794	0.13794	7.250
19	8.613	0.1161	63.440	0.01576	0.13576	7.366
20	9.646	0.1037	72.052	0.01388	0.13388	7.469
21	10.804	0.0926	81.699	0.01224	0.13224	7.562
22	12.100	0.0826	92.503	0.01081	0.13081	7.645
23	13.552	0.0738	104.603	0.00956	0.12956	7.718
24	15.179	0.0659	118.155	0.00846	0.12846	7.784
25	17.000	0.0588	133.334	0.00750	0.12750	7.843
26	19.040	0.0525	150.334	0.00665	0.12665	7.896
27	21.325	0.0469	169.374	0.00590	0.12590	7.943
28	23.884	0.0419	190.699	0.00524	0.12524	7.984
29	26.750	0.0374	214.583	0.00466	0.12466	8.022
30	29.960	0.0334	241.333	0.00414	0.12414	8.055
35	52.800	0.0189	431.663	0.00232	0.12232	8.176
40	93.051	0.0107	767.091	0.00130	0.12130	8.244
45	163.988	0.0061	1358.230	0.00074	0.12074	8.283
50	289.002	0.0035	2400.018	0.00042	0.12042	8.304
55	509.321	0.0020	4236.005	0.00024	0.12024	8.317
60	897.597	0.0011	7471.641	0.00013	0.12013	8.324
65	1581.872	0.0006	13173.937	0.00008	0.12008	8.328
70	2787.800	0.0004	23223.332	0.00004	0.12004	8.330
75	4913.056	0.0002	40933.799	0.00002	0.12002	8.332
80	8658.483	0.0001	72145.692	0.00001	0.12001	8.332

15% Interest Factors

Period n	Single-payment compound-amount (SPCA) — Future value of $1 $(1 + i)^n$	Single-payment present-worth (SPPW) — Present value of $1 $\dfrac{1}{(1 + i)^n}$	Uniform-series compound-amount (USCA) — Future value of uniform series of $1 $\dfrac{(1 + i)^n - 1}{i}$	Sinking-fund payment (SFP) — Uniform series whose future value is $1 $\dfrac{i}{(1 + i)^n - 1}$	Capital recovery (CR) — Uniform series with present value of $1 $\dfrac{i(1 + i)^n}{(1 + i)^n - 1}$	Uniform-series present-worth (USPW) — Present value of uniform series of $1 $\dfrac{(1 + i)^n - 1}{i(1 + i)^n}$
1	1.150	0.8696	1.000	1.00000	1.15000	0.870
2	1.322	0.7561	2.150	0.46512	0.61512	1.626
3	1.521	0.6575	3.472	0.28798	0.43798	2.283
4	1.749	0.5718	4.993	0.20027	0.35027	2.855
5	2.011	0.4972	6.742	0.14832	0.29832	3.352
6	2.313	0.4323	8.754	0.11424	0.26424	3.784
7	2.660	0.3759	11.067	0.09036	0.24036	4.160
8	3.059	0.3269	13.727	0.07285	0.22285	4.487
9	3.518	0.2843	16.786	0.05957	0.20957	4.772
10	4.046	0.2472	20.304	0.04925	0.19925	5.019
11	4.652	0.2149	24.349	0.04107	0.19107	5.234
12	5.350	0.1869	29.002	0.03448	0.18448	5.421
13	6.153	0.1625	34.352	0.02911	0.17911	5.583
14	7.076	0.1413	40.505	0.02469	0.17469	5.724
15	8.137	0.1229	47.580	0.02102	0.17102	5.847
16	9.358	0.1069	55.717	0.01795	0.16795	5.954
17	10.761	0.0929	65.075	0.01537	0.16537	6.047
18	12.375	0.0808	75.836	0.01319	0.16319	6.128
19	14.232	0.0703	88.212	0.01134	0.16134	6.198
20	16.367	0.0611	102.444	0.00976	0.15976	6.259
21	18.822	0.0531	118.810	0.00842	0.15842	6.312
22	21.645	0.0462	137.632	0.00727	0.15727	6.359
23	24.891	0.0402	159.276	0.00628	0.15628	6.399
24	28.625	0.0349	184.168	0.00543	0.15543	6.434
25	32.919	0.0304	212.793	0.00470	0.15470	6.464
26	37.857	0.0264	245.712	0.00407	0.15407	6.491
27	43.535	0.0230	283.569	0.00353	0.15353	6.514
28	50.066	0.0200	327.104	0.00306	0.15306	6.534
29	57.575	0.0174	377.170	0.00265	0.15265	6.551
30	66.212	0.0151	434.745	0.00230	0.15230	6.566
35	133.176	0.0075	881.170	0.00113	0.15113	6.617
40	267.864	0.0037	1779.090	0.00056	0.15056	6.642
45	538.769	0.0019	3585.128	0.00028	0.15028	6.654
50	1083.657	0.0009	7217.716	0.00014	0.15014	6.661
55	2179.622	0.0005	14524.148	0.00007	0.15007	6.664
60	4383.999	0.0002	29219.992	0.00003	0.15003	6.665
65	8817.787	0.0001	58778.583	0.00002	0.15002	6.666

EXHIBIT D-9   Tables of Conversion Factors[9]

Angle	°	′	″	rad	rev
I degree =	I	60	3600	1.745×10^{-2}	2.778×10^{-3}
I minute =	1.667×10^{-2}	I	60	2.909×10^{-4}	4.630×10^{-5}
I second =	2.778×10^{-4}	1.667×10^{-2}	I	4.848×10^{-6}	7.716×10^{-7}
I radian =	57.30	3438	2.063×10^{5}	I	0.1592
I revolution =	360	2.16×10^{4}	1.296×10^{6}	6.283	I

I artillery mil = $\frac{1}{6400}$ rev = 0.0009817 rad = 0°.05625

Length	m	km	i	f	mi
I meter =	I	10^{-3}	39.37	3.281	6.214×10^{-4}
I kilometer =	1000	I	3.937×10^{4}	3281	0.6214
I inch =	0.0254	2.54×10^{-5}	I	0.0833	1.578×10^{-5}
I foot =	0.3048	3.048×10^{-4}	12	I	1.894×10^{-4}
I statute mile =	1609	1.609	6.336×10^{4}	5280	I

I angstrom = 10^{-10} m   I millimicron (mμ) = 10^{-9} m   I fathom = 6 f
I X-unit = 10^{-13} m   I light-year = 9.4600×10^{12} km   I yard (yd) = 3 f
I micron (μ) = 10^{-6} m   I parsec = 3.084×10^{13} km   I rod = 16.5 f
I nautical mile = 1852 m = 1.1508 mi = 6076.10 f   I mil = 10^{-3} i
I astronomical unit = 149.5×10^{6} km   I league = 3 naut miles

Area	m²	cm²	f²	i²
I square meter =	I	10^{4}	10.76	1550
I square centimeter =	10^{-4}	I	1.076×10^{-3}	0.1550
I square foot =	9.290×10^{-2}	929.0	I	144
I square inch =	6.452×10^{-4}	6.452	6.944×10^{-3}	I

I square mile = 27,878,400 f² = 640 acres   I acre = 43,560 f²   I are = 100 m²
I circular mil = 7.854×10^{-7} i²   I barn = 10^{-28} m²   I hectare = 100 are

Volume	m³	cm³	f³	i³
I cubic meter =	I	10^{6}	35.31	6.102×10^{4}
I cubic centimeter =	10^{-6}	I	3.531×10^{-5}	0.06102
I cubic foot =	2.832×10^{-2}	28,320	I	1728
I cubic inch =	1.639×10^{-5}	16.39	5.787×10^{-4}	I

I U.S. fluid gallon = 4 quarts = 8 pints = 128 fluid ounces = 231 i³
I British Imperial gallon = the volume of 10 lb of water at 62° F = 277.42 i³
I liter = the volume of 1 kg of water at its maximum density = 1000.028 cm³

---

[9] George Shortley and Dudley Williams, *Elements of Physics*, 4th edition © 1965. Reprinted by permission of Prentice-Hall, Inc., Englewood Cliffs, N. J., pp. Appendix vii-xi.

Mass	g	kg	lb	sl	ton
I gram =	I	0.001	0.002205	6.852×10^{-5}	1.102×10^{-6}
I kilogram =	1000	I	2.205	6.852×10^{-2}	1.102×10^{-3}
I pound (avoirdupois) =	453.6	0.4536	I	3.108×10^{-2}	0.0005
I slug =	1.459×10^4	14.59	32.17	I	1.609×10^{-2}
I ton-mass =	9.072×10^5	907.2	2000	62.16	I

I avoirdupois pound = 16 avoirdupois ounces = 7000 grains
I troy or apothecaries' pound = 12 troy or apothecaries' ounces
= 0.8229 avoirdupois pound = 5760 grains
I long ton = 2240 lb = 20 cwt   I stone = 14 lb   I hundredweight (cwt) = 112 lb
I metric ton = 1000 kg = 2205 lb   I carat = 0.2 g   I pennyweight (dwt) = 24 grains
I atomic mass unit (amu) = 1.6604×10^{-27} kg

Time	y	d	h	min	s
I year =	I	365.2	8.766×10^3	5.259×10^5	3.156×10^7
I day =	2.738×10^{-3}	I	24	1440	86400
I hour =	1.141×10^{-4}	4.167×10^{-2}	I	60	3600
I minute =	1.901×10^{-6}	6.944×10^{-4}	1.667×10^{-2}	I	60
I second =	3.169×10^{-8}	1.157×10^{-5}	2.778×10^{-4}	1.667×10^{-2}	I

I sidereal day = period of rotation of earth = 86,164 s
I year = period of revolution of earth = 365.24219879 d

Frequency: I hertz = I Hz = I s^{-1}

Density	sl/f³	lb/f³	lb/i³	kg/m³	g/cm³
I slug per f³ =	I	32.17	1.862×10^{-2}	515.4	0.5154
I pound per f³ =	3.108×10^{-2}	I	5.787×10^{-4}	16.02	1.602×10^{-2}
I pound per i³ =	53.71	1728	I	2.768×10^4	27.68
I kg per m³ =	1.940×10^{-3}	6.243×10^{-2}	3.613×10^{-5}	I	0.001
I gram per cm³ =	1.940	62.43	3.613×10^{-2}	1000	I

Speed	f/s	km/h	m/s	mi/h	knot
I foot per second =	I	1.097	0.3048	0.6818	0.5925
I kilometer per hour =	0.9113	I	0.2778	0.6214	0.5400
I meter per second =	3.281	3.6	I .	2.237	1.944
I mile per hour =	1.467	1.609	0.4470	I	0.8689
I knot =	1.688	1.852	0.5144	1.151	I

I knot = I nautical mile/hr

Force	dyne	kgf	N	p	poundal
1 dyne =	1	1.020×10^{-6}	10^{-5}	2.248×10^{-6}	7.233×10^{-5}
1 kilogram-force =	9.807×10^5	1	9.807	2.205	70.93
1 newton =	10^5	0.1020	1	0.2248	7.233
1 pound =	4.448×10^5	0.4536	4.448	1	32.17
1 poundal =	1.383×10^4	1.410×10^{-2}	0.1383	3.108×10^{-2}	1

1 kgf = 9.80665 N   1 p = 32.17398 pdl

Pressure	atm	inch of water	cm Hg	N/m²	p/i²
1 atmosphere =	1	406.8	76	1.013×10^5	14.70
1 inch of water [a] =	2.458×10^{-3}	1	0.1868	249.1	0.03613
1 cm mercury[a] =	1.316×10^{-2}	5.353	1	1333	0.1934
1 newton per m² =	9.869×10^{-6}	0.004105	7.501×10^{-4}	1	1.450×10^{-4}
1 pound per i² =	6.805×10^{-2}	27.68	5.171	6.895×10^3	1

[a] Under standard gravitational acceleration, and temperature of 4° C for water, 0° C for mercury.

1 bar = 10^6 dyne/cm²   1 millibar = 10^3 dyne/cm²
1 cm of water = 98.07 N/m²   1 f of water = 62.43 p/f²

Energy	BTU	fp	J	kcal	kWh
1 British thermal unit =	1	777.9	1055	0.2520	2.930×10^{-4}
1 foot-pound =	1.285×10^{-3}	1	1.356	3.240×10^{-4}	3.766×10^{-7}
1 joule =	9.481×10^{-4}	0.7376	1	2.390×10^{-4}	2.778×10^{-7}
1 kilocalorie =	3.968	3086	4184	1	1.163×10^{-3}
1 kilowatt-hour =	3413	2.655×10^6	3.6×10^6	860.2	1

See also table of relativistic mass-energy equivalents on p. xi.
1 kcal = 2.612×10^{22} eV   1 horsepower-hour = 1.980×10^6 fp
1 erg = 10^{-7} joule   1 therm = 10^5 BTU

Power	BTU/h	fp/s	hp	kcal/s	kW	W
1 BTU/h =	1	0.2161	3.929×10^{-4}	7.000×10^{-5}	2.930×10^{-4}	0.2930
1 fp/s =	4.628	1	1.818×10^{-3}	3.239×10^{-4}	1.356×10^{-3}	1.356
1 horsepower =	2545	550	1	0.1782	0.7457	745.7
1 kcal/s =	1.429×10^4	3087	5.613	1	4.184	4184
1 kilowatt =	3413	737.6	1.341	0.2390	1	1000
1 watt =	3.413	0.7376	1.341×10^{-3}	2.390×10^{-4}	0.001	1

Electric charge	abC	C	statC
I abcoulomb (I EMU) =	I	10	2.998×10^{10}
I coulomb =	0.1	I	2.998×10^{9}
I statcoulomb (I ESU) =	3.336×10^{-11}	3.336×10^{-10}	I

I franklin = I Fr = I statC   I ampere-hour = 3600 C

Electric current	abA	A	statA
I abampere (I EMU) =	I	10	2.998×10^{10}
I ampere =	0.1	I	2.998×10^{9}
I statampere (I ESU) =	3.336×10^{-11}	3.336×10^{-10}	I

I biot = I Bi = I abA

Electric potential	abV	V	statV
I abvolt (I EMU) =	I	10^{-8}	3.336×10^{-11}
I volt =	10^{8}	I	3.336×10^{-3}
I statvolt (I ESU) =	2.998×10^{10}	299.8	I

Electric resistance	abohm	Ω	statohm
I abohm (I EMU) =	I	10^{-9}	1.113×10^{-21}
I ohm =	10^{9}	I	1.113×10^{-12}
I statohm (I ESU) =	8.987×10^{20}	8.987×10^{11}	I

Capacitance	abF	F	μF	statF
I abfarad (I EMU) =	I	10^{9}	10^{15}	8.987×10^{20}
I farad =	10^{-9}	I	10^{6}	8.987×10^{11}
I microfarad =	10^{-15}	10^{-6}	I	8.987×10^{5}
I statfarad (I ESU) =	1.113×10^{-21}	1.113×10^{-12}	1.113×10^{-6}	I

Inductance	abH	H	mH	statH
I abhenry (I EMU) =	I	10^{-9}	10^{-6}	1.113×10^{-21}
I henry =	10^{9}	I	1000	1.113×10^{-12}
I millihenry =	10^{6}	0.001	I	1.113×10^{-15}
I stathenry (I ESU) =	8.987×10^{20}	8.987×10^{11}	8.987×10^{14}	I

Magnetic flux	Mx	kiloline	Wb
I maxwell (I line or I EMU) =	I	0.001	10^{-8}
I kiloline =	1000	I	10^{-5}
I weber =	10^8	10^5	I

I ESU = 299.8 weber

Magnetic intensity ℬ	G	kiloline/i^2	T	mG
I gauss (line per cm^2) =	I	6.452×10^{-3}	10^{-4}	1000
I kiloline per square inch =	155.0	I	1.550×10^{-2}	1.550×10^5
I tesla =	10^4	64.52	I	10^7
I milligauss =	0.001	6.452×10^{-6}	10^{-7}	I

I T = I Wb/m^2 I ESU = 2.998×10^6 T   I gamma = 10^{-2} mG = 10^{-9} T

Magnetomotive force	abA-turn	A-turn	Gi
I abampere-turn =	I	10	12.57
I ampere-turn =	0.1	I	1.257
I gilbert =	7.958×10^{-2}	0.7958	I

I ESU = I stastampere-turn = 3.336×10^{-10} A-turn

Magnetizing force ℋ	abA/cm	A/in	A/m	Oe
I abampere-turn per centimeter =	I	25.40	1000	12.57
I ampere-turn per inch =	3.937×10^{-2}	I	39.37	0.4947
I ampere-turn per meter =	0.001	2.540×10^{-2}	I	1.257×10^{-2}
I oersted =	7.958×10^{-2}	2.021	79.58	I

I oersted = I gilbert/cm   I ESU = 3.336×10^{-8} A-turn/m

Mass-energy equivalents	kg	amu[a]	J	MeV
I kilogram~	I	6.025×10^{26}	8.987×10^{-16}	5.610×10^{29}
I atomic mass unit~	1.660×10^{-27}	I	1.492×10^{-10}	931.5
I joule~	1.113×10^{-17}	6.705×10^9	I	6.242×10^{12}
I million electron-volts~	1.783×10^{-30}	1.074×10^{-3}	1.602×10^{-13}	I

[a] The abbreviation u has been internationally recommended.

Index